Computational Methods

of

Feature Selection

Chapman & Hall/CRC
Data Mining and Knowledge Discovery Series

SERIES EDITOR

Vipin Kumar
University of Minnesota
Department of Computer Science and Engineering
Minneapolis, Minnesota, U.S.A

AIMS AND SCOPE

This series aims to capture new developments and applications in data mining and knowledge discovery, while summarizing the computational tools and techniques useful in data analysis. This series encourages the integration of mathematical, statistical, and computational methods and techniques through the publication of a broad range of textbooks, reference works, and handbooks. The inclusion of concrete examples and applications is highly encouraged. The scope of the series includes, but is not limited to, titles in the areas of data mining and knowledge discovery methods and applications, modeling, algorithms, theory and foundations, data and knowledge visualization, data mining systems and tools, and privacy and security issues.

PUBLISHED TITLES

UNDERSTANDING COMPLEX DATASETS: Data Mining with Matrix Decompositions
David Skillicorn

COMPUTATIONAL METHODS OF FEATURE SELECTION
Huan Liu and Hiroshi Motoda

Chapman & Hall/CRC
Data Mining and Knowledge Discovery Series

Computational Methods

of

Feature Selection

Edited by

Huan Liu • Hiroshi Motoda

Chapman & Hall/CRC
Taylor & Francis Group

Boca Raton   London   New York

Chapman & Hall/CRC is an imprint of the
Taylor & Francis Group, an **informa** business

Chapman & Hall/CRC
Taylor & Francis Group
6000 Broken Sound Parkway NW, Suite 300
Boca Raton, FL 33487-2742

International Standard Book Number-13: 978-1-58488-878-9 (Hardcover)

Library of Congress Cataloging-in-Publication Data

Liu, Huan, 1958-
   Computational methods of feature selection / authors/editors, Huan Liu and
Hiroshi Motoda.
      p. cm. -- (Chapman & Hall/CRC data mining and knowledge
   discovery)
   Includes bibliographical references and index.
   ISBN 978-1-58488-878-9 (alk. paper)
   1. Database management. 2. Data mining. 3. Machine learning. I. Motoda,
Hiroshi. II. Title. III. Series.

QA76.9.D3L5652 2007
005.74--dc22   2007027465

Visit the Taylor & Francis Web site at
http://www.taylorandfrancis.com

and the CRC Press Web site at
http://www.crcpress.com

Preface

It has been ten years since we published our first two books on feature selection in 1998. In the past decade, we witnessed a great expansion of feature selection research in multiple dimensions. We experienced the fast data evolution in which extremely high-dimensional data, such as high-throughput data of bioinformatics and Web/text data, became increasingly common. They stretch the capabilities of conventional data processing techniques, pose new challenges, and stimulate accelerated development of feature selection research in two major ways. One trend is to improve and expand the existing techniques to meet the new challenges. The other is to develop brand new algorithms directly targeting the arising challenges. In this process, we observe many feature-selection-centered activities, such as one well-received competition, two well-attended tutorials at top conferences, and two multi-disciplinary workshops, as well as a special development section in a recent issue of *IEEE Intelligent Systems*, to name a few.

This collection bridges the widening gap between existing texts and the rapid developments in the field, by presenting recent research works from various disciplines. It features excellent survey work, practical guides, exciting new directions, and comprehensive tutorials from leading experts. The book also presents easy-to-understand illustrations, state-of-the-art methodologies, and algorithms, along with real-world case studies ranging from text classification, to Web mining, to bioinformatics where high-dimensional data are pervasive. Some vague ideas suggested in our earlier book have been developed into mature areas with solid achievements, along with progress that could not have been imagined ten years ago. With the steady and speedy development of feature selection research, we sincerely hope that this book presents distinctive and representative achievements; serves as a convenient point for graduate students, practitioners, and researchers to further the research and application of feature selection; and sparks a new phase of feature selection research. We are truly optimistic about the impact of feature selection on massive, high-dimensional data and processing in the near future, and we have no doubt that in another ten years, when we look back, we will be humbled by the newfound power of feature selection, and by its indelible contributions to machine learning, data mining, and many real-world challenges.

Huan Liu and Hiroshi Motoda

Acknowledgments

The inception of this book project was during SDM 2006's feature selection workshop. Randi Cohen, an editor of Chapman and Hall/CRC Press, eloquently convinced one of us that it was a time for a new book on feature selection. Since then, she closely worked with us to make the process easier and smoother and allowed us to stay focused. With Randi's kind and expert support, we were able to adhere to the planned schedule when facing unexpected difficulties. We truly appreciate her generous support throughout the project.

This book is a natural extension of the two successful feature selection workshops held at SDM 2005[1] and SDM 2006.[2] The success would not be a reality without the leadership of two workshop co-organizers (Robert Stine of Wharton School and Leonard Auslender of SAS); the meticulous work of the proceedings chair (Lei Yu of Binghamton University); and the altruistic efforts of PC members, authors, and contributors. We take this opportunity to thank all who helped to advance the frontier of feature selection research.

The authors, contributors, and reviewers of this book played an instrumental role in this project. Given the limited space of this book, we could not include all quality works. Reviewers' detailed comments and constructive suggestions significantly helped improve the book's consistency in content, format, comprehensibility, and presentation. We thank the authors who patiently and timely accommodated our (sometimes many) requests.

We would also like to express our deep gratitude for the gracious help we received from our colleagues and students, including Zheng Zhao, Lei Tang, Quan Nguyen, Payam Refaeilzadeh, and Shankara B. Subramanya of Arizona State University; Kozo Ohara of Osaka University; and William Nace and Kenneth Gorreta of AFOSR/AOARD, Air Force Research Laboratory.

Last but not least, we thank our families for their love and support. We are grateful and happy that we can now spend more time with our families.

Huan Liu and Hiroshi Motoda

---

[1] The 2005 proceedings are at http://enpub.eas.asu.edu/workshop/.
[2] The 2006 proceedings are at http://enpub.eas.asu.edu/workshop/2006/.

Contributors

Jesús S. Aguilar-Ruiz
Pablo de Olavide University,
Seville, Spain

Constantin F. Aliferis
Vanderbilt University, Nashville,
Tennessee

Paolo Avesani
ITC-IRST, Trento, Italy

Susan M. Bridges
Mississippi State University,
Mississippi

Alexander Borisov
Intel Corporation, Chandler,
Arizona

Shane Burgess
Mississippi State University,
Mississippi

Diana Chan
Mississippi State University,
Mississippi

Claudia Diamantini
Universitá Politecnica delle
Marche, Ancona, Italy

Rezarta Islamaj Dogan
University of Maryland, College
Park, Maryland and National
Center for Biotechnology Infor-
mation, Bethesda, Maryland

Carlotta Domeniconi
George Mason University, Fair-
fax, Virginia

Jennifer G. Dy
Northeastern University, Boston,
Massachusetts

André Elisseeff
IBM Research, Zürich, Switzerland

Susana Eyheramendy
Ludwig-Maximilians Universität
München, Germany

George Forman
Hewlett-Packard Labs, Palo
Alto, California

Lise Getoor
University of Maryland, College
Park, Maryland

Dimitrios Gunopulos
University of California, Riverside

Isabelle Guyon
ClopiNet, Berkeley, California

Trevor Hastie
Stanford University, Stanford,
California

Joshua Zhexue Huang
University of Hong Kong, Hong
Kong, China

Mohamed Kamel
University of Waterloo, Ontario,
Canada

Igor Kononenko
University of Ljubljana, Ljubljana, Slovenia

David Madigan
Rutgers University, New Brunswick, New Jersey

Masoud Makrehchi
University of Waterloo, Ontario, Canada

Michael Ng
Hong Kong Baptist University, Hong Kong, China

Emanuele Olivetti
ITC-IRST, Trento, Italy

Domenico Potena
Universitá Politecnica delle Marche, Ancona, Italy

José C. Riquelme
University of Seville, Seville, Spain

Roberto Ruiz
Pablo de Olavide University, Seville, Spain

Marko Robnik Šikonja
University of Ljubljana, Ljubljana, Slovenia

David J. Stracuzzi
Arizona State University, Tempe, Arizona

Yijun Sun
University of Florida, Gainesville, Florida

Wei Tang
Florida Atlantic University, Boca Raton, Florida

Kari Torkkola
Motorola Labs, Tempe, Arizona

Eugene Tuv
Intel Corporation, Chandler, Arizona

Sriharsha Veeramachaneni
ITC-IRST, Trento, Italy

W. John Wilbur
National Center for Biotechnology Information, Bethesda, Maryland

Jun Xu
Georgia Institute of Technology, Atlanta, Georgia

Yunming Ye
Harbin Institute of Technology, Harbin, China

Lei Yu
Binghamton University, Binghamton, New York

Shi Zhong
Yahoo! Inc., Sunnyvale, California

Hui Zou
University of Minnesota, Minneapolis

Contents

III Weighting and Local Methods   167

Part I

Introduction and Background

Chapter 1

Less Is More

Huan Liu

Arizona State University

Hiroshi Motoda

AFOSR/AOARD, Air Force Research Laboratory

As our world expands at an unprecedented speed from the physical into the virtual, we can conveniently collect more and more data in any ways one can imagine for various reasons. Is it "The more, the merrier (better)"? The answer is "Yes" and "No." It is "Yes" because we can at least get what we might need. It is also "No" because, when it comes to a point of too much, the existence of inordinate data is tantamount to non-existence if there is no means of effective data access. More can mean less. Without the processing of data, its mere existence would not become a useful asset that can impact our business, and many other matters. Since continued data accumulation is inevitable, one way out is to devise data selection techniques to keep pace with the rate of data collection. Furthermore, given the sheer volume of data, data generated by computers or equivalent mechanisms must be processed automatically, in order for us to tame the data monster and stay in control.

Recent years have seen extensive efforts in feature selection research. The field of feature selection expands both in depth and in breadth, due to increasing demands for dimensionality reduction. The evidence can be found in many recent papers, workshops, and review articles. The research expands from classic *supervised* feature selection to *unsupervised* and *semi-supervised* feature selection, to selection of different feature types such as *causal* and *structural* features, to different kinds of data like high-throughput, text, or images, to feature selection evaluation, and to wide applications of feature selection where data abound.

No book of this size could possibly document the extensive efforts in the frontier of feature selection research. We thus try to sample the field in several ways: asking established experts, calling for submissions, and looking at the recent workshops and conferences, in order to understand the current developments. As this book aims to serve a wide audience from practitioners to researchers, we first introduce the basic concepts and the essential problems with feature selection; next illustrate feature selection research in parallel to supervised, unsupervised, and semi-supervised learning; then present an overview of feature selection activities included in this collection; and last contemplate some issues about evolving feature selection. The book is organized in five parts: (I) Introduction and Background, (II) Extending Feature Selection, (III) Weighting and Local Methods, (IV) Text Feature Selection, and (V) Feature Selection in Bioinformatics. These five parts are relatively independent and can be read in any order. For a newcomer to the field of feature selection, we recommend that you read Chapters 1, 2, 9, 13, and 17 first, then decide on which chapters to read further according to your need and interest. Rudimentary concepts and discussions of related issues such as feature extraction and construction can also be found in two earlier books [10, 9]. Instance selection can be found in [11].

1.1   Background and Basics

One of the fundamental motivations for feature selection is the *curse of dimensionality* [6]. Plainly speaking, two *close* data points in a 2-d space are likely *distant* in a 100-d space (refer to Chapter 2 for an illustrative example). For the case of classification, this makes it difficult to make a prediction of unseen data points by a hypothesis constructed from a limited number of training instances. The number of features is a key factor that determines the size of the hypothesis space containing all hypotheses that can be learned from data [13]. A hypothesis is a pattern or function that predicts classes based on given data. The more features, the larger the hypothesis space. Worse still, the linear increase of the number of features leads to the exponential increase of the hypothesis space. For example, for N binary features and a binary class feature, the hypothesis space is as big as 2^{2^N}. Therefore, feature selection can efficiently reduce the hypothesis space by removing irrelevant and redundant features. The smaller the hypothesis space, the easier it is to find correct hypotheses. Given a fixed-size data sample that is part of the underlying population, the reduction of dimensionality also lowers the number of required training instances. For example, given M, when the number of binary features $N = 10$ is reduced to $N = 5$, the ratio of $M/2^N$ increases exponentially. In other words, it virtually increases the number of training instances. This helps to better constrain the search of correct hypotheses.

Feature selection is essentially a task to remove irrelevant and/or redundant features. *Irrelevant features* can be removed without affecting learning performance [8]. *Redundant features* are a type of irrelevant feature [16]. The distinction is that a redundant feature implies the co-presence of another feature; individually, each feature is relevant, but the removal of one of them will not affect learning performance. The selection of features can be achieved in two ways: One is to rank features according to some criterion and select the top k features, and the other is to select a minimum subset of features without learning performance deterioration. In other words, subset selection algorithms can automatically determine the number of selected features, while feature ranking algorithms need to rely on some given threshold to select features. An example of feature ranking algorithms is detailed in Chapter 9. An example of subset selection can be found in Chapter 17.

Other important aspects of feature selection include models, search strategies, feature quality measures, and evaluation [10]. The three typical models are filter, wrapper, and embedded. An embedded model of feature selection integrates the selection of features in model building. An example of such a model is the decision tree induction algorithm, in which at each branching node, a feature has to be selected. The research shows that even for such a learning algorithm, feature selection can result in improved learning performance. In a wrapper model, one employs a learning algorithm and uses its performance to determine the quality of selected features. As shown in Chapter 2, filter and wrapper models are not confined to supervised feature selection, and can also apply to the study of unsupervised feature selection algorithms.

Search strategies [1] are investigated and various strategies are proposed including forward, backward, floating, branch-and-bound, and randomized. If one starts with an empty feature subset and adds relevant features into the subset following a procedure, it is called *forward selection*; if one begins with a full set of features and removes features procedurally, it is *backward selection*. Given a large number of features, either strategy might be too costly to work. Take the example of forward selection. Since k is usually unknown *a priori*, one needs to try $\binom{N}{1} + \binom{N}{2} + ... + \binom{N}{k}$ times in order to figure out k out of N features for selection. Therefore, its time complexity is $O(2^N)$. Hence, more efficient algorithms are developed. The widely used ones are sequential strategies. A *sequential forward selection* (SFS) algorithm selects one feature at a time until adding another feature does not improve the subset quality with the condition that a selected feature remains selected. Similarly, a *sequential backward selection* (SBS) algorithm eliminates one feature at a time and once a feature is eliminated, it will never be considered again for inclusion. Obviously, both search strategies are heuristic in nature and cannot guarantee the optimality of the selected features. Among alternatives to these strategies are randomized feature selection algorithms, which are discussed in Chapter 3. A relevant issue regarding exhaustive and heuristic searches is whether there is any reason to perform exhaustive searches if time complexity were not a concern. Research shows that exhaustive search can lead the features that exacerbate data overfitting, while heuristic search is less prone to data overfitting in feature selection, facing small data samples.

The *small sample problem* addresses a new type of "wide" data where the number of features (N) is several degrees of magnitude more than the number of instances (M). High-throughput data produced in genomics and proteomics and text data are typical examples. In connection to the curse of dimensionality mentioned earlier, the wide data present challenges to the reliable estimation of the model's performance (e.g., accuracy), model selection, and data overfitting. In [3], a pithy illustration of the small sample problem is given with detailed examples.

The evaluation of feature selection often entails two tasks. One is to compare two cases: before and after feature selection. The goal of this task is to observe if feature selection achieves its intended objectives (recall that feature selection does not confine it to improving classification performance). The aspects of evaluation can include the number of selected features, time, scalability, and learning model's performance. The second task is to compare two feature selection algorithms to see if one is better than the other for a certain task. A detailed empirical study is reported in [14]. As we know, there is no universally superior feature selection, and different feature selection algorithms have their special edges for various applications. Hence, it is wise to find a suitable algorithm for a given application. An initial attempt to address the problem of selecting feature selection algorithms is presented in [12], aiming to mitigate the increasing complexity of finding a suitable algorithm from many feature selection algorithms.

Another issue arising from feature selection evaluation is *feature selection bias*. Using the *same* training data in both feature selection and classification learning can result in this selection bias. According to statistical theory based on regression research, this bias can exacerbate data over-fitting and negatively affect classification performance. A recommended practice is to use separate data for feature selection and for learning. In reality, however, separate datasets are rarely used in the selection and learning steps. This is because we want to use as much data as possible in both selection and learning. It is against this intuition to divide the training data into two datasets leading to the reduced data in both tasks. Feature selection bias is studied in [15] to seek answers if there is discrepancy between the current practice and the statistical theory. The findings are that the statistical theory is correct, but feature selection bias has limited effect on feature selection for classification.

Recently researchers started paying attention to interacting features [7]. Feature interaction usually defies those heuristic solutions to feature selection evaluating individual features for efficiency. This is because interacting features exhibit properties that cannot be detected in individual features. One simple example of interacting features is the XOR problem, in which both features together determine the class and each individual feature does not tell much at all. By combining careful selection of a feature quality measure and design of a special data structure, one can heuristically handle some feature interaction as shown in [17]. The randomized algorithms detailed in Chapter 3 may provide an alternative. An overview of various additional issues related to improving classification performance can be found in [5]. Since there are many facets of feature selection research, we choose a theme that runs in parallel with supervised, unsupervised, and semi-supervised learning below, and discuss and illustrate the underlying concepts of disparate feature selection types, their connections, and how they can benefit from one another.

1.2 Supervised, Unsupervised, and Semi-Supervised Feature Selection

In one of the early surveys [2], all algorithms are supervised in the sense that data have class labels (denoted as X_l). Supervised feature selection algorithms rely on measures that take into account the class information. A well-known measure is *information gain*, which is widely used in both feature selection and decision tree induction. Assuming there are two features F_1 and F_2, we can calculate feature F_i's information gain as $E_0 - E_i$, where E is entropy. E_0 is the entropy before the data split using feature F_i, and can be calculated as $E_0 = \sum_c p_c \log p_c$, where p is the estimated probability of class c and $c = 1, 2, ..., C$. E_i is the entropy after the data split using F_i. A better feature can result in larger information gain. Clearly, class information plays a critical role here. Another example is the algorithm ReliefF, which also uses the class information to determine an instance's "near-hit" (a neighboring instance having the same class) and "near-miss" (a neighboring instance having different classes). More details about ReliefF can be found in Chapter 9. In essence, supervised feature selection algorithms try to find features that help *separate data of different classes* and we name it *class-based separation*. If a feature has no effect on class-based separation, it can be removed. A good feature should, therefore, help enhance class-based separation.

In the late 90's, research on unsupervised feature selection intensified in order to deal with data without class labels (denoted as X_u). It is closely related to unsupervised learning [4]. One example of unsupervised learning is clustering, where similar instances are grouped together and dissimilar ones are separated apart. Similarity can be defined by the distance between two instances. Conceptually, the two instances are similar if the distance between the two is small, otherwise they are dissimilar. When all instances are connected pair-wisely, breaking the connections between those instances that are far apart will form clusters. Hence, clustering can be thought as achieving *locality-based separation*. One widely used clustering algorithm is k-means. It is an iterative algorithm that categorizes instances into k clusters. Given predetermined k centers (or centroids), it works as follows: (1) Instances are categorized to their closest centroid, (2) the centroids are recalculated using the instances in each cluster, and (3) the first two steps are repeated until the centroids do not change. Obviously, the key concept is distance calculation, which is sensitive to dimensionality, as we discussed earlier about the curse of dimensionality. Basically, if there are many irrelevant or redundant features, clustering will be different from that with only relevant features. One toy example can be found in Figure 1.1 in which two well-formed clusters in a 1-d space (x) become two different clusters (denoted with different shapes, circles vs. diamonds) in a 2-d space after introducing an irrelevant feature y. Unsupervised feature selection is more difficult to deal with than supervised feature selection. However, it also is a very useful tool as the majority of data are unlabeled. A comprehensive introduction and review of unsupervised feature selection is presented in Chapter 2.

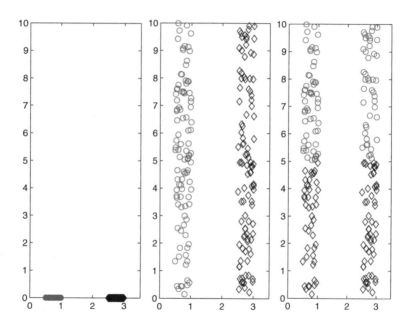

FIGURE 1.1: An illustrative example: left - two well-formed clusters; middle - after an irrelevant feature is added; right - after applying 2-means clustering.

When a small number of instances are labeled but the majority are not, semi-supervised feature selection is designed to take advantage of both the large number of unlabeled instances and the labeling information as in semi-supervised learning. Intuitively, the additional labeling information should help constrain the search space of unsupervised feature selection. In other words, semi-supervised feature selection attempts to align locality-based separation and class-based separations Since there are a large number of unlabeled data and a small number of labeled instances, it is reasonable to use unlabeled data to form some potential clusters and then employ labeled data to find those clusters that can achieve both locality-based and class-based separations. For the two possible clustering results in Figure 1.1, if we are given one correctly labeled instance each for the clusters of circles and diamonds, the correct clustering result (the middle figure) will be chosen. The idea of semi-supervised feature selection can be illustrated as in Figure 1.2 showing how the properties of X_l and X_u complement each other and work together to find relevant features. Two feature vectors (corresponding to two features, f and f') can generate respective cluster indicators representing different clustering results: The left one can satisfy both constraints of X_l and X_u, but the right one can only satisfy X_u. For semi-supervised feature selection, we want to select f over f'. In other words, there are two equally good ways to cluster the data as shown in the figure, but only one way can also attain class-based

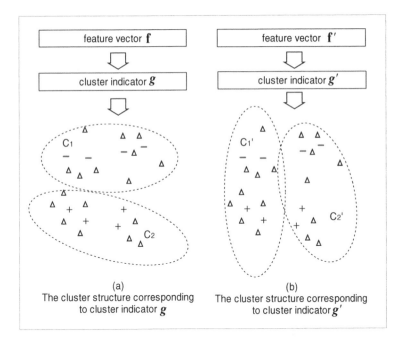

FIGURE 1.2: The basic idea for comparing the fitness of cluster indicators according to both X_l (labeled data) and X_u (unlabeled data) for semi-supervised feature selection. "-" and "+" correspond to instances of negative and positive classes, and "Δ" to unlabeled instances.

separation. A semi-supervised feature selection algorithm sSelect is proposed in [18], and sSelect is effective to use both data properties when locality-based separation and class-based separation do not generate conflicts. We expect to witness a surge of study on semi-supervised feature selection. The reason is two-fold: It is often affordable to carefully label a small number of instances, and it also provides a natural way for human experts to inject their knowledge into the feature selection process in the form of labeled instances.

Above, we presented and illustrated the development of feature selection in parallel to supervised, unsupervised, and semi-supervised learning to meet the increasing demands of labeled, unlabeled, and partially labeled data. It is just one perspective of feature selection that encompasses many aspects. However, from this perspective, it can be clearly seen that as data evolve, feature selection research adapts and develops into new areas in various forms for emerging real-world applications. In the following, we present an overview of the research activities included in this book.

1.3   Key Contributions and Organization of the Book

The ensuing chapters showcase some current research issues of feature selection. They are categorically grouped into five parts, each containing four chapters. The first chapter in Part I is this introduction. The other three discuss issues such as unsupervised feature selection, randomized feature selection, and causal feature selection. Part II reports some recent results of empowering feature selection, including active feature selection, decision-border estimate, use of ensembles with independent probes, and incremental feature selection. Part III deals with weighting and local methods such as an overview of the ReliefF family, feature selection in k-means clustering, local feature relevance, and a new interpretation of Relief. Part IV is about text feature selection, presenting an overview of feature selection for text classification, a new feature selection score, constraint-guided feature selection, and aggressive feature selection. Part V is on Feature Selection in Bioinformatics, discussing redundancy-based feature selection, feature construction and selection, ensemble-based robust feature selection, and penalty-based feature selection. A summary of each chapter is given next.

1.3.1   Part I - Introduction and Background

Chapter 2 is an overview of unsupervised feature selection, finding the smallest feature subset that best uncovers interesting, natural clusters for the chosen criterion. The existence of irrelevant features can misguide clustering results. Both filter and wrapper approaches can apply as in a supervised setting. Feature selection can either be global or local, and the features to be selected can vary from cluster to cluster. Disparate feature subspaces can have different underlying numbers of natural clusters. Therefore, care must be taken when comparing two clusters with different sets of features.

Chapter 3 is also an overview about randomization techniques for feature selection. Randomization can lead to an efficient algorithm when the benefits of good choices outweigh the costs of bad choices. There are two broad classes of algorithms: Las Vegas algorithms, which guarantee a correct answer but may require a long time to execute with small probability, and Monte Carlo algorithms, which may output an incorrect answer with small probability but always complete execution quickly. The randomized complexity classes define the probabilistic guarantees that an algorithm must meet. The major sources of randomization are the input features and/or the training examples. The chapter introduces examples of several randomization algorithms.

Chapter 4 addresses the notion of causality and reviews techniques for learning causal relationships from data in applications to feature selection. Causal Bayesian networks provide a convenient framework for reasoning about causality and an algorithm is presented that can extract causality from data by finding the Markov blanket. Direct causes (parents), direct effects (children), and other direct causes of the direct effects (spouses) are all members of the Markov blanket. Only direct causes are strongly causally relevant. The knowledge of causal relationships can benefit feature selection, e.g., explaining relevance in terms of causal mechanisms, distinguishing between actual features and experimental artifacts, predicting the consequences of actions, and making predictions in a non-stationary environment.

1.3.2   Part II - Extending Feature Selection

Chapter 5 poses an interesting problem of active feature sampling in domains where the feature values are expensive to measure. The selection of features is based on the maximum benefit. A benefit function minimizes the mean-squared error in a feature relevance estimate. It is shown that the minimum mean-squared error criterion is equivalent to the maximum average change criterion. The results obtained by using a mixture model for the joint class-feature distribution show the advantage of the active sampling policy over the random sampling in reducing the number of feature samples. The approach is computationally expensive. Considering only a random subset of the missing entries at each sampling step is a promising solution.

Chapter 6 discusses feature extraction (as opposed to feature selection) based on the properties of the decision border. It is intuitive that the direction normal to the decision boundary represents an informative direction for class discriminability and its effectiveness is proportional to the area of decision border that has the same normal vector. Based on this, a labeled vector quantizer that can efficiently be trained by the Bayes risk weighted vector quantization (BVQ) algorithm was devised to extract the best linear approximation to the decision border. The BVQ produces a decision boundary feature matrix, and the eigenvectors of this matrix are exploited to transform the original feature space into a new feature space with reduced dimensionality. It is shown that this approach is comparable to the SVM-based decision boundary approach and better than the MLP (Multi Layer Perceptron)-based approach, but with a lower computational cost.

Chapter 7 proposes to compare feature relevance against the relevance of its randomly permuted version (or probes) for classification/regression tasks using random forests. The key is to use the same distribution in generating a probe. Feature relevance is estimated by averaging the relevance obtained from each tree in the ensemble. The method iterates over the remaining features by removing the identified important features using the residuals as new target variables. It offers autonomous feature selection taking into account non-linearity, mixed-type data, and missing data in regressions and classifications. It shows excellent performance and low computational complexity, and is able to address massive amounts of data.

Chapter 8 introduces an incremental feature selection algorithm for high-dimensional data. The key idea is to decompose the whole process into feature ranking and selection. The method first ranks features and then resolves the redundancy by an incremental subset search using the ranking. The incremental subset search does not retract what it has selected, but it can decide not to add the next candidate feature, i.e., skip it and try the next according to the rank. Thus, the average number of features used to construct a learner during the search is kept small, which makes the wrapper approach feasible for high-dimensional data.

1.3.3   Part III - Weighting and Local Methods

Chapter 9 is a comprehensive description of the Relief family algorithms. Relief exploits the context of other features through distance measures and can detect highly conditionally-dependent features. The chapter explains the idea, advantages, and applications of Relief and introduces two extensions: ReliefF and RReliefF. ReliefF is for classification and can deal with incomplete data with multi-class problems. RReliefF is its extension designed for regression. The variety of the Relief family shows the general applicability of the basic idea of Relief as a non-myopic feature quality measure.

Chapter 10 discusses how to automatically determine the important features in the k-means clustering process. The weight of a feature is determined by the sum of the within-cluster dispersions of the feature, which measures its importance in clustering. A new step to calculate the feature weights is added in the iterative process in order not to seriously affect the scalability. The weight can be defined either globally (same weights for all clusters) or locally (different weights for different clusters). The latter, called subspace k-means clustering, has applications in text clustering, bioinformatics, and customer behavior analysis.

Chapter 11 is in line with Chapter 5, but focuses on local feature relevance and weighting. Each feature's ability for class probability prediction at each point in the feature space is formulated in a way similar to the weighted χ-square measure, from which the relevance weight is derived. The weight has a large value for a direction along which the class probability is not locally constant. To gain efficiency, a decision boundary is first obtained by an SVM, and its normal vector nearest to the point in query is used to estimate the weights reflected in the distance measure for a k-nearest neighbor classifier.

Chapter 12 gives further insights into Relief (refer to Chapter 9). The working of Relief is proven to be equivalent to solving an online convex optimization problem with a margin-based objective function that is defined based on a nearest neighbor classifier. Relief usually performs (1) better than other filter methods due to the local performance feedback of a nonlinear classifier when searching for useful features, and (2) better than wrapper methods due to the existence of efficient algorithms for a convex optimization problem. The weights can be iteratively updated by an EM-like algorithm, which guarantees the uniqueness of the optimal weights and the convergence. The method was further extended to its online version, which is quite effective when it is difficult to use all the data in a batch mode.

1.3.4 Part IV - Text Classification and Clustering

Chapter 13 is a comprehensive presentation of feature selection for text classification, including feature generation, representation, and selection, with illustrative examples, from a pragmatic view point. A variety of feature generating schemes is reviewed, including word merging, word phrases, character N-grams, and multi-fields. The generated features are ranked by scoring each feature independently. Examples of scoring measures are information gain, χ-square, and bi-normal separation. A case study shows considerable improvement of F-measure by feature selection. It also shows that adding two word phrases as new features generally gives good performance gain over the features comprising only selected words.

Chapter 14 introduces a new feature selection score, which is defined as the posterior probability of inclusion of a given feature over all possible models, where each model corresponds to a different set of features that includes the given feature. The score assumes a probability distribution on the words of the documents. Bernoulli and Poisson distributions are assumed respectively when only the presence or absence of a word matters and when the number of occurrences of a word matters. The score computation is inexpensive, and the value that the score assigns to each word has an appealing Bayesian interpretation when the predictive model corresponds to a naive Bayes model. This score is compared with five other well-known scores.

Chapter 15 focuses on dimensionality reduction for semi-supervised clustering where some weak supervision is available in terms of pairwise instance constraints (must-link and cannot-link). Two methods are proposed by leveraging pairwise instance constraints: pairwise constraints-guided feature projection and pairwise constraints-guided co-clustering. The former is used to project data into a lower dimensional space such that the sum-squared distance between must-link instances is minimized and the sum-squared distance between cannot-link instances is maximized. This reduces to an elegant eigenvalue decomposition problem. The latter is to use feature clustering benefitting from pairwise constraints via a constrained co-clustering mechanism. Feature clustering and data clustering are mutually reinforced in the co-clustering process.

Chapter 16 proposes aggressive feature selection, removing more than 95% features (terms) for text data. Feature ranking is effective to remove irrelevant features, but cannot handle feature redundancy. Experiments show that feature redundancy can be as destructive as noise. A new multi-stage approach for text feature selection is proposed: (1) pre-processing to remove stop words, infrequent words, noise, and errors; (2) ranking features to identify the most informative terms; and (3) removing redundant and correlated terms. In addition, term redundancy is modeled by a term-redundancy tree for visualization purposes.

1.3.5   Part V - Feature Selection in Bioinformatics

Chapter 17 introduces the challenges of microarray data analysis and presents a redundancy-based feature selection algorithm. For high-throughput data like microarrays, redundancy among genes becomes a critical issue. Conventional feature ranking algorithms cannot effectively handle feature redundancy. It is known that if there is a Markov blanket for a feature, the feature can be safely eliminated. Finding a Markov blanket is computationally heavy. The solution proposed is to use an approximate Markov blanket, in which it is assumed that the Markov blanket always consists of one feature. The features are first ranked, and then each feature is checked in sequence if it has any approximate Markov blanket in the current set. This way it can efficiently find all predominant features and eliminate the rest. Biologists would welcome an efficient filter algorithm to feature redundancy. Redundancy-based feature selection makes it possible for a biologist to specify what genes are to be included before feature selection.

Chapter 18 presents a scalable method for automatic feature generation on biological sequence data. The algorithm uses sequence components and domain knowledge to construct features, explores the space of possible features, and identifies the most useful ones. As sequence data have both compositional and positional properties, feature types are defined to capture these properties, and for each feature type, features are constructed incrementally from the simplest ones. During the construction, the importance of each feature is evaluated by a measure that best fits to each type, and low ranked features are eliminated. At the final stage, selected features are further pruned by an embedded method based on recursive feature elimination. The method was applied to the problem of splice-site prediction, and it successfully identified the most useful set of features of each type. The method can be applied to complex feature types and sequence prediction tasks such as translation start-site prediction and protein sequence classification.

Chapter 19 proposes an ensemble-based method to find robust features for biomarker research. Ensembles are obtained by choosing different alternatives at each stage of data mining: three normalization methods, two binning methods, eight feature selection methods (including different combination of search methods), and four classification methods. A total of 192 different classifiers are obtained, and features are selected by favoring frequently appearing features that are members of small feature sets of accurate classifiers. The method is successfully applied to a publicly available Ovarian Cancer Dataset, in which case the original attribute is the m/z (mass/charge) value of mass spectrometer and the value of the feature is its intensity.

Chapter 20 presents a penalty-based feature selection method, *elastic net*, for genomic data, which is a generalization of lasso (a penalized least squares method with L_1 penalty for regression). Elastic net has a nice property that irrelevant features receive their parameter estimates equal to 0, leading to sparse and easy to interpret models like lasso, and, in addition, strongly correlated relevant features are all selected whereas in lasso only one of them is selected. Thus, it is a more appropriate tool for feature selection with high-dimensional data than lasso. Details are given on how elastic net can be applied to regression, classification, and sparse eigen-gene analysis by simultaneously building a model and selecting relevant and redundant features.

1.4 Looking Ahead

Feature selection research has found applications in many fields where large (either row-wise or column-wise) volumes of data present challenges to effective data analysis and processing. As data evolve, new challenges arise and the expectations of feature selection are also elevated, due to its own success. In addition to high-throughput data, the pervasive use of Internet and Web technologies has been bringing about a great number of new services and applications, ranging from recent Web 2.0 applications to traditional Web services where multi-media data are ubiquitous and abundant. Feature selection is widely applied to find topical terms, establish group profiles, assist in categorization, simplify descriptions, facilitate personalization and visualization, among many others.

The frontier of feature selection research is expanding incessantly in answering the emerging challenges posed by the ever-growing amounts of data, multiple sources of heterogeneous data, data streams, and disparate data-intensive applications. On one hand, we naturally anticipate more research on semi-supervised feature selection, unifying supervised and unsupervised feature selection [19], and integrating feature selection with feature extraction. On the other hand, we expect new feature selection methods designed for various types of features like causal, complementary, relational, structural, and sequential features, and intensified research efforts on large-scale, distributed, and real-time feature selection. As the field develops, we are optimistic and confident that feature selection research will continue its unique and significant role in taming the data monster and helping turning data into nuggets.

References

[1] A. Blum and P. Langley. Selection of relevant features and examples in machine learning. *Artificial Intelligence*, 97:245–271, 1997.

[2] M. Dash and H. Liu. Feature selection methods for classifications. *Intelligent Data Analysis: An International Journal*, 1(3):131–156, 1997.

[3] E. Dougherty. Feature-selection overfitting with small-sample classifier design. *IEEE Intelligent Systems*, 20(6):64–66, November/December 2005.

[4] J. Dy and C. Brodley. Feature selection for unsupervised learning. *Journal of Machine Learning Research*, 5:845–889, 2004.

[5] I. Guyon and A. Elisseeff. An introduction to variable and feature selection. *Journal of Machine Learning Research (JMLR)*, 3:1157–1182, 2003.

[6] T. Hastie, R. Tibshirani, and J. Friedman. *The Elements of Statistical Learning*. Springer, 2001.

[7] A. Jakulin and I. Bratko. Testing the significance of attribute interactions. In *ICML '04: Twenty-First International Conference on Machine Learning*. ACM Press, 2004.

[8] G. John, R. Kohavi, and K. Pfleger. Irrelevant feature and the subset selection problem. In W. Cohen and H. H., editors, *Machine Learning: Proceedings of the Eleventh International Conference*, pages 121–129, New Brunswick, NJ: Rutgers University, 1994.

[9] H. Liu and H. Motoda, editors. *Feature Extraction, Construction and Selection: A Data Mining Perspective*. Boston: Kluwer Academic Publishers, 1998. 2nd Printing, 2001.

[10] H. Liu and H. Motoda. *Feature Selection for Knowledge Discovery & Data Mining*. Boston: Kluwer Academic Publishers, 1998.

[11] H. Liu and H. Motoda, editors. *Instance Selection and Construction for Data Mining.* Boston: Kluwer Academic Publishers, 2001.

[12] H. Liu and L. Yu. Toward integrating feature selection algorithms for classification and clustering. *IEEE Trans. on Knowledge and Data Engineering*, 17(3):1–12, 2005.

[13] T. Mitchell. *Machine Learning.* New York: McGraw-Hill, 1997.

[14] P. Refaeilzadeh, L. Tang, and H. Liu. On comparison of feature selection algorithms. In *AAAI 2007 Workshop on Evaluation Methods for Machine Learning II, Vancouver, British Columbia, Canada*, July 2007.

[15] S. Singhi and H. Liu. Feature subset selection bias for classification learning. In *International Conference on Machine Learning*, 2006.

[16] L. Yu and H. Liu. Efficient feature selection via analysis of relevance and redundancy. *Journal of Machine Learning Research (JMLR)*, 5(Oct):1205–1224, 2004.

[17] Z. Zhao and H. Liu. Searching for interacting features. In *Proceedings of IJCAI - International Joint Conference on AI*, January 2007.

[18] Z. Zhao and H. Liu. Semi-supervised feature selection via spectral analysis. In *Proceedings of SIAM International Conference on Data Mining (SDM-07)*, 2007.

[19] Z. Zhao and H. Liu. Spectral feature selection for supervised and unsupervised learning. In *Proceedings of International Conference on Machine Learning*, 2007.

Chapter 2

Unsupervised Feature Selection

Jennifer G. Dy

Northeastern University

2.1 Introduction

Many existing databases are unlabeled, because large amounts of data make it difficult for humans to manually label the categories of each instance. Moreover, human labeling is expensive and subjective. Hence, unsupervised learning is needed. Besides being unlabeled, several applications are characterized by high-dimensional data (e.g., text, images, gene). However, not all of the features domain experts utilize to represent these data are important for the learning task. We have seen the need for feature selection in the supervised learning case. This is also true in the unsupervised case. Unsupervised means there is no teacher, in the form of class labels. One type of unsupervised learning problem is clustering. The goal of clustering is to group "similar" objects together. "Similarity" is typically defined in terms of a metric or a probability density model, which are both dependent on the features representing the data.

In the supervised paradigm, feature selection algorithms maximize some function of prediction accuracy. Since class labels are available in supervised learning, it is natural to keep only the features that are related to or lead to these classes. But in unsupervised learning, we are not given class labels. Which features should we keep? Why not use all the information that we have? The problem is that not all the features are important. Some of the features may be redundant and some may be irrelevant. Furthermore, the existence of several irrelevant features can misguide clustering results. Reducing

19

the number of features also facilitates comprehensibility and ameliorates the problem that some unsupervised learning algorithms break down with high-dimensional data. In addition, for some applications, the goal is not just clustering, but also to find the important features themselves.

A reason why some clustering algorithms break down in high dimensions is due to the curse of dimensionality [3]. As the number of dimensions increases, a fix data sample becomes exponentially sparse. Additional dimensions increase the volume exponentially and spread the data such that the data points would look equally far. Figure 2.1 (a) shows a plot of data generated from a uniform distribution between 0 and 2 with 25 instances in one dimension. Figure 2.1 (b) shows a plot of the same data in two dimensions, and Figure 2.1 (c) displays the data in three dimensions. Observe that the data become more and more sparse in higher dimensions. There are 12 samples that fall inside the unit-sized box in Figure 2.1 (a), seven samples in (b) and two in (c). The sampling density is proportional to $M^{1/N}$, where M is the number of samples and N is the dimension. For this example, a sampling density of 25 in one dimension would require $25^3 = 125$ samples in three dimensions to achieve a similar sample density.

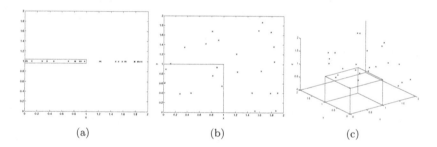

| (a) | (b) | (c) |

FIGURE 2.1: Illustration for the curse of dimensionality. These are plots of a 25-sample data generated from a uniform distribution between 0 and 2. (a) Plot in one dimension, (b) plot in two dimensions, and (c) plot in three dimensions. The boxes in the figures show unit-sized bins in the corresponding dimensions. Note that data are more sparse with respect to the unit-sized volume in higher dimensions. There are 12 samples in the unit-sized box in (a), 7 samples in (b), and 2 samples in (c).

As noted earlier, supervised learning has class labels to guide the feature search. In unsupervised learning, these labels are missing, and in fact its goal is to find these grouping labels (also known as cluster assignments). Finding these cluster labels is dependent on the features describing the data, thus making feature selection for unsupervised learning difficult.

Dy and Brodley [14] define the goal of feature selection for unsupervised learning as:

to find the smallest feature subset that best uncovers "interesting natural" groupings (clusters) from data according to the chosen criterion.

Without any labeled information, in unsupervised learning, we need to make some assumptions. We need to define what "interesting" and "natural" mean in the form of criterion or objective functions. We will see examples of these criterion functions later in this chapter.

Before we proceed with how to do feature selection on unsupervised data, it is important to know the basics of clustering algorithms. Section 2.2 briefly describes clustering algorithms. In Section 2.3 we review the basic components of feature selection algorithms. Then, we present the methods for unsupervised feature selection in Sections 2.4 and 2.5, and finally provide a summary in Section 2.6.

2.2 Clustering

The goal of clustering is to group similar objects together. There are two types of clustering approaches: partitional and hierarchical. Partitional clustering provides one level of clustering. Hierarchical clustering, on the other hand, provides multiple levels (hierarchy) of clustering solutions. Hierarchical approaches can proceed bottom-up (agglomerative) or top-down (divisive). Bottom-up approaches typically start with all instances as clusters and then, at each level, merge clusters that are most similar with each other. Top-down approaches divide the data into k clusters at each level. There are several methods for performing clustering. A survey of these algorithms can be found in [29, 39, 18].

In this section we briefly present two popular partitional clustering algorithms: k-means and finite mixture model clustering. As mentioned earlier, similarity is typically defined by a metric or a probability distribution. K-means is an approach that uses a metric, and finite mixture models define similarity by a probability density.

Let us denote our dataset as $X = \{x_1, x_2, \ldots, x_M\}$. X consists of M data instances x_k, $k = 1, 2, \ldots, M$, and each x_k represents a single N-dimensional instance.

2.2.1 The K-Means Algorithm

The goal of k-means is to partition X into K clusters $\{C_1, \ldots, C_K\}$. The most widely used criterion function for the k-means algorithm is the sum-squared-error (SSE) criterion. SSE is defined as

$$SSE = \sum_{j=1}^{K} \sum_{x_k \in C_j} \|x_k - \mu_j\|^2 \tag{2.1}$$

where μ_j denotes the mean (centroid) of those instances in cluster C_j.

K-means is an iterative algorithm that locally minimizes the SSE criterion. It assumes each cluster has a hyper-spherical structure. "K-means" denotes the process of assigning each data point, x_k, to the cluster with the nearest mean. The k-means algorithm starts with initial K centroids, then it assigns each remaining point to the nearest centroid, updates the cluster centroids, and repeats the process until the K centroids do not change (convergence). There are two versions of k-means: One version originates from Forgy [17] and the other version from Macqueen [36]. The difference between the two is when to update the cluster centroids. In Forgy's k-means [17], cluster centroids are re-computed after all the data points have been assigned to their nearest centroids. In Macqueen's k-means [36], the cluster centroids are re-computed after each data assignment. Since k-means is a greedy algorithm, it is only guaranteed to find a local minimum, the solution of which is dependent on the initial assignments. To avoid local optimum, one typically applies random restarts and picks the clustering solution with the best SSE. One can refer to [47, 4] for other ways to deal with the initialization problem.

Standard k-means utilizes Euclidean distance to measure dissimilarity between the data points. Note that one can easily create various variants of k-means by modifying this distance metric (e.g., other L_p norm distances) to ones more appropriate for the data. For example, on text data, a more suitable metric is the cosine similarity. One can also modify the objective function, instead of SSE, to other criterion measures to create other clustering algorithms.

2.2.2   Finite Mixture Clustering

A finite mixture model assumes that data are generated from a mixture of K component density functions, in which $p(x_k|\theta_j)$ represents the density function of component j for all $j's$, where θ_j is the parameter (to be estimated) for cluster j. The probability density of data x_k is expressed by

$$p(x_k) = \sum_{j=1}^{K} \alpha_j p(x_k|\theta_j) \tag{2.2}$$

where the $\alpha's$ are the mixing proportions of the components (subject to $\alpha_j \geq 0$ and $\sum_{j=1}^{K} \alpha_j = 1$). The log-likelihood of the M observed data points is then given by

$$\mathcal{L} = \sum_{k=1}^{M} \ln\{\sum_{j=1}^{K} \alpha_j p(x_k|\theta_j)\} \tag{2.3}$$

It is difficult to directly optimize (2.3), therefore we apply the Expectation-Maximization (EM) [10] algorithm to find a (local) maximum likelihood or maximum a posteriori (MAP) estimate of the parameters for the given data set. EM is a general approach for estimating the maximum likelihood or MAP estimate for missing data problems. In the clustering context, the missing or hidden variables are the class labels. The EM algorithm iterates between an Expectation-step (E-step), which computes the expected complete data log-likelihood given the observed data and the model parameters, and a Maximization-step (M-step), which estimates the model parameters by maximizing the expected complete data log-likelihood from the E-step, until convergence. In clustering, the E-step is similar to estimating the cluster membership and the M-step estimates the cluster model parameters. The clustering solution that we obtain in a mixture model is what we call a "soft"-clustering solution because we obtain an estimated cluster membership (i.e., each data point belongs to all clusters with some probability weight of belonging to each cluster). In contrast, k-means provides a "hard"-clustering solution (i.e., each data point belongs to only a single cluster).

Analogous to metric-based clustering, where one can develop different algorithms by utilizing other similarity metric, one can design different probability-based mixture model clustering algorithms by choosing an appropriate density model for the application domain. A Gaussian distribution is typically utilized for continuous features and multinomials for discrete features. For a more thorough description of clustering using finite mixture models, see [39] and a review is provided in [18].

2.3 Feature Selection

Feature selection algorithms has two main components: (1) feature search and (2) feature subset evaluation.

2.3.1 Feature Search

Feature search strategies have been widely studied for classifications. Generally speaking, search strategies used for supervised classifications can also be used for clustering algorithms. We repeat and summarize them here for completeness. An exhaustive search would definitely find the optimal solution; however, a search on 2^N possible feature subsets (where N is the number of features) is computationally impractical. More realistic search strategies have been studied. Narendra and Fukunaga [40] introduced the branch and bound algorithm, which finds the optimal feature subset if the criterion function used is monotonic. However, although the branch and bound algorithm makes problems more tractable than an exhaustive search, it becomes impractical for feature selection problems involving more than 30 features [43]. Sequential search methods generally use greedy techniques and hence do not guarantee global optimality of the selected subsets, only local optimality. Examples of sequential searches include sequential forward selection, sequential backward elimination, and bidirectional selection [32, 33]. Sequential forward/backward search methods generally result in an $O(N^2)$ worst case search. Marill and Green [38] introduced the sequential backward selection (SBS) [43] method, which starts with all the features and sequentially eliminates one feature at a time (eliminating the feature that contributes least to the criterion function). Whitney [50] introduced sequential forward selection (SFS) [43], which starts with the empty set and sequentially adds one feature at a time. A problem with these hill-climbing search techniques is that when a feature is deleted in SBS, it cannot be re-selected, while a feature added in SFS cannot be deleted once selected. To prevent this effect, the Plus-l-Minus-r (l-r) search method was developed by Stearns [45]. Indeed, at each step the values of l and r are pre-specified and fixed. Pudil et al. [43] introduced an adaptive version that allows l and r values to "float." They call these methods floating search methods: sequential forward floating selection (SFFS) and sequential backward floating selection (SBFS) based on the dominant search method (i.e., either in the forward or backward direction). Random search methods such as genetic algorithms and random mutation hill climbing add some randomness in the search procedure to help to escape from a local optimum. In some cases when the dimensionality is very high, one can only afford an individual search. Individual search methods evaluate each feature individually according to a criterion or a condition [24]. They then select features, which either satisfy the condition or are top-ranked.

2.3.2   Feature Evaluation

Not all the features are important. Some of the features may be irrelevant and some of the features may be redundant. Each feature or feature subset needs to be evaluated based on importance by a criterion. Different criteria may select different features. It is actually deciding the evaluation criteria that makes feature selection in clustering difficult. In classification, it is natural to keep the features that are related to the labeled classes. However, in clustering, these class labels are not available. Which features should we keep? More specifically, how do we decide which features are relevant/irrelevant, and which are redundant?

Figure 2.2 gives a simple example of an irrelevant feature for clustering. Suppose data have features F_1 and F_2 only. Feature F_2 does not contribute to cluster discrimination, thus, we consider feature F_2 to be irrelevant. We want to remove irrelevant features because they may mislead the clustering algorithm (especially when there are more irrelevant features than relevant ones). Figure 2.3 provides an example showing feature redundancy. Observe

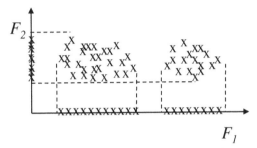

FIGURE 2.2: In this example, feature F_2 is irrelevant because it does not contribute to cluster discrimination.

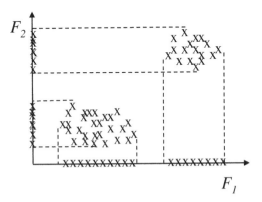

FIGURE 2.3: In this example, features F_1 and F_2 have redundant information, because feature F_1 provides the same information as feature F_2 with regard to discriminating the two clusters.

that both features F_1 and F_2 lead to the same clustering results. Therefore, we consider features F_1 and F_2 to be redundant.

2.4   Feature Selection for Unlabeled Data

There are several feature selection methods for clustering. Similar to supervised learning, these feature selection methods can be categorized as either filter or wrapper approaches [33] based on whether the evaluation methods depend on the learning algorithms[1].

As Figure 2.4 shows, the wrapper approach wraps the feature search around the learning algorithms that will ultimately be applied, and utilizes the learned results to select the features. On the other hand, as shown in Figure 2.5, the filter approach utilizes the data alone to decide which features should be kept,

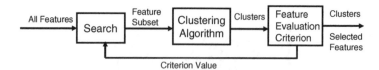

FIGURE 2.4: Wrapper approach for feature selection for clustering.

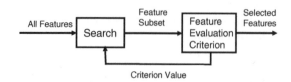

FIGURE 2.5: Filter approach for feature selection for clustering.

without running the learning algorithm. Usually, a wrapper approach may lead to better performance compared to a filter approach for a particular learning algorithm. However, wrapper methods are more computationally expensive since one needs to run the learning algorithm for every candidate feature subset.

In this section, we present the different methods categorized into filter and wrapper approaches.

2.4.1   Filter Methods

Filter methods use some intrinsic property of the data to select features without utilizing the clustering algorithm that will ultimately be applied. The basic components in filter methods are the feature search method and the feature selection criterion. Filter methods have the challenge of defining feature relevance (interestingness) and/or redundancy without applying clustering on the data.

Talavera [48] developed a filter version of his wrapper approach that selects features based on feature dependence. He claims that irrelevant features are features that do not depend on the other features. Manoranjan et al. [37] introduced a filter approach that selects features based on the entropy of distances between data points. They observed that when the data are clustered, the distance entropy at that subspace should be low. He, Cai, and Niyogi [26] select features based on the Laplacian score that evaluates features based on their locality preserving power. The Laplacian score is based on the premise that two data points that are close together probably belong to the same cluster.

These three filter approaches try to remove features that are not relevant.

Another way to reduce the dimensionality is to remove redundancy. A filter approach primarily for reducing redundancy is simply to cluster the features. Note that even though we apply clustering, we consider this as a filter method because we cluster on the feature space as opposed to the data sample space. One can cluster the features using a k-means clustering [36, 17] type of algorithm with feature correlation as the similarity metric. Instead of a cluster mean, represent each cluster by the feature that has the highest correlation among features within the cluster it belongs to.

Popular techniques for dimensionality reduction without labels are principal components analysis (PCA) [30], factor analysis, and projection pursuit [20, 27]. These early works in data reduction for unsupervised data can be thought of as filter methods, because they select the features prior to applying clustering. But rather than selecting a subset of the features, they involve some type of feature transformation. PCA and factor analysis aim to reduce the dimension such that the representation is as faithful as possible to the original data. As such, these techniques aim at reducing dimensionality by removing redundancy. Projection pursuit, on the other hand, aims at finding "interesting" projections (defined as the directions that are farthest from Gaussian distributions and close to uniform). In this case, projection pursuit addresses relevance. Another method is independent component analysis (ICA) [28]. ICA tries to find a transformation such that the transformed variables are statistically independent. Although the goals of ICA and projection pursuit are different, the formulation in ICA ends up being similar to that of projection pursuit (i.e., they both search for directions that are farthest from the Gaussian density). These techniques are filter methods, however, they apply transformations on the original feature space. We are interested in subsets of the original features, because we want to retain the original meaning of the features. Moreover, transformations would still require the user to collect all the features to obtain the reduced set, which is sometimes not desired.

2.4.2 Wrapper Methods

Wrapper methods apply the clustering algorithm to evaluate the features. They incorporate the clustering algorithm inside the feature search and selection. Wrapper approaches consist of: (1) a search component, (2) a clustering algorithm, and (3) a feature evaluation criterion. See Figure 2.4.

One can build a feature selection wrapper approach for clustering by simply picking a favorite search method (any method presented in Section 2.3.1), and apply a clustering algorithm and a feature evaluation criterion. However, there are issues that one must take into account in creating such an algorithm. In [14], Dy and Brodley investigated the issues involved in creating a general wrapper method where any feature selection, clustering, and selection criteria can be applied. The first issue they observed is that it is not a good idea to use the same number of clusters throughout the feature search because different feature subspaces have different underlying numbers of "natural" clusters. Thus, the clustering algorithm should also incorporate finding the number of clusters in feature search. The second issue they discovered is that various selection criteria are biased with respect to dimensionality. They then introduced a cross-projection normalization scheme that can be utilized by any criterion function.

Feature subspaces have different underlying numbers of clusters. When we are searching for the best feature subset, we run into a new problem: The value of the number of clusters depends on the feature subset. Figure 2.6 illustrates this point. In two dimensions $\{F_1, F_2\}$ there are three clusters, whereas in one dimension (the projection of the data only on F_1) there are only two clusters. It is not a good idea to use a fixed number of clusters in feature search, because different feature subsets require different numbers of clusters. And, using a fixed number of clusters for all feature sets does not model the data in the respective subspace correctly. In [14], they addressed finding the number of clusters by applying a Bayesian information criterion penalty [44].

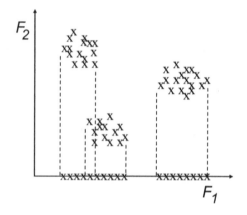

FIGURE 2.6: The number of cluster components varies with dimension.

Feature evaluation criterion should not be biased with respect to dimensionality. In a wrapper approach, one searches in feature space, applies clustering in each candidate feature subspace, S_i, and then evaluates the results (clustering in space S_i) with other cluster solutions in other subspaces, S_j, $j \neq i$, based on an evaluation criterion. This can be problematic especially when S_i and S_j have different dimensionalities. Dy and Brodley [14] examined two feature selection criteria: maximum likelihood and scatter separability. They have shown that the scatter separability criterion prefers higher dimensionality. In other words, the criterion value monotonically increases as features are added (i.e., the dimension is increased) assuming identical clustering assignments [40]. However, the separability criterion may not be monotonically increasing with respect to dimension when the clustering assignments change. Scatter separability or the trace criterion prefers higher dimensions, intuitively, because data are more scattered in higher dimensions, and mathematically, because adding features means adding more terms in the trace function. Ideally, one would like the criterion value to remain the same if the discrimination information is the same. Maximum likelihood, on the other hand, prefers lower dimensions. The problem occurs when we compare feature set A with feature set B wherein set A is a subset of B. The problem is that the joint probability of a single point $\{x, y\}$ is less than or equal to its marginal probability x when the conditional probability is less than one. For sequential searches, this can lead to the trivial result of selecting only a single feature.

To ameliorate this bias, Dy and Brodley [14] suggest a cross-projection scheme that can be applied with any feature evaluation criterion. The idea is to project the cluster solution to the subspaces that we are comparing, because the ultimate goal is to find the subspace that yields good clustering. Given two feature subsets, S_1 and S_2, with different dimensions, clustering our data using subset S_1 produces cluster C_1. In the same way, we obtain the clusters C_2 using the features in S_2. Which feature subset, S_1 or S_2, enables us to discover better clusters? Let $CRIT(S_i, C_j)$ be the feature selection criterion value using feature subset F_i to represent the data and C_j as the clustering assignment. $CRIT(\cdot)$ represents any criterion (e.g., maximum likelihood, scatter separability). Normalize the criterion value for S_1, C_1 as

$$normalizedValue(S_1, C_1) = CRIT(S_1, C_1) \cdot CRIT(S_2, C_1)$$

and the criterion value for S_2, C_2 as

$$normalizedValue(S_2, C_2) = CRIT(S_2, C_2) \cdot CRIT(S_1, C_2).$$

If $normalizedValue(S_i, C_i) > normalizedValue(S_j, C_j)$, we choose feature subset S_i. When the normalized criterion values are equal for S_i and S_j, we favor the subset with the lower cardinality. Another way to normalize the bias of a feature evaluation criterion with respect to dimensionality is to measure the criterion function of the clustering solution obtained by any subset S_i onto the set of all of the original features. This way, one can compare any candidate subset.

Now, one can build any feature selection wrapper approach for unlabeled data, by performing any favorite feature search, clustering, and evaluation criterion, and take these two issues into account.

For wrapper approaches, the clustering method deals with defining a "similarity" metric or defines what "natural" means. The feature selection criterion defines what "interestingness" means. These two criteria need not be the same. Typically one should choose an appropriate clustering algorithm (which is defined by a clustering objective function and a similarity metric) based on a problem domain. For example, an appropriate metric for text data might be the cosine similarity or a mixture of multinomial model for clustering. For data described by continuous features, one might define Gaussian clusters as the "natural" groups. The feature evaluation criterion should quantify what type of features the user is interested in. If the user wishes to find features that optimize the clustering algorithm in finding the natural clusters, then an appropriate criterion for feature evaluation is the same criterion as the objective function for the clustering algorithm. If the user is interested in features that find clusters that are well-separated, then criteria such as scatter separability are appropriate. Unlike supervised learning, which has class labels to guide the feature search, unsupervised feature selection relies on criterion functions and would thus require domain knowledge to choose the appropriate objective functions.

Dy and Brodley [14] examined two feature selection criteria: maximum likelihood and scatter separability, for a wrapper method that applies a sequential forward search wrapped around Gaussian mixture model clustering. Recall that to cluster data, we need to make assumptions and define what "natural" grouping means. Note that with this model, the assumption is that each of the "natural" groups is Gaussian. To evaluate the feature subset, they tried maximum likelihood and scatter separability. Here, they tried to define what "interestingness" means. Maximum likelihood (ML) is the same criterion used in the clustering algorithm. ML prefers the feature subspace that can be modeled best as a Gaussian mixture. They also explored scatter separability, because it can be used with many clustering algorithms. Scatter separability is similar to the criterion function used in discriminant analysis. It measures how far apart the clusters are from each other normalized by their within cluster distance. High values of ML and scatter separability are desired. The conclusion was that no one criterion is best for all applications. For an image retrieval application, Dy et al. [15] applied a sequential forward search wrapped around Gaussian mixture model clustering and the scatter separability for feature evaluation. The features were continuous valued image features; hence, the choice of the Gaussian mixture model for clustering, and since the goal was to retrieve similar images from the same cluster, the separability criterion was chosen for selecting the features.

Gennari [22] incorporated feature selection (they call "attention") to CLASSIT (an incremental concept formation hierarchical clustering algorithm). The attention algorithm inspects the features starting with the most salient ("per-attribute contribution to category utility") attribute to the least salient attribute, and stops inspecting features if the remaining features do not change the current clustering decision. The purpose of this attention mechanism is to increase efficiency without loss of prediction accuracy. Devaney and Ram [11] applied both sequential forward and backward selection to search the feature space and hierarchically clustered the data using COBWEB as the induction algorithm for each candidate feature subset, and evaluated these feature subsets using the category utility metric (COBWEB's cluster criterion) as the feature selection criterion function. To improve the efficiency of the feature subset search, they introduced AICC, which is an attribute-incremental concept learner for COBWEB that learns an $n + 1$ (or $n - 1$) descriptor concept hierarchy using the existing n-descriptor hierarchy and the new feature to add (or remove). Talavera [48] applied filter and wrapper approaches to COBWEB, and used a feature dependence measure to select features. Vaithyanathan and Dom [49] proposed a probabilistic objective function for both feature selection and clustering, and applied it to text. They modeled the text data as a mixture of multinomials and used a Bayesian approach to estimate the parameters. To search the feature space, they applied distributional clustering to pre-select candidate subsets and then picked the candidate subset that led to the largest value in the objective function. Vaithyanathan and Dom [49] incorporated finding the number of clusters in their Bayesian formulation. They address dimensionality bias by formulating the objective function as the integrated likelihood of the joint distribution of the relevant and irrelevant features and assumed the relevant and irrelevant features as conditionally independent given the class. The dimensionality of the objective function will be equal to the original number of features no matter how many relevant features there are. Kim, Street, and Menczer [31] applied an evolutionary local selection algorithm (ELSA) to search the feature subset and number of clusters on two clustering algorithms: k-means and Gaussian mixture clustering (with diagonal covariances), and a Pareto front to combine multiple objective evaluation functions. Law, Figueiredo, and Jain [34] added feature saliency, a measure of feature relevance, as a missing variable to a probabilistic objective function. The objective function was similar to that in [49] (i.e., the objective function modeled relevant features as conditionally independent given the cluster component label, and irrelevant features with a probability density identical for all components). To add feature saliency, they utilized the conditional feature independence assumption to build their model. Then, they derived an Expectation-Maximization (EM) [10] algorithm to estimate the feature saliency for a mixture of Gaussians. Law, Figueiredo, and Jain's [34] method is able to find the features and clusters simultaneously through a single EM run. They also developed a wrapper approach that selects features using Kullback-Leibler divergence and entropy. They address finding the number of clusters with a minimum message length criterion and dimensionality bias by formulating the objective function as the likelihood of the data for both the relevant and irrelevant features similar to [49].

2.5   Local Approaches

Unsupervised feature selection algorithms can be categorized as filter or wrapper approaches. Another way to group the methods are based on whether the approach is global or local. Global methods select a single set of features for all the clusters. Local methods select subsets of features associated with each cluster. The feature subsets for each cluster can be different. All the methods presented earlier are global methods. In this section, we present two types of local unsupervised feature selection approaches: subspace clustering and co-clustering.

2.5.1   Subspace Clustering

As local methods, subspace clustering evaluates features only from each cluster, as opposed to global methods that evaluate features from all the data (all clusters). Typical subspace clustering approaches measure the existence of a cluster in a feature subspace based on density. They take advantage of the downward closure property of density to reduce the search space. The downward closure property of density states that if there are dense units in d dimensions, then there are dense units in all $d-1$ dimensional projections. One can start from one dimension going up until no more dense units are found. When no more dense units are found, the algorithm combines adjacent dense units to form clusters. Density is measured by creating histograms in each dimension and measuring the density within each bin. A unit is considered dense if its density is higher than a user-defined threshold. Thus, the quality of clusters found is dependent on tuning the density thresholds and grid size, which can be difficult to set.

One of the first subspace clustering algorithm is CLIQUE [1]. Here is where the term subspace clustering was coined. CLIQUE proceeds level-by-level from one feature to the highest dimension or until no more feature subspaces with clusters (regions with high density points) are generated. The idea is that dense clusters in dimensionality d should remain dense in $d-1$. Once the dense units are found, CLIQUE keeps the units with the high coverage (fraction of the dataset covered by the dense units). Then, clusters are found by combining adjacent dense and high-coverage units. By combining adjacent units, CLIQUE is capable of discovering irregular-shaped clusters, and points can belong to multiple clusters. CLIQUE allows one to discover different clusters from various subspaces and combine the results.

Several new subspace clustering methods were developed after CLIQUE. ENCLUS [7] is similar to CLIQUE except that it measures entropy rather than density. A subspace with clusters typically has lower entropy than those without clusters. MAFIA [23] is an extension of CLIQUE that enables the grid-size to be adaptive. Other approaches that adaptively determine the grid-size are CBF [6], CLTree[35], and DOC [42]. To learn more about subspace clustering, there is a survey in [41].

2.5.2   Co-Clustering/Bi-Clustering

As mentioned earlier, one can perform feature selection by clustering in feature space to reduce redundancy. An approach called co-clustering initially inspired by Hartigan [25] has become recently popular due to research on microarray analysis. Co-clustering tries to find the coherence exhibited by the subset of instances on the subset of features. In microarray analysis, one may want to find the genes that respond similarly to the environment conditions; in text clustering, one may wish to find the co-occurence of words and documents. Co-clustering, also known as bi-clustering, is simply the clustering of both the row (sample space) and column (feature space) simultaneously. The algorithms for performing co-clustering typically quantify the quality of a co-clustering as a measure of the approximation error between the original data matrix and the reconstructed matrix from a co-clustering. And the techniques to solve this problem alternate clustering the rows and the columns to find the co-clusters. Dhillon, Mallela, and Modha [12] introduced an information theoretic formulation for co-clustering. The objective is to find the clusters that minimizes the loss in mutual information subject to the constraints that the numbers of row and column clusters are held fix. Banerjee et al. [2] provide a generalized approach for co-clustering such that any Bregman divergence [5] can be used in the objective function. Bregman divergence covers a large class of divergence measures, which include the Kullback-Liebler divergence and the squared Euclidean distance. They show that the update steps that alternately update the row and column cluster and the minimum Bregman solution will progressively decrease the matrix approximation error and lead to a locally optimal co-clustering solution. The general method is; (1) Start with an arbitrary row and column clustering, compute the approximation matrix; (2) hold the column clustering fixed and update the row clusters, then compute a new approximation matrix; (3) hold the row clustering fixed and update the column clusters, then compute a new approximation matrix, and repeat steps (2) and (3) until convergence.

Cheng and Church [8] and Cho et al. [9] developed bi-clustering algorithms that utilize the squared Euclidean distance. The δ-cluster algorithm [51] is another bi-clustering algorithm. It swaps attributes and data points iteratively to find a solution that leads to the highest coherence that a particular attribute or instance brings to the cluster, where coherence is measured by the Pearson correlation. Friedman and Meulman [21] designed a distance measure for attribute-value data for clustering on subsets of attributes, and allows feature subsets for each cluster to be different. Their algorithm, COSA, starts by initializing the weights for the features; it then clusters the data based on these weights and recompute the weights until the solution stabilizes. The weight update increases the weight on attributes with smaller dispersion within each group, where the degree of this increase is controlled by a parameter λ. The cluster update minimizes a criterion that minimizes the inverse exponential mean with separate attribute weighting within each cluster.

2.6   Summary

For a fixed amount of data samples, the higher the dimension, the more sparse the data space is. The data points in high dimensions would look equally far. Because of this, many clustering algorithms break down in high dimensions. In addition, usually not all the features are important – some are redundant and some are irrelevant. Data with several irrelevant features can misguide the clustering results. There are two ways to reduce the dimensionality: feature transformation and feature selection. Feature transformation reduces the dimension by applying some type of linear or non-linear function on the original features, whereas feature selection selects a subset of the original features. One may wish to perform feature selection rather than transformation because one may wish to keep the original meaning of the features. Furthermore, after feature selection, one does not need to measure the features that are not selected. Feature transformation, on the other hand, still needs all the features to extract the reduced dimensions.

This chapter presents a survey of methods to perform feature selection on unsupervised data. One can select a global set of features or a local set. Global means that one selects a single subset of features that clusters the data. Local means that different sets of features are chosen for each cluster. Global feature selection methods can be classified as a filter or a wrapper approach. Filter does not take into account the final clustering algorithm in evaluating features whereas wrapper incorporates the clustering inside the feature search and selection. Local feature selection methods include subspace clustering and co-clustering approaches. Subspace clustering tries to find the clusters hidden in high-dimensional data by proceeding from one dimension going up to higher dimensions and searching for high density regions. Subspace clustering can find clusters in overlapping subspaces, the points can also belong to multiple clusters, and, for the methods presented here, because they connect adjacent regions to form clusters, they can also discover irregularly shaped clusters. Co-clustering simultaneously finds feature subsets and clusters by alternating clustering the rows and the columns.

The key to feature selection in clustering is defining what feature relevance and redundancy mean. Different researchers introduced varying criteria for feature selection. To define interestingness and relevance, measures such as scatter separability, entropy, category utility, maximum likelihood, density, and consensus have been proposed. Redundancy is implicitly handled by the search process (e.g., when adding new features do not change the evaluation criterion), or explicitly through feature correlation, or through compression techniques. Defining interestingness is really difficult because it is a relative concept. Given the same data, what is interesting to a physician will be different from what is interesting to an insurance company. Thus, no single criterion is best for all applications. This led to research work on visualization as a guide to feature search [13]. This led Kim, Street, and Menczer [31] to optimize multi-objective criteria. This difficulty of defining interestingness also led to work in looking at ensembles of clusters from different projections (or feature subspaces) and applying a consensus of solutions to provide the final clustering [16, 46, 19]. Another avenue for research, to aid in defining interestingness, is semi-supervised feature selection. Knowing a few labeled points or constrained must-link and cannot-link pairs can help guide the feature search.

Acknowledgment

This research was supported by NSF CAREER IIS-0347532.

Notes

1 When discussing filters and wrappers, approach, method, and model are used exchangeably.

References

[1] R. Agrawal, J. Gehrke, D. Gunopulos, and P. Raghavan. Automatic subspace clustering of high dimensional data for data mining applications. In *Proceedings ACM SIGMOD International Conference on Management of Data*, pages 94–105, Seattle, WA, ACM Press, June 1998.

[2] A. Banerjee, I. S. Dhillon, J. Ghosh, S. Merugu, and D. S. Modha. A generalized maximum entropy approach to bregman co-clustering and matrix approximations. In *Proceedings of the Tenth ACM SIGKDD International Conference on Knowledge Discovery and Data Mining*, pages 509–514, August 2004.

[3] R. E. Bellman. *Adaptive Control Processes*. Princeton University Press, Princeton, NJ, 1961.

[4] P. S. Bradley and U. M. Fayyad. Refining initial points for K-means clustering. In *Proceedings of the Fifteenth International Conference on Machine Learning*, pages 91–99, San Francisco, CA, Morgan Kaufmann, 1998.

[5] Y. Censor and S. Zenios. *Parallel Optimization: Theory, Algorithms, and Applications.* Oxford University Press, 1998.

[6] J.-W. Chang and D.-S. Jin. A new cell-based clustering method for large, high-dimensional data in data mining applications. In *Proceedings of the 2002 ACM Symposium on Applied Computing*, pages 503–507. ACM Press, 2002.

[7] C. H. Cheng, A. W. Fu, and Y. Zhang. Entropy-based subspace clustering for mining numerical data. In *Proceedings of the fifth ACM SIGKDD International Conference on Knowledge Discovery and Data Mining*, pages 84–93. ACM Press, August 1999.

[8] Y. Cheng and G. M. Church. Biclustering of expression data. In *Proceedings of the eighth International Conference on Intelligent Systems for Molecular Biology*, pages 93–103, 2000.

[9] H. Cho, I. S. Dhillon, Y. Guan, and S. Sra. Minimum sum-squared residue co-clustering of gene expression data. In *Proceedings of the Fourth SIAM International Conference on Data Mining*, pages 114–125, 2004.

[10] A. P. Dempster, N. M. Laird, and D. B. Rubin. Maximum likelihood from incomplete data via the em algorithm. *Journal Royal Statistical Society, Series B*, 39(1):1–38, 1977.

[11] M. Devaney and A. Ram. Efficient feature selection in conceptual clustering. In *Proceedings of the Fourteenth International Conference on Machine Learning*, pages 92–97, Nashville, TN, Morgan Kaufmann, 1997.

[12] I. S. Dhillon, S. Mallela, and D. S. Modha. Information-theoretic co-clustering. In *Proceedings of the ninth ACM SIGKDD International Conference on Knowledge Discovery and Data Mining (KDD)*, pages 89–98, August 2003.

[13] J. G. Dy and C. E. Brodley. Interactive visualization and feature selection for unsupervised data. In *Proceedings of the Sixth ACM SIGKDD International Conference on Knowledge Discovery and Data Mining*, pages 360–364, Boston, MA, ACM Press, August 2000.

[14] J. G. Dy and C. E. Brodley. Feature selection for unsupervised learning. *Journal of Machine Learning Research*, 5:845–889, August 2004.

[15] J. G. Dy, C. E. Brodley, A. Kak, L. S. Broderick, and A. M. Aisen. Unsupervised feature selection applied to content-based retrieval of lung images. *IEEE Transactions on Pattern Analysis and Machine Intelli-*

gence, 25(3):373–378, March 2003.

[16] X. Fern and C. E. Brodley. Solving cluster ensemble problems by bipartite graph partitioning. In *Proceedings of the 21st International Conference on Machine Learning*, pages 281–288, 2004.

[17] E. Forgy. Cluster analysis of multivariate data: Efficiency vs. interpretability of classifications. *Biometrics*, 21:768, 1965.

[18] C. Fraley and A. E. Raftery. Model-based clustering, discriminant analysis, and density estimation. *Journal of the American Statistical Association*, 97(458):611–631, June 2002.

[19] A. Fred and A. K. Jain. Combining multiple clustering using evidence accumulation. *IEEE Transactions on Pattern Analysis and Machine Intelligence*, 27(6):835–850, 2005.

[20] J. H. Friedman. Exploratory projection pursuit. *Journal American Statistical Association*, 82:249–266, 1987.

[21] J. H. Friedman and J. J. Meulman. Clustering objects on subsets of attributes. *Journal Royal Statistical Society B*, 2004.

[22] J. H. Gennari. Concept formation and attention. In *Proceedings of the Thirteenth Annual Conference of the Cognitive Science Society*, pages 724–728, Chicago, IL, Lawrence Erlbaum, 1991.

[23] S. Goil, H. Nagesh, and A. Choudhary. Mafia: Efficient and scalable subspace clustering for very large data sets. Technical report, Northwestern University, IL, June 1999.

[24] I. Guyon and A. Elisseeff. An introduction to variable and feature selection. *Journal of Machine Learning Research*, 3:1157–1182, 2003.

[25] J. A. Hartigan. Direct clustering of a data matrix. *Journal of the American Statistical Association*, 67(337):123–129, March 1972.

[26] X. He, D. Cai, and P. Niyogi. Laplacian score for feature selection. In Y. Weiss, B. Schölkopf, and J. Platt, editors, *Advances in Neural Information Processing Systems 18*, pages 507–514. MIT Press, Cambridge, MA, 2006.

[27] P. Huber. Projection pursuit. *The Annals of Statistics*, 13(2):435–475, 1985.

[28] A. Hyvärinen. Survey on independent component analysis. *Neural Computing Surveys*, 2:94–128, 1999.

[29] A. K. Jain, M. N. Murty, and P. J. Flynn. Data clustering: A review. *ACM Computing Surveys*, 31(3):264–323, 1999.

[30] I. Jolliffe. *Principal Component Analysis*. Springer, New York, Second

edition edition, 2002.

[31] Y. S. Kim, N. Street, and F. Menczer. Evolutionary model selection in unsupervised learning. *Intelligent Data Analysis*, 6:531–556, 2002.

[32] J. Kittler. Feature set search algorithms. In *Pattern Recognition and Signal Processing*, pages 41–60, 1978.

[33] R. Kohavi and G. H. John. Wrappers for feature subset selection. *Artificial Intelligence, special issue on relevance*, 97(1-2):273–324, 1997.

[34] M. H. Law, M. Figueiredo, and A. K. Jain. Feature selection in mixture-based clustering. In *Advances in Neural Information Processing Systems 15*, Vancouver, December 2002.

[35] B. Liu, Y. Xia, and P. S. Yu. Clustering through decision tree construction. In *Proceedings of the ninth International Conference on Information and Knowledge Management*, pages 20–29. ACM Press, 2000.

[36] J. Macqueen. Some methods for classifications and analysis of multivariate observations. *Proc. Symp. Mathematical Statistics and Probability, 5th, Berkeley*, 1:281–297, 1967.

[37] D. Manoranjan, K. Choi, P. Scheuermann, and H. Liu. Feature selection for clustering - a filter solution. In *Proceedings of the Second IEEE International Conference on Data Mining*, pages 115–122, December 2002.

[38] T. Marill and D. M. Green. On the effectiveness of receptors in recognition systems. *IEEE Transactions on Information Theory*, 9:11–17, 1963.

[39] G. McLachlan and D. Peel. *Finite Mixture Models*. Wiley, New York, 2000.

[40] P. M. Narendra and K. Fukunaga. A branch and bound algorithm for feature subset selection. *IEEE Transactions on Computers*, C-26(9):917–922, September 1977.

[41] L. Parsons, E. Haque, and H. Liu. Subspace clustering for high dimensional data: A review. *SIGKDD Explorations*, 6(1):90–105, June 2004.

[42] C. M. Procopiuc, M. Jones, P. K. Agarwal, and T. M. Murali. A monte carlo algorithm for fast projective clustering. In *Proceedings of the 2002 ACM SIGMOD International Conference on Management of Data*, pages 418–427. ACM Press, 2002.

[43] P. Pudil, Novovičová, and J. Kittler. Floating search methods in feature selection. *Pattern Recognition Letters*, 15:1119–1125, 1994.

[44] G. Schwarz. Estimating the dimension of a model. *The Annals of Statistics*, 6(2):461–464, 1978.

[45] S. D. Stearns. On selecting features for pattern classifiers. In *Third*

International Conf. on Pattern Recognition, pages 71–75, 1976.

[46] A. Strehl and J. Ghosh. Cluster ensembles - a knowledge reuse framework for combining multiple partitions. *Journal on Machine Learning Research*, 3(583-617), 2002.

[47] T. Su and J. G. Dy. In search of deterministic methods for initializing k-means andgaussian mixture clustering. *Intelligent Data Analysis*, 11(4), 2007.

[48] L. Talavera. Feature selection as a preprocessing step for hierarchical clustering. In *Proceedings of the Sixteenth International Conference on Machine Learning*, pages 389–397, Bled, Slovenia, Morgan Kaufmann, 1999.

[49] S. Vaithyanathan and B. Dom. Model selection in unsupervised learning with applications to document clustering. In *Proceedings of the Sixteenth International Conference on Machine Learning*, pages 433–443, Bled, Slovenia, Morgan Kaufmann, 1999.

[50] A. W. Whitney. A direct method of nonparametric measurement selection. *IEEE Transactions on Computers*, 20:1100–1103, 1971.

[51] J. Yang, W. Wang, H. Wang, and P. Yu. δ-clusters: Capturing subspace correlation in a large data set. In *Proceedings of the 18th International Conference on Data Engineering*, pages 517–528, 2002.

Chapter 3

Randomized Feature Selection

David J. Stracuzzi

Arizona State University

3.1   Introduction

Randomization is an algorithmic technique that has been used to produce provably efficient algorithms for a wide variety of problems. For many applications, randomized algorithms are either the simplest or the fastest algorithms available, and sometimes both [16]. This chapter provides an overview of randomization techniques as applied to feature selection. The goal of this chapter is to provide the reader with sufficient background on the topic to stimulate both new applications of existing randomized feature selection methods, and research into new algorithms. Motwani and Raghavan [16] provide a more broad and widely applicable introduction to randomized algorithms.

Learning algorithms must often make choices during execution. Randomization is useful when there are many ways available in which to proceed but determining a guaranteed good way is difficult. Randomization can also lead to efficient algorithms when the benefits of good choices outweigh the costs of bad choices, or when good choices occur more frequently than bad choices. In the context of feature selection, randomized methods tend to be useful when the space of possible feature subsets is prohibitively large. Likewise, randomization is often called for when deterministic feature selection algorithms are prone to becoming trapped in local optima. In these cases, the ability of randomization to sample the feature subset space is of particular value.

In the next section, we discuss two types of randomizations that may be applied to a given problem. We then provide an overview of three complexity classes used in the analysis of randomized algorithms. Following this brief theoretical introduction, we discuss explicit methods for applying randomization to feature selection problems, and provide examples. Finally, the chapter concludes with a discussion of several advanced issues in randomization, and a summary of key points related to the topic.

3.2   Types of Randomizations

Randomized algorithms can be divided into two broad classes. *Las Vegas* algorithms always output a correct answer, but may require a long time to execute with small probability. One example of a Las Vegas algorithm is the randomized quicksort algorithm (see Cormen, Lieserson, and Rivest [4], for example). Randomized quicksort selects a pivot point at random, but always produces a correctly sorted output. The goal of randomization is to avoid degenerate inputs, such as a pre-sorted sequence, which produce the worst-case $O(n^2)$ runtime of the deterministic (pivot point always the same) quicksort algorithm. The effect is that randomized quicksort achieves the *expected* runtime of $O(n \log n)$ with high probability, regardless of input.

Monte Carlo algorithms may output an incorrect answer with small probability, but always complete execution quickly. As an example of a Monte Carlo algorithm, consider the following method for computing the value of π, borrowed from Krauth [11]. Draw a circle inside a square such that the sides of the square are tangent to the circle. Next, toss pebbles (or coins) randomly in the direction of the square. The ratio of pebbles inside the circle to those inside the entire square should be approximately $\frac{\pi}{4}$. Pebbles that land outside the square are ignored.

Notice that the longer the algorithm runs (more pebbles tossed) the more accurate the solution. This is a common, but not required, property of randomized algorithms. Algorithms that generate initial solutions quickly and then improve them over time are also known as *anytime algorithms* [22]. Anytime algorithms provide a mechanism for trading solution quality against computation time. This approach is particularly relevant to tasks, such as feature selection, in which computing the optimal solution is infeasible.

Some randomized algorithms are neither guaranteed to execute efficiently nor to produce a correct output. Such algorithms are typically also labeled as Monte Carlo. The type of randomization used for a given problem depends on the nature and needs of the problem. However, note that a Las Vegas algorithm may be converted into a Monte Carlo algorithm by having it output a random (possibly incorrect) answer whenever the algorithm requires more than a specified amount of time to complete. Similarly, a Monte Carlo algorithm may be converted into a Las Vegas algorithm by executing the algorithm repeatedly with independent random choices. This assumes that the solutions produced by the Monte Carlo algorithm can be verified.

3.3   Randomized Complexity Classes

The probabilistic behavior that gives randomized algorithms their power also makes them difficult to analyze. In this section, we provide a brief introduction to three complexity classes of practical importance for randomized algorithms. Papadimitriou [18] provides a rigorous and detailed discussion of these and other randomized complexity classes. For simplicity, we focus on decision algorithms, or those that output "yes" and "no" answers, for the remainder of this section.

Randomized algorithms are related to nondeterministic algorithms. Nondeterministic algorithms choose, at each step, among zero or more possible next steps, with no specification of which choice should be taken. Contrast this to deterministic algorithms, which have exactly one next step available at each step of the algorithm. Note the difference between nondeterministic choices and conditional control structures, such as *if . . . then* statements, which are fully determined by the input to the algorithm. A nondeterministic algorithm accepts its input if there exists some sequence of choices that result in a "yes" answer. The well-known class $\mathcal{NP}$ therefore includes languages accepted by nondeterministic algorithms in a polynomial number of steps, while class $\mathcal{P}$ does the same for languages accepted by deterministic algorithms.

Randomized algorithms differ from nondeterministic algorithms in that they accept inputs probabilistically rather than existentially. The randomized complexity classes therefore define probabilistic guarantees that an algorithm must meet. For example, consider the class $\mathcal{RP}$, for *randomized polynomial time*. $\mathcal{RP}$ encompasses algorithms that accept good inputs (members of the underlying language) with non-trivial probability, always reject bad inputs (non-members of the underlying language), and always execute in polynomial time. More formally, a language $L \in \mathcal{RP}$ if some randomized algorithm R accepts string $s \in L$ with probability $\frac{1}{\epsilon}$ for any ϵ that is polynomial in $|s|$, rejects $s' \notin L$ with probability 1, and requires a polynomial number of steps in $|s|$.

Notice that the definition of $\mathcal{RP}$ corresponds to the set of Monte Carlo algorithms that can make mistakes only if the input string is a member of the target language. The complement of this class, co-$\mathcal{RP}$, then corresponds to the set of algorithms that can make mistakes only if the input string is *not* a member of the target language. Furthermore, the intersection of these two classes, $\mathcal{RP}\cap$ co-$\mathcal{RP}$, corresponds to the set of Las Vegas algorithms that execute in worst-case polynomial time.

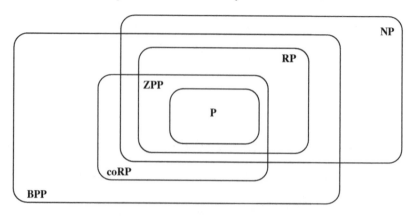

FIGURE 3.1: Illustration of the randomized complexity classes in relation to each other and the deterministic classes P and NP.

To see why, first note that each problem in the intersection has two Monte Carlo algorithms. One algorithm never outputs a false positive, while the other never outputs a false negative. By conducting many repeated and independent executions of both algorithms, we are guaranteed to eventually arrive at the correct output. (Recall that Las Vegas algorithms always output the correct answer, but may take a long time to do so.) This intersection is also known as the class $\mathcal{ZPP}$, for *polynomial randomized algorithms with zero probability of error*.

In practice we can use algorithms in $\mathcal{RP}$ to construct Monte Carlo algorithms that produce the correct output with high probability simply by running them polynomially many times. If any execution accepts the input, then we return "yes." Since algorithms in $\mathcal{RP}$ never produce false positive results, we can guarantee that the probability of a false negative becomes small. Here, that probability is $(1 - \frac{1}{\epsilon})^k$ for k executions of the algorithm.

The third and largest complexity class of practical importance is $\mathcal{BPP}$, for *polynomial time algorithms with bounded probability of error*. Unlike $\mathcal{RP}$ and $\mathcal{ZPP}$, $\mathcal{BPP}$ allows a randomized algorithm to commit both false positive and false negative errors. This class encompasses algorithms that accept good inputs a majority of the time and rejects bad inputs a majority of the time. More formally, a language $L \in \mathcal{BPP}$ if some randomized algorithm R accepts $s \in L$ with probability $\frac{1}{2} + \frac{1}{\epsilon}$ and accepts $s \notin L$ with probability $\frac{1}{2} - \frac{1}{\epsilon}$ for any ϵ polynomial in $|s|$. Like $\mathcal{RP}$ and $\mathcal{ZPP}$, we can create an algorithm that produces the correct result with high probability simply by executing repeatedly an algorithm that meets the stated minimums.

Figure 3.1 illustrates the relationships among the randomized classes, and shows how the randomized classes are related to the deterministic classes P and NP. Note that the figure assumes that $P \neq NP$, which is an open problem. If this assumption turns out to be false, then the complexity classes will collapse into one or just a few classes.

Finally, note that the randomized complexity classes are semantic as opposed to syntactic classes such as $\mathcal{P}$ and $\mathcal{NP}$. Semantic class membership depends on the *meaning* of a specific algorithm instead of the *format* of the algorithm. For example, we can determine whether an algorithm is a member of class $\mathcal{P}$ by counting the number of times the input is processed. Conversely, we must consider the probability that a given input is *accepted* to determine membership in the class $\mathcal{RP}$. Thus, there is no simple way to check whether a given randomized algorithm fits into a given randomized complexity class. There can be no complete problems for such classes [18].

3.4 Applying Randomization to Feature Selection

A critical step in constructing a randomized algorithm is to decide which aspect of the target problem to randomize. In some cases there may be only one clear option. For example, in the deterministic quicksort algorithm, the pivot is typically chosen arbitrarily as the first element of the current array. However, any fixed choice of pivot would work equally well, so randomizing the selection in an effort to protect against degenerate inputs is successful. Other problems may offer several candidates for randomization.

We formulate the specific feature selection problem considered here as follows. Given a set of supervised training examples described by a set of input features or variables **x** and a target concept or function y, produce a subset of the original input variables that predicts best the target concept or function when combined into a hypothesis by a learning algorithm. The term "predicts best" may be defined in a variety of ways, depending on the specific application. In this context, there are at least two possible sources of randomization.

The first source is the set of input variables. A feature selection algorithm may choose at random which variables to include in a subset. The resulting algorithm searches for the best variable subset by sampling the space of possible subsets. This approach to randomization carries an important advantage. As compared to the popular greedy stepwise search algorithms [1, 8], which add or remove a single variable at a time, randomization protects against local minima. A broad sampling of subsets is unlikely to concentrate effort on any one portion of the search space. Conversely, if many subsets have equally high quality, then a randomized approach will also tend to find a solution quickly.

Randomizing over the set of variables is less likely to be effective if one or a few of the variable subsets is much better than all of the others. The probability of selecting one particular subset at random out of all possible subsets is simply too small. A second issue with this type of randomization is that there is no clear choice of when to stop sampling. A parameter must be set arbitrarily within the algorithm, or the algorithm can be run until the available computation time expires (as an anytime algorithm).

The second possible source of randomization is the set of training examples, often known as the prototype selection problem. If the number of available examples is very large, an algorithm can select at random which examples to include in a given subset evaluation. The resulting algorithm may conduct a traditional deterministic search through the space of feature subsets, but evaluates those subsets based on a random sample of data. This option is particularly useful when the number of examples available is intractably large, or the available computation time is short.

Notice that as a side effect, randomization reduces the confidence with which the feature selection algorithm produces results. By sampling only a small portion of the space of variable subsets, we lose confidence that the algorithm's final output is actually the best possible subset. Likewise, when we sample the set of available training data, we lose confidence in the accuracy of our evaluation of a given feature subset. Such effects are of particular concern for algorithms that randomize on both the set of input variables and the set of examples. The approach offers the possibility of combining the advantages of both randomization methods, but it also reduces confidence in two ways. Concerns about confidence must be balanced carefully against any reductions in computation.

3.5 The Role of Heuristics

A fundamental goal of computer science is to find correct or optimal problem solutions using a minimum of computation. For many problems, no known algorithm can produce such a solution efficiently. Heuristics are therefore used to relax one or both of these demands on optimality and efficiency.

Randomization itself is a problem solving heuristic. A randomized algorithm may trade optimality for efficiency by searching only a sampled portion of the state space, instead of the entire state space. In many cases there is no guarantee that the best possible solution will be found, but often a relatively good solution is found with an acceptable amount of computation.

Many algorithms employ multiple heuristics. One type of heuristic appropriate to a randomized algorithm is a sampling bias. In the context of feature selection, an algorithm that always samples uniformly from the entire space of feature subsets to obtain its next candidate solution uses randomization as its only heuristic. However, algorithms that bias their samples, for example by sampling only in the neighborhood of the current best solution, employ a second heuristic in conjunction with randomization.

A variety of sampling biases are possible for feature and prototype selection algorithms. We illustrate several examples in the following section. However, not all sampling biases are appropriate to all selection problems. A sampling bias that quickly focuses the search on a small set of features may not be appropriate if there are several disjoint feature sets capable of producing good learner performance. Likewise, an approach that samples the space broadly throughout the search may not be appropriate if the number of features is large but few are relevant. As noted above, randomization may not be a good choice of heuristic if there is some reason to believe that only a very small number of feature subsets produce desirable results, while all other subsets produce undesirable results. In this case, random sampling is unlikely to uncover the solution efficiently.

Successful application of a randomized (or deterministic) selection algorithm requires some understanding of the underlying feature space. The heuristics and sampling biases used must be appropriate to the given task. Viewed oppositely, successful application of a randomized algorithm *implies* that the underlying feature space exhibits particular characteristics, and these characteristics depend on the specific heuristics used to solve the problem.

3.6 Examples of Randomized Selection Algorithms

We now consider specific examples of randomized feature and prototype selection algorithms. The goal is to illustrate ways in which randomization can be applied to the feature selection problem. We consider both Las Vegas and Monte Carlo methods, and a variety of performance guarantees, along with the strengths and weaknesses of each approach. The algorithms discussed here also illustrate a variety of heuristics and sampling biases. As is often the case, no one algorithm uniformly dominates another. The goal of this section is to familiarize readers with existing methods for randomized selection, and to provide the background necessary to make informed choices.

3.6.1 A Simple Las Vegas Approach

The key characteristic of a Las Vegas algorithm is that it must eventually produce the correct solution. In the case of feature selection, this means that the algorithm must produce a minimal subset of features that optimizes some criteria, such as classification accuracy. The Las Vegas Filter (LVF) algorithm discussed by Liu and Setino [12] achieves this goal, albeit under specific conditions.

LVF searches for a minimal subset of features to describe a given set of supervised training examples $\mathbf{X} = < \mathbf{x}_1, y_1 >, \ldots, < \mathbf{x}_M, y_M >$, where $|\mathbf{x}_i| = N$.

Given:
    Examples $\mathbf{X} = <\mathbf{x}_1, y_1>, \ldots, <\mathbf{x}_M, y_M>$
    Maximum allowable inconsistancy γ
    Number of attributes N
    Number of iterations t_{max}

Algorithm:
    $S_{best} \leftarrow$ all N attributes
    $c_{best} \leftarrow N$
    for $i \leftarrow 1$ **to** t_{max} **do**
        $c \leftarrow$ random number between 0 and c_{best}
        $S \leftarrow$ random selection of c features to include
        if $Inconsistancy(S, \mathbf{X}) \leq \gamma$ **then**
            $S_{best} \leftarrow S$
            $c_{best} \leftarrow c$
    return(S_{best})

FIGURE 3.2: The Las Vegas Filter algorithm [12].

The subsets are selected uniformly at random with respect to the set of all possible subsets. They are then evaluated according to an inconsistency criterion, which tests the extent to which the reduced-dimension data can still separate the class labels. If the newly selected subset is both smaller in size and has an equal or lesser inconsistency rate, then the subset is retained. LVF performs this simple sampling procedure repeatedly, stopping after a predetermined number of iterations, t_{max}. Figure 3.2 shows the LVF algorithm.

There are two important caveats to the LVF algorithm. First, the algorithm can only be labeled as a Las Vegas algorithm if it is allowed to run sufficiently long to find the optimal solution. For training data described by N input features, we expect to need approximately 2^N iterations. In the case where $t_{max} \ll 2^N$, the algorithm should be considered Monte Carlo. Notice that the Monte Carlo version of the algorithm may be used in an anytime format by returning the current best feature subset at any point during execution.

The second caveat to LVF regards the allowable inconsistency rate, γ. This parameter controls the trade-off between the size of the returned feature subset and the ability of that subset to distinguish among examples. If we set γ equal to the inconsistency rate of the full data set $\mathbf{X}_{()}$, then LVF is guaranteed to find the optimal solution under the conditions described above for t_{max}. However, a larger inconsistency rate allows LVF to reach smaller feature subsets more quickly. The algorithm then effectively becomes a greedy local search and is susceptible to local minima. LVF ignores any subset that is selected with size larger than the current best. If a larger subset exists that has a lower inconsistency rate, then the algorithm will not find it. Thus, given an inconsistency rate larger than that of the full data set, LVF must be

Given:

Examples $\mathbf{X} = < \mathbf{x}_1, y_1 >, < \mathbf{x}_M, y_M >, \ldots$

Number of iterations t_{max}

Number of prototypes p

Algorithm:

$\mathbf{X}_{best} \leftarrow$ random selection of p examples from $\mathbf{X}$

for $i \leftarrow 1$ to t_{max} do

$\quad \mathbf{X}' \leftarrow$ random selection of p examples from $\mathbf{X}$

$\quad$ if $k\mathrm{NN}(\mathbf{X}', \mathbf{X}) > k\mathrm{NN}(\mathbf{X}_{best}, \mathbf{X})$ then

$\quad\quad\quad \mathbf{X}_{best} \leftarrow \mathbf{X}'$

return($\mathbf{X}_{best}$)

FIGURE 3.3: The MC1 algorithm [19].

considered as a Monte Carlo algorithm, regardless of the number of iterations performed.

3.6.2 Two Simple Monte Carlo Approaches

Consider next two applications of Monte Carlo randomization to feature selection. The goal of the first is to reduce the computational requirements of the nearest neighbor learner by sampling over the set of available training examples. The algorithm, called MC1 [19], repeatedly samples the data set in an attempt to find a small subset of prototypes (training examples) that allow nearest neighbor to generalize well to unseen examples.

The MC1 procedure begins by selecting p prototypes at random from the available examples, where p is chosen in advance by the user. Classification accuracy for nearest neighbor is then computed over the entire training set. If the selected set of examples leads to higher accuracy than the previous best subset, then the new subset is retained. This procedure is repeated t_{max} times, where t_{max} is also specified in advance by the user. The example subset, which yields the highest accuracy, is then returned at the end of the procedure and used on test data. Figure 3.3 summarizes the MC1 algorithm.

Notice that if we set t_{max} sufficiently large, then we are virtually guaranteed to find the best possible set of prototypes for a given value of p. Thus, like the LVF algorithm, MC1 behaves like a Las Vegas algorithm in the limit. Unlike LVF, which attempts to find the minimum number of features, MC1 does not necessarily find the minimum number of prototypes, p. Notice also that MC1 makes no assumptions particular to the nearest neighbor learner. The selection algorithm can therefore be adapted as a general purpose wrapper and can be used with any classification learning algorithm.

Skalak's experiments [19] show that MC1 performs best when the training and test data exhibit well-defined class boundaries. Put another way, MC1 performs well when there is little overlap between examples from different classes. This may be an artifact of the nearest neighbor algorithm and not of Monte Carlo randomization in general. Nevertheless, the result reinforces the notion that we cannot expect randomized techniques to find a single specific solution within a large search space.

The Relief algorithm [9] demonstrates a different use of Monte Carlo randomization. Relief is a basic, two-class filtering algorithm that ranks variables according to a statistical measure of how well individual features separate the two classes. In an effort to reduce the computational cost of calculating these statistics, Relief selects examples at random for the computation.

Given:
     Examples $\mathbf{X} = <\mathbf{x}_1, y_1 >, \ldots < \mathbf{x}_m, y_m >$
     Relevancy cut-off τ
     Number of iterations t_{max}

Algorithm:
     Partition $\mathbf{X}$ by class into $\mathbf{X}^+$ and $\mathbf{X}^-$
     Initialize $\mathbf{w} = (0, 0, \ldots, 0)$
     for $i \leftarrow 1$ **to** t_{max} **do**   *//compute relevance*
         $\mathbf{x}_i \leftarrow$ random example $\mathbf{x} \in \mathbf{X}$
         $\mathbf{x}_i^+ \leftarrow$ nearest $\mathbf{x}^+ \in \mathbf{X}^+$ to $\mathbf{x}_i$
         $\mathbf{x}_i^- \leftarrow$ nearest $\mathbf{x}^- \in \mathbf{X}^-$ to $\mathbf{x}_i$
         if $\mathbf{x}_i \in \mathbf{X}^+$ **then**
                 update$(\mathbf{w}, \mathbf{x}_i, \mathbf{x}_i^+, \mathbf{x}_i^-)$
         else
                 update$(\mathbf{w}, \mathbf{x}_i, \mathbf{x}_i^-, \mathbf{x}_i^+)$
     for $i \leftarrow 1$ **to** N **do**   *//select most relevant*
         if $\frac{\mathbf{w}_i}{t_{max}} \geq \tau$ **then**
             feature i is relevant

Procedure update$(\mathbf{w}, \mathbf{x}, \mathbf{x}^+, \mathbf{x}^-)$ *//update relevance values*
     for $i \leftarrow 1$ **to** N **do**
         $\mathbf{w}_i \leftarrow \mathbf{w}_i - \text{diff}(\mathbf{x}, \mathbf{x}^+)^2 + \text{diff}(\mathbf{x}, \mathbf{x}^-)^2$

FIGURE 3.4: The Relief algorithm [9].

Briefly, the algorithm operates by calculating a weight value for each of the N available features. These weights are calculated using a random sample of examples from the full set of supervised examples $\mathbf{X}$. Relief selects a training example $\mathbf{x}_i$ at random and then finds, according to Euclidean distance, the nearest same-class example $\mathbf{x}_i^+$ and the nearest different-class example $\mathbf{x}_i^-$. These examples are then used to update the weight value for each feature according to the difference between $\mathbf{x}_i$, $\mathbf{x}_i^+$, and $\mathbf{x}_i^-$. Here, the difference for nominal feature k is defined as 1 if $\mathbf{x}_{i,k}$ and $\mathbf{x}_{j,k}$ have different nominal values, and is defined as 0 if they are the same. For numerical features, the difference is simply $\mathbf{x}_{i,k} - \mathbf{x}_{j,k}$ normalized into the range $[0, 1]$. This procedure is repeated t_{max} times for some preset value of t_{max}. Features with weights greater than a specified value τ are considered relevant to the target output variable. Figure 3.4 summarizes the Relief algorithm.

Notice the similarities and differences between MC1 and Relief. Both algorithms use randomization to avoid evaluating all M available training examples. MC1 achieves this goal by evaluating many hypotheses on different random example subsets, while Relief simply selects one random subset of examples on which to perform evaluations. Relief's approach is faster computationally but cannot provide the user with any confidence that the selection of examples is representative of the sample space. In particular, the fewer examples selected, the less likely the random subset will provide a representative sample of the space. MC1 mitigates this problem by searching for the most beneficial, and presumably representative, example subset.

3.6.3   Random Mutation Hill Climbing

Skalak [19] discusses a feature selection approach based on randomized local search, called random mutation hill climbing (RMHC). As with the MC1 algorithm, the goal is to reduce the computational cost of the nearest neighbor learner while maximizing classification accuracy. Unlike MC1, which samples the space of possible prototype subsets, the RMHC algorithm conducts a more localized search by changing only one included prototype per iteration.

RMHC uses a single bit vector to encode the index of each of the p selected prototypes. This bit vector is initialized randomly, and the algorithm proceeds by flipping one randomly selected bit on each iteration. This has the effect of replacing exactly one prototype with another. The new set of prototypes is then evaluated on the entire training set using nearest neighbor and is retained if it produces higher accuracy than the current set. Otherwise the change is discarded. The algorithm terminates after a fixed number of iterations, t_{max}. Figure 3.5 summarizes the RMHC algorithm. Note that, like MC1, RMHC can be adapted for use with learning algorithms other than nearest neighbor.

Skalak also describes a variant of the algorithm in which the bit vector also encodes which of the features are selected for use. Here, when a bit is selected for flipping, it may either change the set of included prototypes or the set of included features. No control over the relative probability of these changes is considered. Experimental results, though limited, suggest that RMHC does improve both the computational requirements and the classification performance of k-nearest neighbor. Notice however, that because the random selections are embedded in a greedy local search, RMHC does not necessarily avoid falling into local extrema. Thus, RMHC is a Monte Carlo algorithm that cannot be converted into a Las Vegas algorithm simply by

Given:
 Examples $\mathbf{X} =< \mathbf{x}_1, y_1 >, \ldots, < \mathbf{x}_m, y_m >$
 Number of iterations t_{max}
 Number of prototypes p

Algorithm:
 $\mathbf{X}_{best} \leftarrow$ random selection of p examples from $\mathbf{X}$
 $\mathbf{b} \leftarrow$ random bit vector encoding p prototype indicies
 for $i \leftarrow 1$ **to** t_{max} **do**
  $j \leftarrow$ random number between $0 \ldots |\mathbf{b}|$
  flip bit $\mathbf{b}_j$
  $\mathbf{X}' \leftarrow$ set of prototypes from $\mathbf{X}$ included by $\mathbf{b}$
  if $k\text{NN}(\mathbf{X}', \mathbf{X}) > k\text{NN}(\mathbf{X}_{best}, \mathbf{X})$ **then**
   $\mathbf{X}_{best} \leftarrow \mathbf{X}'$
 return$(\mathbf{X}_{best})$

FIGURE 3.5: The random mutation hill climbing algorithm [19].

increasing the number of iterations, t_{max}. We can still convert RMHC to a Las Vegas algorithm by running the algorithm many times, however.

3.6.4 Simulated Annealing

Simulated annealing [10, 2] is a general purpose stochastic search algorithm inspired by a process used in metallurgy. The heating and slow cooling technique of annealing allows the initially excited and disorganized atoms of a metal to find strong, stable configurations. Likewise, simulated annealing seeks solutions to optimization problems by initially manipulating the solution at random (high temperature), and then slowly increasing the ratio of greedy improvements taken (cooling) until no further improvements are found.

To apply simulated annealing, we must specify three parameters. First is an annealing schedule, which consists of an initial and final temperature, T_0 and T_{final}, along with an annealing (cooling) constant ΔT. Together these govern how the search will proceed over time and when the search will stop. The second parameter is a function used to evaluate potential solutions (feature subsets). The goal of simulated annealing is to optimize this function. For this discussion, we assume that higher evaluation scores are better. In the context of feature selection, relevant evaluation functions include the accuracy of a given learning algorithm using the current feature subset (creating a wrapper algorithm) or a variety of statistical scores (producing a filter algorithm).

The final parameter for simulated annealing is a neighbor function, which takes the current solution and temperature as input and returns a new, "nearby" solution. The role of the temperature is to govern the size of the neighborhood. At high temperature the neighborhood should be large, allow-

Given:

Examples $\mathbf{X} = \langle \mathbf{x}_1, y_1 \rangle, \ldots \langle \mathbf{x}_m, y_m \rangle$

Annealing schedule, T_0, T_{final} and ΔT with $0 < \Delta T < 1$

Feature subset evaluation function $Eval(\cdot, \cdot)$

Feature subset neighbor function $Neighbor(\cdot, \cdot)$

Algorithm:

$S_{best} \leftarrow$ random feature subset

while $T_i > T_{final}$ **do**

    $S_i \leftarrow Neighbor(S_{best}, T_i)$

    $\Delta E \leftarrow Eval(S_{best}, \mathbf{X}) - Eval(S_i, \mathbf{X})$

    if $\Delta E < 0$ **then**   *//if new subset better*

        $S_{best} \leftarrow S_i$

    else   *//if new subset worse*

        $S_{best} \leftarrow S_i$ with probability $\exp(\frac{-\Delta E}{T_i})$

    $T_{i+1} \leftarrow \Delta T \times T_i$

return(S_{best})

FIGURE 3.6: A basic simulated annealing algorithm.

ing the algorithm to explore broadly. At low temperature, the neighborhood should be small, forcing the algorithm to explore locally. For example, suppose we represent the set of available features as a bit vector, such that each bit indicates the presence or absence of a particular feature. At high temperature, the neighbor function may flip many bits to produce the next solution, while at low temperature the neighbor function may flip just one bit.

The simulated annealing algorithm, shown in Figure 3.6, attempts to iteratively improve a randomly generated initial solution. On each iteration, the algorithm generates a neighboring solution and computes the difference in quality (energy, by analogy to metallurgy process) between the current and candidate solutions. If the new solution is better, then it is retained. Otherwise, the new solution is retained with a probability that is dependent on the quality difference, ΔE, and the temperature. The temperature is then reduced for the next iteration.

Success in simulated annealing depends heavily on the choice of the annealing schedule. If ΔT is too large (near one), the temperature decreases slowly, resulting in slow convergence. If ΔT is too small (near zero), then the temperature decreases quickly and convergence will likely reach a local extrema. Moreover, the range of temperatures used for an application of simulated annealing must be scaled to control the probability of accepting a low-quality candidate solution. This probability, computed as $\exp(\frac{-\Delta E}{T_i})$, should be large at high temperature and small at low temperature to facilitate exploration early in the search and greedy choices later in the search.

In spite of the strong dependence on the cooling schedule, simulated annealing is guaranteed to converge provided that the schedule is sufficiently long [6]. From a theoretical point of view, simulated annealing is therefore a Las Vegas algorithm. However, in practice, the convergence guarantee requires intractably long cooling schedules, resulting in a Monte Carlo algorithm. Although the literature contains relatively few examples of simulated annealing applications to feature selection, the extent to which the algorithm can be customized (annealing schedule, neighbor function, evaluation function) makes it a good candidate for future work. As noted above, simulated annealing supports both wrapper and filter approaches to feature selection.

3.6.5   Genetic Algorithms

Like simulated annealing, genetic algorithms are a general purpose mechanism for randomized search. There are four key aspects to their use: encoding, population, operators, and fitness. First, the individual states in the search space must be encoded into some string-based format, typically bit-strings, similar to those used by RMHC. Second, an initial population of individuals (search states, such as feature subsets) must be selected at random. Third, one or more operators must be defined as a method for exchanging information among individuals in the population. Operators define how the search proceeds through state space. Typical operators include crossover, which pairs two individuals for the exchange of substrings, and mutation, which changes a randomly selected bit in an individual string with low probability. Finally, a fitness function must be defined to evaluate the quality of states in the population. The goal of genetic algorithms is to optimize the population with respect to the fitness function.

The search conducted by a genetic algorithm proceeds iteratively. Individuals in the population are first selected probabilistically with replacement based on their fitness scores. Selected individuals are then paired and crossover is performed, producing two new individuals. These are next mutated with low probability and finally injected into the next population. Figure 3.7 shows a basic genetic algorithm.

Genetic algorithms have been applied to the feature selection problem in several different ways. For example, Vafaie and De Jong [21] describe a straightforward use of genetic algorithms in which individuals are represented by bit-strings. Each bit marks the presence or absence of a specific feature. The fitness function then evaluates individuals by training and then testing a specified learning algorithm based on only the features that the individual specifies for inclusion.

In a similar vein, SET-Gen [3] uses a fitness function that includes both the accuracy of the induced model and the comprehensibility of the model. The learning model used in their experiments was a decision tree, and comprehensibility was defined as a combination of tree size and number of features used. The FSS-EBNA algorithm [7] takes a more complex approach to crossover by using a Bayesian network to mate individuals.

Given:

Examples $\mathbf{X} = < \mathbf{x}_1, y_1 >, \ldots < \mathbf{x}_m, y_m >$
Fitness function $f(\cdot, \cdot)$
Fitness threshold τ
Population size p

Algorithm:

$P_0 \leftarrow$ population of p random individuals
for $k \leftarrow 0$ **to** ∞ **do**
    $sum \leftarrow 0$
    for each individual $i \in P_k$ **do**   //*compute pop fitness*
        $sum \leftarrow sum + f(i, \mathbf{X})$
        if $f(i, \mathbf{X}) \geq \tau$ **then**
            return(i)
    for each individual $i \in P_k$ **do**   //*compute selection probs*
        $\mathrm{Pr}_k[i] \leftarrow \frac{f(i, \mathbf{X})}{sum}$
    for $j \leftarrow 1$ **to** $\frac{p}{2}$ **do**   //*select and breed*
        select $i_1, i_2 \in P_k$ according to Pr_k with replacement
        $i_1, i_2 \leftarrow crossover(i_1, i_2)$
        $i_1 \leftarrow mutate(i_1)$
        $i_2 \leftarrow mutate(i_2)$
        $P_{k+1} \leftarrow P_{k+1} + \{i_1, i_2\}$

FIGURE 3.7: A basic genetic algorithm.

Two well-known issues with genetic algorithms relate to the computational cost of the search and local minima in the evaluation function. Genetic algorithms maintain a population (100 is a common size) of search space states that are mated to produce offspring with properties of both parents. The effect is an initially broad search that targets more specific areas of the space as the search progresses. Thus, genetic algorithms tend to drift through the search space based on the properties of individuals in the population. A wide variety of states, or feature subsets in this case, are explored. However, the cost of so much exploration can easily exceed the cost of a traditional greedy search.

The second problem with genetic algorithms occurs when the evaluation function is non-monotonic. The population may quickly focus on a local maximum in the search space and become trapped. The mutation operator, a broad sampling of the state space in the initial population, and several other tricks are known to mitigate this effect. Goldberg [5] and Mitchell [15] provide detailed discussions of the subtleties and nuances involved in setting up a genetic search. Nevertheless, there is no guarantee that genetic algorithms will produce the best, or even a good, result. This issue may arise with any probabilistic algorithm, but some are more prone to becoming trapped in suboptimal solutions than others.

3.6.6   Randomized Variable Elimination

Each of the algorithms considered so far uses a simple form of randomization to explore the space of feature or example subsets. MC1 and LVF both sample the space of possible subsets globally, while RMHC samples the space in the context of a greedy local search. Simulated annealing and genetic algorithms, meanwhile, conduct initially broad searches that incrementally target more specific areas over time. The next algorithm we consider samples the search space in a more directed manner.

Randomized variable elimination (RVE) [20] is a wrapper method motivated by the idea that, in the presence of many irrelevant variables, the probability of selecting several irrelevant variables simultaneously at random is high. RVE searches backward through the space of variable subsets, attempting to eliminate one or more variables per step. Randomization allows for the selection of irrelevant variables with high probability, while selecting multiple variables allows the algorithm to move through the space without incurring the cost of evaluating the intervening points in the search space. RVE conducts its search along a very narrow trajectory, sampling variable subsets sparsely, rather than broadly and uniformly. This more structured approach allows RVE to reduce substantially the total cost of identifying relevant variables.

A backward search serves two purposes for this algorithm. First, backward elimination eases the problem of recognizing irrelevant or redundant variables. As long as a core set of relevant variables remains intact, removing other variables should not harm the performance of a learning algorithm. Indeed, the learner's performance may increase as irrelevant features are removed from consideration. In contrast, variables whose relevance depends on the presence of other variables may have no noticeable effect when selected in a forward manner. Thus, mistakes should be recognized immediately via backward elimination, while good selections may go unrecognized by a forward selection algorithm.

The second purpose of backward elimination is to ease the process of selecting variables for removal. If most variables in a problem are irrelevant, then a random selection of variables is likely to uncover them. Conversely, a random selection is unlikely to turn up relevant variables in a forward search. Thus, forward search must work harder to find each relevant variable than backward search does for irrelevant variables.

RVE begins by executing the learning algorithm $\mathcal{L}$ on data that include all N variables. This generates an initial hypothesis h. Next, the algorithm selects k input variables at random for removal. To determine the value of k, RVE computes a cost (with respect to a given learning algorithm) of attempting to remove k input variables out of n remaining variables given that r are relevant. Note that knowledge of r is required here, although the assumption is later removed. A table $k_{opt}(n, r)$ of values for k given n and

Given:

Examples $\mathbf{X} = <\mathbf{x}_1, y_1>, \ldots <\mathbf{x}_m, y_m>$
Learning algorithm $\mathcal{L}$
Number of input features N
Number of relevant features r

Algorithm:

$n \leftarrow N$
$h \leftarrow$ hypothesis produced by $\mathcal{L}$ on all N inputs
compute schedule $k_{opt}(i, r)$ for $r < i \leq N$ by dynamic programming
while $n > r$ **do**
    select $k_{opt}(n, r)$ variables at random and remove them
    $h' \leftarrow$ hypothesis produced by $\mathcal{L}$ on $n - k_{opt}(n, r)$ inputs
    if $error(h', \mathbf{X}) < error(h, \mathbf{X})$ **then**
        $n \leftarrow n - k_{opt}(n, r)$
        $h \leftarrow h'$
    else
        replace the $k_{opt}(n, r)$ selected variables
return(h)

FIGURE 3.8: The randomized variable elimination algorithm [20].

r is then computed via dynamic programming by minimizing the aggregate cost of removing all $N - r$ irrelevant variables. Note that n represents the number of remaining variables, while N denotes the total number of variables in the original problem.

On each iteration, RVE selects $k_{opt}(n, r)$ variables at random for removal. The learning algorithm is then trained on the remaining $n - k_{opt}(n, r)$ inputs, and a hypothesis h' is produced. If the error $e(h')$ is less than the error of the previous best hypothesis $e(h)$, then the selected inputs are marked as irrelevant and are all simultaneously removed from future consideration. If the learner was unsuccessful, meaning the new hypothesis had larger error, then at least one of the selected inputs must have been relevant. The removed variables are replaced, a new set of $k_{opt}(n, r)$ inputs is selected, and the process repeats. The algorithm terminates when all $N - r$ irrelevant inputs have been removed. Figure 3.8 shows the RVE algorithm.

Analysis of RVE [20] shows that the algorithm expects to evaluate only $O(r \log(N))$ variable subsets to remove all irrelevant variables. This is a striking result, as it implies that a randomized backward selection wrapper algorithm evaluates fewer subsets and requires less total computation than forward selection wrapper algorithms. Stracuzzi and Utgoff provide a detailed formal analysis of randomized variable elimination [20].

The assumption that the number of relevant variables r is known in advance plays a critical role in the RVE algorithm. In practice, this is a strong assumption that is not typically met. Stracuzzi and Utgoff [20] therefore provide a version of the algorithm, called RVE*r*S (pronounced "reverse"), that conducts a binary search for r during RVE's search for relevant variables.

Experimental studies suggest RVE*r*S evaluates a sublinear number of variable subsets for problems with sufficiently many variables. This conforms to the performance predicted by the analysis of RVE. Experiments also show that for problems with hundreds or thousands of variables, RVE*r*S typically requires less computation than a greedy forward selection algorithm while producing competitive accuracy results. In practice, randomized variable elimination is likely to be effective for any problem that contains many irrelevant variables.

3.7   Issues in Randomization

The preceding sections in this chapter covered the basic use of randomization as an algorithmic technique, specifically as applied to feature selection. We now consider more advanced issues in applying randomization. Of particular interest and importance are sampling techniques, and the source of the random numbers used in the algorithms.

3.7.1   Pseudorandom Number Generators

Randomized algorithms necessarily depend on the ability to produce a sequence of random numbers. However, deterministic machines such as modern computers are not capable of producing sequences of truly random numbers. John von Neumann once stated that, "Anyone who considers arithmetical methods of producing random digits is, of course, in a state of sin" [17]. In practice, we must rely on pseudorandom number generators to provide sequences of numbers that exhibit statistical properties similar to those of genuinely random numbers.

The main property of pseudorandom numbers that differs from true random numbers is periodicity. No matter how sophisticated a pseudorandom number generating algorithm may be, it will eventually revisit a past state and begin repeating the number sequence. Other possible problems with pseudorandom number generators include non-uniform distribution of the output sequence, correlation of successive values (predictability), and short periods for certain starting points. The presence of any of these properties can cause poor or unexpected performance in a randomized algorithm.

The primary defense against such undesirable results is to select a good pseudorandom number generator prior to running any experiments. For example, the Mersenne twister algorithm [14] has proved useful for statistical simulations and generative modeling purposes. The algorithm has a very long period of 2^{19937}, provides a provably good distribution of values, and is computationally inexpensive. A variety of other suitable, but less complex algorithms are also available, particularly if the user knows in advance that the length of the required sequence of pseudorandom numbers is short.

3.7.2 Sampling from Specialized Data Structures

A second possible pitfall in the use of randomized algorithms stems from sampling techniques. For small databases, such as those that can be stored in a simple table or matrix, examples and features (rows and columns, respectively) may be selected by simply picking an index at random. However, many large databases are stored in more sophisticated, non-linear data structures. Uniformly distributed, random samples of examples cannot be extracted from such databases via simple, linear sampling methods.

An improperly extracted sample is unlikely to be representative of the larger database. The results of a feature selection or other learning algorithm run on such a sample may not extrapolate well to the rest of the database. In other words, the error achieved by feature selection and/or learning algorithm on a sampled test database will be overly optimistic. In general, different data structures will require different specialized sampling methods.

One example of a specialized sampling algorithm is discussed by Makawita, Tan, and Liu [13]. Here, the problem is to sample uniformly from a search tree that has a variable number of children at internal nodes. The naive approach of simply starting at the root and then selecting random children at each step until reaching a leaf (known as a random walk) will tend to oversample from paths that have few children at each internal node. This is an artifact of the data structure and not the data themselves, and so is unacceptable. The presented solution is to bias the acceptance of the leaf node into the sample by keeping track of the fanout at each internal node along the path. Leaves from paths with low fanout are accepted with lower probability than those from paths with high fanout. The sampling bias of the naive algorithm is thus removed.

3.8 Summary

The feature selection problem possesses characteristics that are critical to successful applications of randomization. First, the space of all possible feature subsets is often prohibitively large. This means that there are many possible choices available at each step in the search, such as which feature to include or exclude next. Second, those choices are often difficult to evaluate, because learning algorithms are expensive to execute and there may be complex interdependencies among features. Third, deterministic selection algorithms are often prone to becoming trapped in local optimal, also due to interdependencies among features. Finally, there are often too many examples available for an algorithm to consider each one deterministically. A randomized approach of sampling feature subset space helps to mitigate all of these circumstances.

In practice, there are several important issues to consider when constructing a randomized algorithm for feature selection. First is the decision of which aspect of the problem will be randomized. One option is to randomize over the set of input variables, causing the resulting algorithm to search for the best variable subset by sampling from the space of all subsets. A second approach is to randomize over the set of training examples, creating an algorithm that considers only a portion of the available training data. Finally, one may also randomize over both the input variables and the training data. In any case, the achieved reduction in computational cost must be balanced against a loss of confidence in the solution.

The second issue to consider in randomization relates to the performance of the resulting algorithm. Some tasks may demand discovery of the best possible feature subset, necessitating the use of a Las Vegas algorithm. Other tasks may sacrifice solution quality for speed, making a Monte Carlo algorithm more appropriate. A third option is to generate an initial solution, and then improve the solution over time, as in anytime algorithms [22]. Many more specific guarantees on performance are also possible.

Other issues in the application of randomization include the quality of the pseudorandom number generator used and the sampling technique that is used. Both of these can impact the performance of the randomized algorithm. The feature selection literature contains examples of the different randomizations methods (randomization over features versus examples), a variety of performance guarantees, and special purpose sampling methods, as discussed throughout the chapter. Although far from a complete exposition, this chapter should provide sufficient information to launch further study of randomized algorithms in the context of feature selection.

References

[1] R. Caruana and D. Freitag. Greedy attribute selection. In *Machine Learning: Proceedings of the Eleventh International Conference*, pages 121–129, New Brunswick, NJ, Morgan Kaufmann, 1994.

[2] V. Černý. A thermodynamical approach to the traveling salesman problem: An efficient simulation algorithm. *Journal of Optimization Theory*

and Applications, 45:41–51, 1985.

[3] K. Cherkauer and J. Shavlik. Growing simpler decision trees to facilitate knowledge discovery. In E. Simoudis, J. Han, and U. M. Fayyad, editors, *Proceedings of the Second International Conference on Knowledge Discovery and Data Mining*, Portland, OR, AAAI Press, 1996.

[4] T. H. Cormen, C. E. Leiserson, and R. L. Rivest. *Introduction to Algorithms*. MIT Press, Cambridge, MA, 1990.

[5] D. Goldberg. *Genetic Algorithms in Search, Optimization, and Machine Learning*. Addison-Wesley, Reading, MA, 1989.

[6] H. M. Hastings. Convergence of simulated annealing. *ACM SIGACT News*, 17(2):52–63, 1985.

[7] I. Inza, P. Larranaga, E. R., and B. Sierra. Feature subset selection by Bayesian network-based optimization. *Artificial Intelligence*, 123(1–2):157–184, 2000.

[8] G. H. John, R. Kohavi, and K. Pfleger. Irrelevant features and the subset selection problem. In *Machine Learning: Proceedings of the Eleventh International Conference*, pages 121–129, New Brunswick, NJ, Morgan Kaufmann, 1994.

[9] K. Kira and L. Rendell. A practical approach to feature selection. In S. D. H. and P. Edwards, editors, *Machine Learning: Proceedings of the Ninth International Conference*, Aberdeen, Scotland, UK, Morgan Kaufmann, 1992.

[10] S. Kirkpatrick, C. D. Gelatt, and M. P. Vecchi. Optimization by simulated annealing. *Science*, 220(4598):671–680, 1983.

[11] W. Krauth. Introduction to monte carlo algorithms. In J. Kertesz and I. Kondor, editors, *Advances in Computer Simulation*, Lecture Notes in Physics. SpringerVerlag, New York, 1998.

[12] H. Liu and R. Setino. A probabilistic approach to feature selection. In L. Saitta, editor, *Machine Learning: Proceedings of the Thirteenth International Conference on Machine Learning*, pages 319–327, Bari, Italy, Morgan Kaufmann, 1996.

[13] D. P. Makawita, K.-L. Tan, and H. Liu. Sampling from databases using B^{+}-trees. In *Proceedings of the 2000 ACM CIKM International Conference on Information and Knowledge Management*, pages 158–164, McLean, VA, ACM, 2000.

[14] M. Matsumoto and T. Nishimura. Mersenne twister: A 623-dimensionally equidistributed uniform pseudorandom number generator. *ACM Transactions on Modeling and Computer Simulation*, 8(1):3 – 30, 1998.

[15] M. Mitchell. *An Introduction to Genetic Algorithms*. MIT Press, Cambridge, MA, 1996.

[16] R. Motwani and P. Raghavan. *Randomized Algorithms*. Cambridge University Press, Cambridge, UK, 1995.

[17] J. von Neumann. Various techniques used in connection with random digits. In *Applied Mathematics Series, no. 12*. 1951.

[18] C. H. Papadimitriou. *Computational Complexity*. Addison-Wesley, Reading, MA, 1994.

[19] D. B. Skalak. Prototype and feature selection by sampling and random mutation hill climbing. In W. W. Cohen and H. Hirsh, editors, *Machine Learning: Proceedings of the Eleventh International Conference*, pages 293–301, New Brunswick, NJ, Morgan Kaufmann, 1994.

[20] D. J. Stracuzzi and P. E. Utgoff. Randomized variable elimination. *Journal of Machine Learning Research*, 5:1331–1364, 2004.

[21] H. Vafaie and K. DeJong. Genetic algorithms as a tool for restructuring feature space representations. In *Proceedings of the Seventh International Conference on Tools with AI*, Herndon, VA, IEEE Computer Society Press, New York, 1995.

[22] S. Zilberstein. Using anytime algorithms in intelligent systems. *AI Magazine*, 17(3), 1996.

Chapter 4

Causal Feature Selection

Isabelle Guyon

Clopinet, California

Constantin Aliferis

Vanderbilt University, Tennessee

André Elisseeff

IBM Zürich, Switzerland

4.1   Introduction

The present chapter makes an argument in favor of understanding and utilizing the notion of causality for feature selection: from an algorithm design perspective, to enhance interpretation, build robustness against violations of the i.i.d. assumption, and increase parsimony of selected feature sets; from the perspective of method characterization, to help uncover superfluous or artifactual selected features, missed features, and features not only predictive but also causally informative.

Determining and exploiting causal relationships is central in human reasoning and decision-making. The goal of determining causal relationships is to predict the consequences of given actions or manipulations. This is fundamentally different from making predictions from observations. Observations imply no experimentation, no interventions on the system under study, whereas actions disrupt the natural functioning of the system.

Confusing observational and interventional predictive tasks yields classical paradoxes [22]. Consider for instance that there seems to be a correlation between being in bed and dying. Should we conclude that we should better not spend time in bed to reduce our risk of dying? No, because arbitrarily *forcing* people to spend time in bed does not normally increase death rate. A plausible causal model is that *disease* causes both an increase in *time spent in bed* and in *death rate*. This example illustrates that a correlated feature (*time spent in bed*) may be predictive of an outcome (*death rate*), if the system is stationary (no change in the distribution of all the variables) and no interventions are made; yet it does not allow us to make predictions if an intervention is made (e.g., forcing people to spend more or less time in bed regardless of their disease condition). This example outlines the fundamental distinction between correlation and causation.

Policy making in health care, economics, or ecology are examples of interventions, of which it is desirable to know the consequences ahead of time (see our application section, Section 4.6). The goal of causal modeling is to provide coarse descriptions of mechanisms, at a level sufficient to predict the result of interventions. The main concepts are reviewed in Sections 4.3 and 4.4. The most established way of deriving causal models is to carry out *randomized controlled experiments* to test hypothetical causal relationships. Yet such experiments are often costly, unethical, or infeasible. This prompted a lot of recent research in learning causal models from observational data [8, 22, 25, 15], which we briefly review in Section 4.5.

Most feature selection algorithms emanating from the machine learning literature (see, e.g., [18, 11], which we briefly review for comparison in Section 4.2) do not seek to model mechanisms: They do not attempt to uncover cause-effect relationships between feature and target. This is justified because uncovering mechanisms is unnecessary for making good predictions in a purely observational setting. In our *death rate* prediction example, classical feature selection algorithms may include without distinction: features that cause the target (like *disease*), features that are consequences of a common cause (like *time spent in bed*, which is a consequence of *disease*, not of *death rate*), or features that are consequences of the target (like *burial rate*). But, while acting on a cause (like *disease*) can influence the outcome (*death rate*), acting on consequences (*burial rate*) or consequences of common causes (*time spent in bed*) cannot. Thus non-causality-aware feature selection algorithms do not lend themselves to making predictions of the results of actions or interventions. Additionally, feature selection algorithms ignoring the data-generating process may select features for their effectiveness to predict the target, regardless of whether such predictive power is characteristic of the system under study or the result of experimental artifacts. To build robustness against changes in measuring conditions, it is important to separate the effects of measurement error from those of the process of interest, as outlined in Section 4.4.2.

On the strong side of feature selection algorithms developed recently [18, 11], relevant features may be spotted among hundreds of thousands of distracters, with less than a hundred examples, in some problem domains. Research in this field has effectively addressed both computational and statistical problems that related to uncovering significant dependencies in such adverse conditions. In contrast, causal models [8, 22, 25, 15] usually deal with just a few variables and a quasi-perfect knowledge of the variable distribution, which implies an abundance of training examples. Thus there are opportunities for cross-fertilization of the two fields: Causal discovery can benefit from feature selection to cut down dimensionality for computational or statistical reasons (albeit with the risk of removing causally relevant features); feature selection can benefit from causal discovery by getting closer to the underlying mechanisms and reveal a more refined notion of relevance (albeit at computational price). This chapter aims to provide material to stimulate research in both directions. More details and examples are found in a technical report [10].

4.2 Classical "Non-Causal" Feature Selection

In this section, we give formal definitions of irrelevance and relevance from the point of view of classical (non-causal) feature selection. We review examples of feature selection algorithms. In what follows, the feature set is a random vector $\mathbf{X} = [X_1, X_2, ... X_N]$ and the target a random variable Y. Training and test data are drawn according to a distribution $P(\mathbf{X}, Y)$. We use the following definitions and notations for independence and conditional independence: Two random variables A and B are *conditionally independent* given a set of random variables $\mathbf{C}$, denoted $A \perp B | \mathbf{C}$, *iff* $P(A, B | \mathbf{C}) = P(A|\mathbf{C})P(B|\mathbf{C})$, for all assignments of values to A, B, and $\mathbf{C}$. If $\mathbf{C}$ is the empty set, then A and B are *independent*, denoted $A \perp B$.

A simple notion of relevance can be defined by considering only the dependencies between the target and individual variables:

DEFINITION 4.1 Individual feature irrelevance. *A feature X_i is individually irrelevant to the target Y iff X_i is independent of Y (denoted $X_i \perp Y$): $P(X_i, Y) = P(X_i)P(Y)$.*

From that definition it should simply follow that all non-individually irrelevant features are individually relevant (denoted $X_i \not\perp Y$). However, when a finite training sample is provided, the statistical significance of the relevance must be assessed. This is done by carrying out a statistical test with null hypothesis "H_0: the feature is individually irrelevant" (that is X_i and Y are statistically independent). For a review, see e.g. [11], Chapters 2 and 3.

Feature selection based on individual feature relevance (Def. 4.1) is called univariate. In the context of other variables, a variable X_i individually relevant to Y may become irrelevant, or vice versa. This renders necessary the notion of multivariate feature selection; see the scenarios of Figure 4.1:

– **Falsely irrelevant variables.** Figure 4.1 (a) shows a classification problem with two input variables X_1 and X_2 and a binary target Y (represented by the star and circle symbols). Variables X_2 and Y are independent $(X_2 \perp Y)$, yet, in the context of variable X_1, they are dependent $(X_2 \not\perp Y | X_1)$. The class separation with X_1 is improved by adding the individually irrelevant variable X_2. Univariate feature selection methods fail to discover the usefulness of variable X_2. Figure 4.1 (b) shows a trickier example in which both variables X_1 and X_2 are independent of Y $(X_1 \perp Y$ and $X_2 \perp Y)$. Each variable taken separately does not separate the target at all, while taken jointly they provide a perfect non-linear separation $(\{X_1, X_2\} \not\perp Y)$. This problem is known in machine learning as the "chessboard problem" and bears resemblance with the XOR and parity problems. Univariate feature selection methods fail to discover the usefulness of variables for such problems.

– **Falsely relevant variables.** Figures 4.1(c) and (d) show an example of the opposite effect, using a regression problem. The continuous variables X_2 and Y are dependent when taken out of the context of the binary variable X_1. However, conditioned on any value of X_1 (represented by the star and circle symbols), they are independent $(X_2 \perp Y | X_1)$. This problem is known in statistics as Simpson's paradox. In this case, univariate feature selection methods might find feature X_2 relevant, even though it is redundant with X_1. If X_1 were unknown (unobserved), the observed dependency between X_2 and Y may be spurious, as it vanishes when the "confounding factor" X_1 is discovered (see Section 4.4.2).

Multivariate feature selection may be performed by searching in the space of possible feature subsets for an optimal subset. The techniques in use have been classified into filters, wrappers, and embedded methods [14, 5]. They differ in the choice of three basic ingredients [18]: search algorithm, objective function, and stopping criterion. Wrappers use the actual risk functional of the machine learning problem at hand to evaluate feature subsets. They must train one learning machine for each feature subset investigated. Filters use an evaluation function other than the actual risk functional, which is usually computationally advantageous. Often no learning machine is involved in the feature selection process. For embedded methods, the feature selection space and the learning machine parameter space are searched simultaneously. For a review of filter, wrapper, and embedded methods, see [11].

From the multivariate perspective, it is useful to generalize the notion of individual relevance (Def. 4.1) to that of relevance in the context of other features. This allows us to rank features in a total order rather than assessing the relevance of feature subsets. We first introduce irrelevance as a consequence of random variable independence and then define relevance by contrast. For simplicity, we provide only asymptotic definitions, which assume the full knowledge of the data distribution. For a discussion of the finite sample case, see the introductory chapter of [11]. In what follows, $\mathbf{X} = [X_1, X_2, ..., X_i, ..., X_N]$ denotes the set of all features, $\mathbf{X}^{\backslash i}$ is the set of all features except X_i, and $\mathbf{V}^{\backslash i}$ is any subset of $\mathbf{X}^{\backslash i}$ (including $\mathbf{X}^{\backslash i}$ itself).

DEFINITION 4.2 Feature irrelevance.   *A feature X_i is irrelevant to*

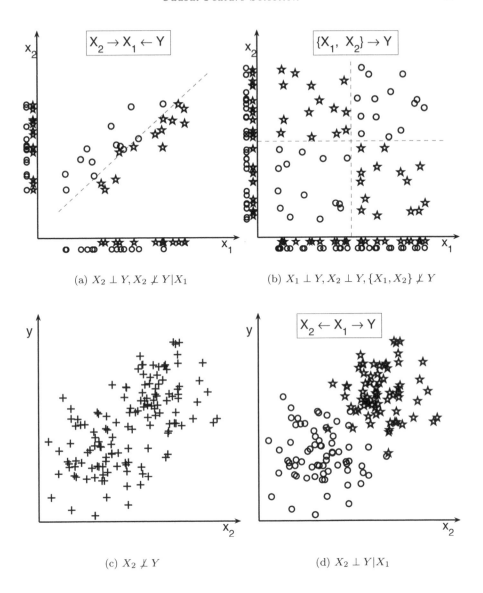

(a) $X_2 \perp Y, X_2 \not\perp Y|X_1$

(b) $X_1 \perp Y, X_2 \perp Y, \{X_1, X_2\} \not\perp Y$

(c) $X_2 \not\perp Y$

(d) $X_2 \perp Y|X_1$

FIGURE 4.1: Multivariate dependencies. (a) **Spouse problem:** Feature X_2 (a spouse of Y having the common child X_1) is individually irrelevant to Y ($X_2 \perp Y$), but it becomes relevant in the context of feature X_1 ($X_2 \not\perp Y|X_1$). (b) **Chessboard problem:** Two individually irrelevant features ($X_1 \perp Y$ and $X_2 \perp Y$) become relevant when taken jointly ($\{X_1, X_2\} \not\perp Y$). (c - d) **Simpson's paradox and the confounder problem:** (c) X_2 is correlated to Y, it is not independent of Y ($X_2 \not\perp Y$). It is individually relevant, but it may become irrelevant in the context of another feature, see case (d). (d) For any value of X_1 (star or circle), X_2 is independent of Y ($X_2 \perp Y|X_1$). Note: We show at the top of each scatter plot the causal structure of the models, which generated the data. In some cases, the same data can be explained by several alternative causal models (see Section 4.3).

the target Y iff *for all subset of features* $\mathbf{V}^{\setminus i}$*, and for all assignments of values,* X_i *is conditionally independent of* Y *given* $\mathbf{V}^{\setminus i}$ *(denoted by* $X_i \perp Y | \mathbf{V}^{\setminus i}$*):* $P(X_i, Y | \mathbf{V}^{\setminus i}) = P(X_i | \mathbf{V}^{\setminus i}) P(Y | \mathbf{V}^{\setminus i})$.

Kohavi and John define a notion of strong and weak relevance [14]. Intuitively and in many practical cases (but not always, as shown below), a strongly relevant feature is needed on its own and cannot be removed without performance degradation, while a weakly relevant feature is redundant with other relevant features and can be omitted if similar features are retained.

DEFINITION 4.3 Strong relevance.   *A feature* X_i *is strongly relevant to the target* Y *iff there exist some values* x, y, *and* $\mathbf{v}$ *with* $P(X_i = x, \mathbf{X}^{\setminus i} = \mathbf{v}) > 0$ *such that:* $P(Y = y | X_i = x, \mathbf{X}^{\setminus i} = \mathbf{v}) \neq P(Y = y | \mathbf{X}^{\setminus i} = \mathbf{v})$.

DEFINITION 4.4 Weak relevance.   *A feature* X_i *is weakly relevant to the target* Y *iff it is not strongly relevant and if there exist a subset of features* $\mathbf{V}^{\setminus i}$ *for which there exist some values* x, y, *and* $\mathbf{v}$ *with* $P(X_i = x, \mathbf{V}^{\setminus i} = \mathbf{v}) > 0$ *such that:* $P(Y = y | X_i = x, \mathbf{V}^{\setminus i} = \mathbf{v}) \neq P(Y = y | \mathbf{V}^{\setminus i} = \mathbf{v})$.

From the above definitions, and noting that $P(Y | X_i, \mathbf{V}^{\setminus i}) = P(Y | \mathbf{V}^{\setminus i})$ implies that $P(X_i, Y | \mathbf{V}^{\setminus i}) = P(X_i | \mathbf{V}^{\setminus i}) P(Y | \mathbf{V}^{\setminus i})$, one can easily see that a feature is either irrelevant, strongly relevant, or weakly relevant.

The issue of relevance has been the subject of much debate (see, e.g., the special issue of *Artificial Intelligence* on relevance [1] and the recent discussion of Tsamardinos and Aliferis [27] challenging the universality of any particular notion of relevance or usefulness of features). Although much remains to be said about such non-causal relevance, we wish now to introduce the concept of causality and show how it will shed light on the notion of feature relevance.

4.3   The Concept of Causality

Formal, widely-accepted definitions of causality have eluded philosophers of science for centuries. However, from an engineering point of view, causality is a very goal-oriented notion, which can simply be defined as finding modes of action on a system, which will result in a desired outcome (for example, taking a drug to cure an illness). Thus, even though causality may not find a perfect definition regrouping all the notions it encompasses in philosophy, psychology, history, law, religion, statistics, physics, and engineering, we can devise tests of causality that satisfy our engineering-oriented goal by assessing the effect of actual or hypothetical manipulations performed on the system [8, 22, 25, 21, 15].

4.3.1 Probabilistic Causality

Our everyday-life concept of causality is very much linked to the time dependency of events. However, such a temporal notion of causality is not always necessary or convenient. In particular, many machine learning problems are concerned with "cross-sectional studies," which are studies where many samples of the state of a system are drawn at a given point in time. Thus, we will drop altogether the reference to time and replace it by the notion of causal ordering. Causal ordering can be understood as fixing a particular time scale and considering only causes happening at time t and effects happening at time $t + \Delta t$, where Δt can be made as small as we want.

In this chapter, causes and consequences will be identified to random variables (RV) rather than events. Borrowing from Glymour and Cooper [8], we adopt here an operational criterion of causality: Given a closed system of interdependent RV, an RV C may be called a cause of another RV E, called its effect or consequence, if imposing changes in the distribution of C (by means of manipulation performed by an agent external to the system) results in changes in the distribution of E. For example, C may be the choice of one of two available treatments for a patient with lung cancer and E may represent 5-year survival. If we randomly assign patients to the two treatments by flipping a fair coin and observe that the probability distribution for 5-year survival differs between the two treatment groups, we can conclude that the choice of treatment causally determines survival in patients with lung cancer.

There is a parallel between the operational test of causality and the notion of individual feature relevance of Definition 4.1. A feature X is individually irrelevant to the target Y *iff* $P(X,Y) = P(X)P(Y)$, that is, assuming that $P(X) > 0$, if $P(Y|X) = P(Y)$. Hence, individual relevance defined by contrast occurs if for some assignment of values $P(Y|X) \neq P(Y)$. In the test of causality, we must first define a manipulation. Borrowing the notation of [22], we will denote by $P(Y|do(X))$ and $P(do(X))$ the distributions resulting from the manipulation of variable X called "$do(X)$". In the test of causality, individual causal relevance occurs if $P(Y|do(X)) \neq P(Y)$.

The definitions of strong and weak feature relevance (Def. 4.3 and 4.4) can also be modified by replacing X_i by $do(X_i)$, yielding a notion of strong and weak causal relevance. Although these definitions are formally interesting in that they establish a parallel with the feature selection framework, they have little practical value. First, they have the same drawback as their non-causal feature relevance counterpart that they require exploring all possible subsets of features and assignment of values to features. Second, the fact that they require exploring all possible manipulations to establish the absence of a causal relationship with certainty is also unrealistic. When we may establish a causal relationship using a manipulation on X_i, thereafter any other manipulation that affects X_i will potentially affect Y. But the converse is not true. We must in principle try "all possible" manipulations to establish with certainty that there is no causal relationship. Practically, however, planned experiments have been devised as canonical manipulations and are commonly relied upon to rule out causal relationships (see, e.g., [23]). Nonetheless, they require conducting experiments, which may be costly, impractical, unethical, or even infeasible. The purpose of the following sections is to introduce the reader to the discovery of causality in the absence of experimentation. Experimentation will be performed punctually, when absolutely needed.

4.3.2   Causal Bayesian Networks

Causal Bayesian networks provide a convenient framework for reasoning about causality between random variables. Causal Bayesian networks implement a notion of causal ordering and do not model causal time dependencies in detail (although they can be extended to do so if desired). Even though other frameworks exist (like structural equation modeling [12, 13]), we limit ourselves in this chapter to Bayesian networks to illustrate simply the connections between feature selection and causality we are interested in.

Recall that in a directed acyclic graph (DAG), a node A is the parent of B (B is the child of A) if there is a direct edge from A to B, and A is the ancestor of B (B is the descendant of A) if there is a direct path from A to B. "Nodes" and "variables" will be used interchangeably. As in previous sections, we denote random variables with uppercase letters X, Y, Z; realizations (values) with lowercase letters, x, y, z; and sets of variables or values with boldface uppercase $\mathbf{X} = [X_1, X_2, ..., X_N]$ or lowercase $\mathbf{x} = [\mathbf{x}_1, \mathbf{x}_2, ..., \mathbf{x}_N]$, respectively. A "target" variable is denoted as Y.

We begin by formally defining a discrete Bayesian network:

DEFINITION 4.5 (Discrete) Bayesian network.   *Let* $\mathbf{X}$ *be a set of discrete random variables and P be a joint probability distribution over all possible realizations of* $\mathbf{X}$*. Let $\mathcal{G}$ be a directed acyclic graph (DAG) and let all nodes of $\mathcal{G}$ correspond one-to-one to members of* $\mathbf{X}$*. We require that for every node $A \in \mathbf{X}$, A is probabilistically independent of all non-descendants of A, given the parents of A (Markov Condition). Then we call the triplet* $\{\mathbf{X}, \mathcal{G}, P\}$ *a (discrete) Bayesian Network or, equivalently, a Belief Network or Probabilistic Network (see, e.g., [21]).*

Discrete Bayesian networks can be generalized to networks of continuous random variables and distributions are then replaced by densities. To simplify our presentation, we limit ourselves to discrete Bayesian networks. A causal Bayesian network is a Bayesian Network $\{\mathbf{X}, \mathcal{G}, P\}$ with the additional semantics that $(\forall A \in \mathbf{X})$ and $(\forall B \in \mathbf{X})$, if there is an edge from A to B in $\mathcal{G}$, then A directly causes B (see, e.g., [25]).

Using the notion of *d-separation* (see, e.g., [22]), it is possible to read from a graph $\mathcal{G}$ if two sets of nodes $\mathbf{A}$ and $\mathbf{B}$ are independent, conditioned on a third set $\mathbf{C}$: $\mathbf{A} \perp_{\mathcal{G}} \mathbf{B}|\mathbf{C}$. Furthermore, in a causal Bayesian network, the existence of a directed path between two nodes indicates a causal relationship. It is usually assumed that in addition to the Markov condition, which is part of the definition of Bayesian networks, another condition called "faithfulness" is also fulfilled (see, e.g., [22]). The faithfulness condition entails *dependencies* in the distribution from the graph while the Markov condition entails that *independencies* in the distribution are represented in the graph. Formally:

DEFINITION 4.6 Faithfulness. *A directed acyclic graph $\mathcal{G}$ is faithful to a joint probability distribution P over a set of variables $\mathbf{X}$ iff every independence present in P is entailed by $\mathcal{G}$ and the Markov condition, that is, $(\forall A \in \mathbf{X}, \forall B \in \mathbf{X}$ and $\forall \mathbf{C} \subset \mathbf{X})$, $A \not\perp_{\mathcal{G}} B|\mathbf{C} \Rightarrow A \not\perp_{P} B|\mathbf{C}$. A distribution P over a set of variables $\mathbf{X}$ is said to be* faithful *iff there exists a DAG $\mathcal{G}$ satisfying the faithfulness condition.*

Together, the Markov and faithfulness conditions guarantee that the Bayesian network will be an accurate map of dependencies and independencies of the represented distribution. Both the Markov condition and the faithfulness conditions can be trivially specialized to causal Bayesian networks.

Bayesian networks are fully defined by their graph and the conditional probabilities $P(X_i|DirectCauses(X_i))$. Those may be given by experts or trained from data (or a combination of both). Once trained, a Bayesian network may be used to compute the joint probability of all variables, by applying the Markov condition $P(X_1, X_2, ..., X_N) = \sum_i P(X_i|DirectCauses(X_i))$, as well as any joint or conditional probabilities involving a subset of variables, using the chain rule $P(A, B, C, D, E) = P(A|B, C, D, E) \, P(B|C, D, E) \, P(C|D, E) \, P(E)$ and marginalization $P(A, B) = \sum_{C,D,E} P(A, B, C, D, E)$. Such calculations are referred to as inference in Bayesian networks. In the worst cases, inference in Bayesian networks is intractable. However, many very efficient algorithms have been described for exact and approximate inference [22, 21].

To learn the structure of a causal Bayesian network, we can test for causal relationships with manipulations. A manipulation in a causal Bayesian network is defined as "clamping" variables to given values, while the other ones are let free to assume values according to the rules of the graph. However, the structure of a causal graph can, to some extent, be determined from observational data (i.e., without manipulation). One method consists in making statistical **tests of conditional independence** between variables, which allows us to determine the causal structure up to Markov equivalence classes (see Section 4.5).

4.4 Feature Relevance in Bayesian Networks

In this section we relate notions of non-causal feature relevance introduced in Section 4.2 with Bayesian networks introduced in Section 4.3.2. Strongly relevant features in the Kohavi-John sense are found in the Bayesian network

DAG in the immediate neighborhood of the target, but they are not necessarily strongly causally relevant. These considerations will allow us in Section 4.4.2 to characterize various cases of features called relevant according to different definitions.

4.4.1   Markov Blanket

Pearl [22] introduced the notion of Markov blanket in a Bayesian network as the set of nodes shielding a given node from the influence of the other nodes (see Figure 4.2). Formally, let $\mathbf{X} \cup Y$ ($Y \notin \mathbf{X}$) be the set of all variables under consideration and $\mathbf{V}$ a subset of $\mathbf{X}$. We denote by "\" the set difference.

DEFINITION 4.7 Markov blanket.   *A subset* $\mathbf{M}$ *of* $\mathbf{X}$ *is a Markov blanket of* Y *iff for any subset* $\mathbf{V}$ *of* $\mathbf{X}$, *and any assignment of values,* Y *is independent of* $\mathbf{V}\backslash\mathbf{M}$ *given* $\mathbf{M}$ *(i.e.,* $Y \perp \mathbf{V}\backslash\mathbf{M}|\mathbf{M}$, *that is,* $P(Y, \mathbf{V}\backslash\mathbf{M}|\mathbf{M}) = P(Y|\mathbf{M})P(\mathbf{V}\backslash\mathbf{M}|\mathbf{M})$ *or for* $P(\mathbf{V}\backslash\mathbf{M}|\mathbf{M}) > 0$, $P(Y|\mathbf{V}\backslash\mathbf{M}, \mathbf{M}) = P(Y|\mathbf{M})$) [21].*

Markov blankets are not unique in general and may vary in size. But, importantly, any given faithful causal Bayesian network (see Section 4.3.2) has a unique Markov blanket, which includes its direct causes (parents), direct effects (children), and direct causes of direct effects (spouses) (see, e.g., [22, 21]). The Markov blanket does not include direct consequences of direct causes (siblings) and direct causes of direct causes (grandparents). To understand the intuition behind Markov blankets, consider the example of Figure 4.2 in which we are looking at the Markov blanket of the central node "lung cancer":

- **Direct causes** (parents): Once all the direct causes have been given, an indirect cause (e.g., "anxiety") does not bring any additional information. In Figure 4.2(e), for instance, increased "anxiety" will eventually increase "smoking" but not influence directly "lung cancer," so it suffices to use "smoking" as a predictor, and we do not need to know about "anxiety." Similarly, any consequence of a direct cause (like "other cancers" in Figure 4.2(d), which is a consequence of "genetic factor 1") brings only indirect evidence, but no additional information once the direct cause "genetic factor 1" is known. Direct causes in faithful distributions are individually predictive, but they may otherwise need to be known jointly to become predictive (see, e.g., Figure 4.1(b): the chessboard/XOR problem, which is an example of unfaithfulness).

- **Direct effects** (children) and **direct causes of direct effects** (spouses): In faithful distributions, direct effects are always predictive of the target. But their predictive power can be enhanced by knowing other possible causes of these direct effects. For instance, in Figure 4.2(a), "allergy" may cause "coughing" independently of whether we have "lung cancer." It is important to know of any "allergy" problem, which would eventually *explain away* that "coughing" might be the result of "lung cancer." Spouses, which do not have a direct connecting path to the target (like "allergy"), are not individually predictive of the target ("lung cancer"), they need a common child ("coughing") to become predictive. However, in unfaithful distributions, children are not necessarily predictive without the help of spouses, for example in the case of the chessboard/XOR problem.

Following [27], we interpret the notion of Markov blankets in faithful distributions in terms Kohavi-John feature relevance as follows:

1. **Irrelevance:** A feature is irrelevant if it is disconnected from Y in the graph.
2. **Strong relevance:** Strongly relevant features form a Markov blanket **M** of Y.
3. **Weak relevance:** Features having a connecting path to Y, but not belonging to **M**, are weakly relevant.

The first statement interprets Definition 4.2 (irrelevance) in terms of disconnection to Y in the graph. It follows directly from the Markov properties of the graph. The second statement casts Definition 4.3 (strong relevance) into the Markov blanket framework. Only strongly relevant features cannot be omitted without changing the predictive power of **X**. Therefore, non-strongly relevant features can be omitted without changing the predictive power of **X**. Hence the set **M** of all strongly relevant features should be sufficient to predict Y, regardless of the values **v** assumed by the other features in $\mathbf{X}\backslash\mathbf{M}$: $P(Y|\mathbf{M}) = P(Y|\mathbf{M}, \mathbf{X}\backslash\mathbf{M} = \mathbf{v})$. Therefore, following Definition 4.7, **M** is a Markov blanket. Markov blankets are unique for faithful distributions (see [22]), which ensures the uniqueness of the set of strongly relevant features for faithful distributions.

The interpretation of the Markov blanket as the set of strongly relevant variables, which is valid for all Bayesian networks, extends to *causal* Bayesian networks. This means that strongly relevant in the Kohavi-John sense includes direct causes (parents), direct effects (children), and direct causes of the direct effects (spouses). Yet only direct causes are *strongly causally relevant* according to our definition (Section 4.3.1). Consequently, *weakly causally relevant* features coincide with indirect causes, which are ancestors in the graph (excluding the parents).

4.4.2 Characterizing Features Selected via Classical Methods

In this section, we analyze in terms of causal relationships several non-trivial cases of multivariate dependencies with artificial examples. This sheds a new light on the notion of feature relevancy. We limit ourselves to the analysis of variables, which are in the immediate proximity of the target, including members of the Markov blanket (MB) and variables in the vicinity of the MB (Figure 4.2). We analyze scenarios illustrating some basic three-variable

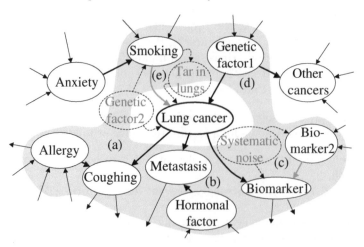

FIGURE 4.2: Markov blanket. The central node "lung cancer" represents a disease of interest, which is our target of prediction. The solid oval nodes in the shaded area include members of the Markov blanket. Given these nodes, the target is independent of the other nodes in the network. The letters identify local three-variable causal templates: (a), (b), and (c) colliders, (d) fork, and (e) chain. The dashed lines/nodes indicate hypothetical unobserved variables providing alternative explanations to the gray arrows.

causal templates: chains $A \to B \to C$, forks $A \leftarrow B \to C$, and colliders $A \to B \leftarrow C$. This allows us to refine the notion of feature relevance into:

- Direct cause (parent)
- Unknown direct cause (absent parent called *confounder*, which may result in mistaking a sibling for a parent)
- Direct effect (child)
- Unknown direct effect (which may cause *sampling bias* and result in mistaking a spouse for a parent)
- Other truly relevant MB members (spouses)
- Nuisance variable members of the MB (also spouses).

We touch upon the problem of causal sufficiency: In the presence of variables, which are unobserved or unknown, the MB does not necessarily include all the strong variables and may include weakly relevant variables. Our examples also include illustrations of the problem of experimental artifacts: We caution against the fact that feature selection algorithms may find nuisance variables (or effects of nuisance variables) "relevant."

Upstream of the Target: Chain and Fork Patterns

Let us first examine the roles played by variables (denoted X_2) directly connected to parents (denoted X_1) of the target Y, including grandparents and siblings. Those are involved in *chain* patterns, $X_2 \to X_1 \to Y$, and *fork* patterns, $X_2 \leftarrow X_1 \rightarrow Y$, both featuring the independence relations: $Y \not\perp X_2$ and $Y \perp X_2|X_1$. An example of a distribution satisfying such independence relations is depicted in Figure 4.1(d). We have already examined this example in Section 4.2: An apparent dependency between X_2 and Y may vanish if we introduce a new variable X_1 (Simpson's paradox). Importantly, the pattern of dependencies does not allow us to determine whether we have a chain or a fork, which prevents us from distinguishing grandparents from siblings; this can sometimes be resolved using dependencies with other variables, higher order moments of the distribution, experiments, or prior knowledge (see Section 4.5). Only parents are part of the Markov blanket and should, in principle, be considered "strongly relevant." The study of some examples allows us to understand the relevance of grandparents and siblings, as well as potential confusions between siblings or grandparents and parents:

- **Relevance of grandparents and siblings: Controllability and specificity of parents.** In our "lung cancer" example of Figures 4.2(d) and (e) the direct causes ("smoking" and "genetic factor 1") are *strongly relevant* (in the Markov blanket). Indirect causes and consequences of causes are only *weakly relevant* (outside the Markov blanket). We argue that siblings and grandparents are nevertheless worthy of attention. In particular, if the direct causes are not controllable (cannot be acted upon), it may be interesting to look at indirect causes (e.g., reducing "anxiety" might indirectly reduce the risk of "lung cancer"). Consequences of direct causes are also interesting for a different reason: They might weaken the relevance of strongly relevant features. In our example, the fact that "genetic factor 1" causes not only "lung cancer" but also "other cancers" makes it a non-specific marker of "lung cancer."

- **Ambiguity between grandparents and siblings: Unknown parents as confounders.** The ambiguity between forks and chains is at the heart of the correlation vs. causation problem. If the "true" parents are not known, grandparents become the most direct identifiable causes. However, if one cannot distinguish between sibling and grandparents, we may falsely think that siblings are causes. This is illustrated by the hypothetical alternative scenarios in Figure 4.2(e). The question is whether "smoking" is a cause of "lung cancer," given that there may be other unknown factors. First scenario: The existence of a more direct cause: "tar in lungs." "Smoking" remains a cause and may still be the most direct controllable cause, retaining its importance even if its no longer member of the MB. Second scenario: The existence of an unknown "genetic factor 2." In the recent years, new restrictions on smoking in public places have been imposed, based on the correlation between smoking and lung cancer. Some tobacco companies have been arguing that there may be a common genetic factor causing both craving for nicotine (and therefore smoking) and a predisposition to get lung cancer. Such a confounding factor, the hypothetical "genetic factor 2," has not been identified to date. But the possibility that it exists outlines the difficulty of drawing conclusions about causal relationships when human experimentation is not possible (in this case for ethical reasons).

Downstream of the Target: Colliders

Patterns of dependencies ($X_2 \perp Y$, $X_2 \not\perp Y|X_1$) are characteristic of unshielded *colliders*: $Y \rightarrow X_1 \leftarrow X_2$. Both children (denoted X_1) and spouses (denoted X_2) are involved in such patterns, which are found downstream of the target Y. As explained before, both children and spouses are members of the Markov blanket, and as such they are "strongly relevants" in the Kohavi-John sense for faithful distributions. Two cases of distributions corresponding to colliders are shown in Figures 4.1(a) and (b). One corresponds to a faithful case (consistent only with: $Y \rightarrow X_1 \leftarrow X_2$) and the other to an unfaithful case (chessboard problem, consistent with several possible graphs: $Y \rightarrow X_1 \leftarrow X_2$, $X_1 \rightarrow Y \leftarrow X_2$, and $Y \rightarrow X_2 \leftarrow X_1$). In either case, spouses can be useful complements of children to improve prediction power. Nonetheless, we must caution against two types of problems that may be encountered: sampling bias and artifacts. We illustrate these various cases:

- **Relevance of a spouse: Explaining away the effect of Y.** Children and spouses are not "causally" relevant, in the sense that manipulating them does not affect the target. Yet, they may be used to make predictions for stationary systems, or as predictors of the effect of manipulations of the target (e.g., the effect of a treatment of "lung cancer"). We previously noted that "allergy" is a useful complement of "coughing" to predict "lung cancer," because knowing about allergy problems allows the doctor to "explain away" the fact that coughing may be the symptom of "lung cancer." Now, after a patient receives a treatment for "lung cancer" (manipulation), a reduction in "coughing" may be an indication of success of the treatment.

- **False causal relevance of a spouse: Sampling bias.** In Figure 4.2(b) we show a scenario of "sampling bias" in the subgraph: *Lungcancer* $\rightarrow$ *Metastasis* $\leftarrow$ *Hormonal factor*. The presence of metastases may be unknown. It may turn out that all the patients showing up in the doctor's office are more likely to have late stage cancer with metastases because only then do they experience alarming fatigue symptoms. In this situation, the sample of patients seen by the doctor is biased. If, from that sample a correlation between a certain hormonal factor and lung cancer is observed, it may be misinterpreted as causal. In reality, the dependency may only be due to the sampling bias. "Hormonal factor" (playing the role of X_2) cannot be used as a predictive factor without knowing about the "metastasis" factor (playing the role of X_1), and wrong results could be inferred if that factor is unknown.

- **False relevance of a spouse: Artifacts.** In Figure 4.2(c), we show an example of an experimental artifact. Assume that we are using an instrument to measure the abundance of some medical diagnosis biomarkers in blood serum (e.g., proteins or metabolites), and we have identified two promising complementary biomarkers (numbered 1 and 2), which have a distribution similar to that of X_1 and X_2 in Figure 4.1(a). The simple model $Y \rightarrow X_1 \leftarrow X_2$ explains the observed distribution, in support of the relevance of "Biomarker2," which can be assumed to be part of the MB. However, the relevance of "Biomarker2" may be challenged if we suspect the existence of some unknown "Systematic noise" variable S due to the measuring instrument. A model $Y \rightarrow X_1 \leftarrow S \rightarrow X_2$ could also explain the observed data. Then S may be part of the MB, not X_2. In that case, feature X_2 is an indirect measurement of the noise S useful to correct the measurement error, but not relevant to the system under study (human disease).

4.5 Causal Discovery Algorithms

In previous sections, we have motivated the introduction of the concept of causality in feature selection. It has long been thought that causal relationships can only be evidenced by "manipulations," as summarized by the motto commonly attributed to Paul Holland and Don Rubin: "No causation without manipulation." For an introduction on manipulation methods of inferring causation, see, for instance, [23]. Yet, in the recent years much fruitful research has been devoted to inferring causal relationships from "observational data," that is, data collected on a system of interest, without planned experiments or intervention. Current books exploring these techniques include [8, 22, 25]. We collectively refer to the algorithms as "causal discovery machine learning methods."

Learning a Bayesian network $\{\mathbf{X}, \mathcal{G}, P\}$ from data consists in two subtasks, sometimes performed jointly, sometimes in sequence: learning the structure of the graph G and learning the probability distribution P. From the point of view of causal discovery and feature selection, learning the structure of the graph is the subtask of interest.

In what follows, we will make the following set of "causal discovery assumptions": (i) Causal sufficiency: The set of observable variables $\mathbf{X}$ is self-*sufficient* to characterize all causal relationships of interest, which imposes that direct common causes of all pairs of variables are observed. (ii) Statistical sufficiency: The learner has access to a sufficiently large training set and reliable statistical tests for determining conditional dependencies and independencies in the original distribution where the data are sampled from.

(iii) Faithfulness: The process that generated the data having the distribution $P(\mathbf{X}, Y)$ can be faithfully represented by the family of models under consideration (here causal Bayesian networks).

4.5.1   A Prototypical Causal Discovery Algorithm

We outline in this section the fundamental operation of the Peter-Clark (PC) algorithm (barring speed-up techniques and implementation details in order to simplify the presentation; see [25] for a complete description). Under the causal discovery assumptions stated above, this algorithm is provably sound in the large sample limit [25], in the sense that it can recover the structure of a Bayesian network (BN) that generated the data, up to a Markov equivalence class (that is, a class of BN sharing the same set of conditional independence conditions).

The algorithm begins with a fully connected unoriented graph and has three phases:

Algorithm: PC

Let A, B, and C be variables in $\mathbf{X}$ and $\mathbf{V}$ any subset of $\mathbf{X}$. Initialize with a fully connected un-oriented graph.

1. Find unoriented edges by using the criterion that variable A shares a direct edge with variable B *iff* no subset of other variables $\mathbf{V}$ can render them conditionally independent ($A \perp B | \mathbf{V}$).

2. Orient edges in "collider" triplets (i.e., of the type $A \rightarrow C \leftarrow B$) using the criterion that if there are direct edges between A, C and between C, B, but not between A and B, then $A \rightarrow C \leftarrow B$, *iff* there is no subset $\mathbf{V}$ containing C such that $A \perp B | \mathbf{V}$.

3. Further orient edges with a constraint-propagation method by adding orientations until no further orientation can be produced, using the two following criteria: (i) If $A \rightarrow B \rightarrow ... \rightarrow C$, and $A - C$ (i.e., there is an undirected edge between A and C), then $A \rightarrow C$. (ii) If $A \rightarrow B - C$, then $B \rightarrow C$.

Without going into details, we note that all of the causal discovery assumptions can be relaxed via a variety of approaches. For example, if the causal sufficiency property does not hold for a pair of variables A and B, and there is at least one common parent C of the pair that is not measured, the PC algorithm might wrongly infer a direct edge between A and B. The FCI algorithm addresses this issue by considering all possible graphs including hidden nodes (latent variables) representing potential unmeasured "confounders," which are consistent with the data. It returns which causal relationships are guaranteed to be unconfounded and which ones cannot be determined by the observed data alone. The FCI algorithm is described in detail in [25]. The PC algorithms and their derivatives remain limited to discovering causal structures up to Markov equivalence classes. For instance, since the two graphs $X \to Y$ and $X \leftarrow Y$ are Markov equivalent, the direction of the arrow cannot be determined from observational data with such methods. Other methods have been proposed to address this problem; see, e.g., [26]. Designed experiments; or "active learning" may be use instead or in conjunction with observational methods to resolve ambiguous cases; see, e.g., [20].

4.5.2 Markov Blanket Induction Algorithms

From our previous discussion it follows that one can apply the PC algorithm (or other algorithms that can learn high-quality causal Bayesian networks) and extract the Markov blanket of a target variable of interest Y. However, when the dataset has tens or hundreds of thousands of variables, or when at least some of them are highly interconnected, applying standard causal discovery algorithms that learn the full network becomes impractical. In those cases, local causal discovery algorithms can be used, which focus on learning the structure of the network only in the immediate neighborhood of Y.

The first two algorithms for Markov blanket induction by Koller and Sahami and Cooper et al. [16, 6] contained many promising ideas, and the latter was successfully applied in the real-life medical problem of predicting community acquired pneumonia mortality; however, they were not guaranteed to find the actual Markov blanket, nor could they be scaled to thousands of variables. Margaritis and Thrun [19] subsequently invented a sound algorithm, Grow-Shrink (GS), however, it required samples at least exponential to the size of the Markov blanket and would not scale to thousands of variables in most real datasets with limited samples sizes. Tsamardinos et al. [27] introduced several improvements to GS with the Iterative Associative Markov Blanket (IAMB) algorithms, while Aliferis et al. [3] introduced HITON ("hiton" means "blanket" in Greek). Both types of algorithms scale well (100,000 variables in a few CPU-hours), but the latter is more sample efficient.

For illustration, we describe HITON in some detail. The same induction criterion as PC is used to find edges (i.e., X_i shares a direct edge with the target Y *iff* there is no subset $\mathbf{V}$ of the variables set $\mathbf{X}$ such that $X_i \perp Y | \mathbf{V}$). However, while PC starts with a fully connected graph, HITON starts with an empty graph. Accordingly, for PC, conditioning sets include large numbers of variables, while, for HITON, they include small dynamically-changing subsets of "neighbors" of Y (direct causes and effects of Y). Spouses are identified by first finding the neighborhood of depth two (i.e., by recursive application of the direct-edge induction step) and then by eliminating non-spouses. This reduces errors from incorrect orientations. As more and more variables are scanned, the algorithm converges to the Markov blanket of Y. Aside from limiting the search in a neighborhood of Y, which already represents a significant computational speedup compared to building an entire Bayesian network, HITON accelerates the search with a number of heuristics, including limiting conditioning sets to sizes permitting the sound estimation of conditional probabilities and prioritizing candidate variables.

An important novelty of local methods is circumventing non-uniform graph connectivity. A network may be non-uniformly dense (or sparse). In a global learning framework, if a region is particularly dense, that region cannot be discovered fast and, when learning with a small sample, it will produce many errors. These errors propagate to remote regions in the network (including those that are learnable accurately and fast with local methods). On the contrary, local methods will be both fast and accurate in the less dense regions. Thus local methods are also competitive for learning full Bayesian networks.

Localizing the search for direct edges is desirable according to the previous explanation, but far from obvious algorithmically [28]. A high-level explanation is that, when building the parents/children sets around Y in a localized manner, we occasionally omit variables X_i not connected to Y but connected to other variables X_j, which are not parents or children of Y. This happens because variables such as X_i act as "hidden variables" insofar as the localized criterion for independence is concerned. It turns out, however, that (i) the configuration in which this problem can occur is rare in real data, and (ii) the problem can be detected by running the localized criterion in the opposite direction (i.e., seeking the parents/children of X_j in a local fashion). This constitutes the symmetry correction of localized learning of direct edges.

The Causal Explorer software package, including HITON and many other useful causal discovery algorithms, is available from the Internet [4].

4.6   Examples of Applications

Causality and feature selection as described in this chapter have been used to achieve various objectives in different areas such as bio-informatics, econometrics, and engineering. We present below one example from each of these fields that illustrates the use of causal and feature techniques in practice.

With the advent of the DNA microarrays technology [24], biologists have collected the expression of thousands of genes under several conditions. Xing et al. were among the first to use a Markov blanket discovery algorithm for feature selection [29] in DNA microarray data, to diagnose disease (two kinds of leukemia). Friedman and colleagues [7] applied a causal discovery technique on microarray data to build a causal network representing the potential dependencies between the regulations of the genes. If the expression level of one gene causes the up or down regulation of another gene, an edge should link them. A simple feature selection technique based on correlation is first applied to select a set of potential causes for each gene. A causal discovery method is then run on the reduced set of potential causes to refine the causal structure. More recently, the Markov blanket discovery algorithm HITON [3] has been applied with success to clinical, genomic, structural, and proteomic data, and mining the medical literature, achieving significantly better reduction in feature set size without classification degradation compared to a wide range of alternative feature selection methods. Other applications include understanding physician decisions and guideline compliance in the diagnosis of melanomas, discovering biomarkers in human cancer data using microarrays and mass spectrometry, and selecting features in the domain of early graft failure in patients with liver transplantations (see [2] for reports and comparisons with other methods).

In biology and medicine, causal discovery aims at guiding scientific discovery, but the causal relationships must then be validated by experiments. The original problem, e.g., the infeasibility of an exhaustive experimental approach to detect and model gene interactions, is addressed using causality by defining a limited number of experiments that should be sufficient to extract the gene regulatory processes. This use of causality is in contrast with our second example, economy and sociology, where experiments in a closed environment are usually not possible, i.e., there is no possible laboratory validation before using the treatment in real situations. Causality has been used by economists for more than 40 years. Some years before artificial intelligence started to address the topic, Clive Granger [9] defined a notion of temporal causality that is still in use today. In 1921, Wright introduced Structure Equation Modeling (SEM) [12, 13], a model widely known by sociologists and economists. It is therefore singular to see that marketing research – a field close to economy and sociology – does not contain much work involving causality. The defense of SEM by Pearl [22] might change the status, though, and causality appears slowly as to be a subject of interest in marketing. From a practical perspective, causality can be directly used to addresses one of the key questions that marketers ask: how to assess the impact of promotions on sales? It is known that many potential factors come into play when computing the effect of promotions: weather, word of mouth, availability, special days (e.g., Valentine's Day), etc. Understanding how these factors influence the sales is interesting from a theoretical point of view but is not the primary objective: What practically matters is what to do next, that is, what will be the effect of promotions versus no promotions next month. This is typically a problem of causal discovery and parameter estimation. Finding the causal link is not enough. It is necessary to know whether the promotion will have a positive effect and how positive it will be in order to compute the expected profit. A promotion that has a small positive effect but costs a lot to implement might not be worth launching. Extracting the true causal structure is also less critical than estimating $P(\text{sales}|\text{do}(\text{promotions}))$.

Failure diagnosis is the last application we shall consider. In diagnosing a failure, engineers are interested in detecting the cause of defect as early as possible to save cost and to reduce the duration of service breach. Bayesian networks and their diagnostic capabilities, which are of particular relevance when the links are causal, have been used to quickly perform a root cause analysis and to design a series of tests minimizing the overall cost of diagnosis and repair. Kraaijeveld et al. [17] present an approach that relies on a user-defined causal structure to infer the most probable causes based on a description of the symptoms.

These three applications show that causality techniques can be used in different settings with different requirements.

4.7   Summary, Conclusions, and Open Problems

Feature selection focuses on uncovering subsets of variables $X_1, X_2, \ldots$ predictive of a target Y. In light of causal relationships, the notion of variable relevance can be refined. In particular, causes are better targets of action of external agents than effects: If X_i is a cause of Y, manipulating it will have an effect on Y, not if X_i is a consequence (or effect). In the language of Bayesian networks, direct causes (parents), direct effects (children), and other direct causes of the direct effects (spouses) are all members of the Markov blanket. The members of the Markov blanket are strongly relevant in the Kohavi-John sense, for faithful distributions. Direct causes are strongly causally relevant. Spouses are not individually relevant, but both parents and children are, in faithful distributions. Both causes and consequences of Y are predictive of Y, but consequences can sometimes be "explained away" by other causes of the consequences of Y. So the full predictive power of children cannot be harvested without the help of spouses. Causes and consequences have different predictive powers when the data distribution changes between training and utilization time, depending on the type of change. In particular, causal features should be more predictive than consequential features, if new unknown "noise" is added to the variables $X_1, X_2, \ldots$ (the co-variate shift problem). If new unknown noise is added to Y, however, consequential variables are a better choice. Unknown features, including possible artifacts or confounders, may cause the whole scaffold of causal feature discovery to fall apart if their possible existence is ignored. Causal feature selection methods can assist the design of new experiments to disambiguate feature relevance.

Connecting feature selection and causality opens many avenues of interesting future research, including:

1. Characterizing theoretically and/or empirically existing and novel feature selection methods in terms of causal validity
2. Developing appropriate metrics, research designs, benchmarks, etc., to empirically study the performance and pros and cons of causal vs. non-causal feature selection methods
3. Studying the concept of relevancy and its relationship with causality beyond faithful distributions and beyond Kohavi-John relevancy
4. Improving computational performance and accuracy of causal feature selection methods for large dimensional problems and small samples

5. Developing a theory of the statistical complexity of learning causal relationships

6. Developing powerful and versatile software environments for causally-oriented feature selection

7. Examining the validity of and relaxing assumptions motivated by efficiency or convenience (e.g., faithfulness, causal sufficiency, normality of distributions, linearity of relationships) when applied to real-world feature selection situations.

The interested reader is encouraged to pursue his reading, starting perhaps with [8, 22, 25, 21, 15].

Acknowledgments

We are grateful to Alexander Statnikov, Philippe Leray, and the reviewers for their helpful comments. Part of this work was initiated while Isabelle Guyon was visiting Prof. Joachim Buhmann at ETH Zürich. His support is gratefully acknowledged. Constantin Aliferis acknowledges support from grants 1R01 LM007948-01, 1U01 HL081332-01, and 1 U24 CA126479-01.

References

[1] Special issue on relevance. *Artificial Intelligence*, 97(1-2), Dec. 1997.

[2] Dicovery systems laboratory bibliography, 2007, http://discover.mc.vanderbilt.edu/discover/public/publications.html.

[3] C. F. Aliferis, I. Tsamardinos, and A. Statnikov. HITON, a novel Markov blanket algorithm for optimal variable selection. In *American Medical Informatics Association (AMIA) Annual Symposium*, pages 21–25, 2003.

[4] C. F. Aliferis, I. Tsamardinos, A. Statnikov, and L. Brown. Causal explorer: A probabilistic network learning toolkit for biomedical discovery. In *2003 International Conference on Mathematics and Engineering Techniques in Medicine and Biological Sciences (METMBS)*, Las Vegas, Nevada, USA, June 23-26 2003, http://discover1.mc.vanderbilt.edu/discover/public/causal_explorer/. CSREA Press.

[5] A. Blum and P. Langley. Selection of relevant features and examples in machine learning. *Artificial Intelligence*, 97(1-2):245–271, Dec. 1997.

[6] G. Cooper and E. Herskovits. A Bayesian method for the induction of probabilistic networks from data. *Mach. Learning*, 9(4):309–347, 1992.

[7] N. Friedman, M. Linial, I. Nachman, and D. Pe'er. Using bayesian networks to analyze expression data. In *RECOMB*, pages 127–135, 2000.

[8] C. Glymour and G. C. Editors. *Computation, Causation, and Discovery.* AAAI Press/The MIT Press, Menlo Park, CA; Cambridge, MA; London, England, 1999.

[9] C. Granger. Investigating causal relations by econometric models and cross-spectral methods. *Econometrica*, 37:424–438, 1969.

[10] I. Guyon, C. Aliferis, and A. Elisseeff. Causal feature selection. Technical report, Berkeley, CA, March 2007, http://clopinet.com/isabelle/Papers/causalFS.pdf.

[11] I. Guyon, S. Gunn, M. Nikravesh, and L. Zadeh, Editors. *Feature Extraction, Foundations and Applications.* Studies in Fuzziness and Soft Computing. Physica-Verlag, Springer, Heidelberg, 2006.

[12] D. Kaplan. *Structural Equation Modeling: Foundations and Extensions*, volume 10 of *Advanced Quantitative Techniques in the Social Sciences series.* Sage, 2000.

[13] R. B. Kline. *Principles and Practice of Structural Equation Modeling.* The Guilford Press, 2005.

[14] R. Kohavi and G. John. Wrappers for feature selection. *Artificial Intelligence*, 97(1-2):273–324, December 1997.

[15] D. Koller and N. Friedman. *Structured Probabilistic Models: Principles and Techniques.* MIT Press, Cambridge, MA, 2007, to appear.

[16] D. Koller and M. Sahami. Toward optimal feature selection. In *13th International Conference on Machine Learning*, pages 284–292, July 1996.

[17] P. C. Kraaijeveld and M. J. Druzdzel. Genierate: An interactive generator of diagnostic bayesian network models. In *16th International Workshop on Principles of Diagnosis*, Monterey, CA, 2005.

[18] H. Liu and H. Motoda. *Feature Extraction, Construction and Selection: A Data Mining Perspective.* Kluwer Academic, 1998.

[19] D. Margaritis and S. Thrun. Bayesian network induction via local neighborhoods. Technical Report CMU-CS-99-134, Carnegie Mellon University, August 1999.

[20] K. Murphy. Active learning of causal bayes net structure, 2001.

[21] R. E. Neapolitan. *Learning Bayesian Networks.* Prentice Hall series in Artificial Intelligence. Prentice Hall, Eaglewood Cliffs, NJ, 2003.

[22] J. Pearl. *Causality: Models, Reasoning and Inference.* Cambridge University Press, Cambridge, UK, March 2000.

[23] D. Rubin. Estimating causal effects of treatments in randomized and nonrandomized studies. *Journal of Educational Psychology,* 66(5):688–701, 1974.

[24] M. Schena, D. Shalon, R. W. Davis, and P. O. Brown. Quantitative monitoring of gene expression patterns with a complementary dna microarray. *Science,* 270(5235):467–470, October 1995.

[25] P. Spirtes, C. Glymour, and R. Scheines. *Causation, Prediction, and Search.* MIT Press, Cambridge, MA, 2000.

[26] X. Sun, D. Janzing, and B. Schölkopf. Causal inference by choosing graphs with most plausible Markov kernels. In *Ninth International Symposium on Artificial Intelligence and Mathematics,* 2006.

[27] I. Tsamardinos and C. Aliferis. Towards principled feature selection: Relevance, filters, and wrappers. In *Ninth International Workshop on Artificial Intelligence and Statistics,* Florida, USA, January 2003.

[28] A. C. Tsamardinos I, Brown LE. The max-min hill-climbing bayesian network structure learning algorithm. *Machine Learning,* 651:31–78, 2006.

[29] E. P. Xing, M. I. Jordan, and R. M. Karp. Feature selection for high-dimensional genomic microarray data. In *Proc. 18th International Conf. on Machine Learning,* pages 601–608. Morgan Kaufmann, San Francisco, CA, 2001.

Part II

Extending Feature Selection

Chapter 5

Active Learning of Feature Relevance

Emanuele Olivetti

SRA Division, ITC-IRST

Sriharsha Veeramachaneni

SRA Division, ITC-IRST

Paolo Avesani

SRA Division, ITC-IRST

5.1 Introduction

This chapter deals with active feature value acquisition for feature relevance estimation in domains where feature values are expensive to measure. The following two examples motivate our work.

Example 1: Molecular reagents called biomarkers are studied for cancer characterization by testing them on biological (e.g., tissue) samples from patients who have been monitored for several years and labeled according to their cancer relapse and survival status. New biomarkers are tested on these biological samples with the goal of obtaining a subset of biomarkers that characterize the disease. In addition to the relapse and survival information, for each patient, information such as grade of the disease, tumor dimensions, and lymphonode status is also available. That is, the samples are class labeled as well as described by some existing features. The goal is to choose the best subset of new features (biomarkers) among many that are most informative about the class label given the existing features. Since each time a biomarker is tested on a biological sample the sample cannot be used for testing other biomarkers, it is desirable to evaluate the biomarkers by testing them on as few samples as possible. Once some of the biomarkers are determined to be informative, they can be tested on all the samples. A detailed description of this problem is presented in [20, 8, 12].

Example 2: In the agricultural domain, biologists study the symptoms of a certain disease by monitoring a controlled collection of trees affected by the disease. A data archive is arranged with each record describing a single tree. All the records are labeled as infected or not infected. The biologists then propose candidate features (e.g., color of leaves, altitude of the tree, new chemical tests, etc.) that could be extracted (or measured) to populate the archive, so as to ultimately arrive at a set of most predictive symptoms. Since the data collection on the field is usually very expensive or time consuming, there is a need for a data acquisition plan that is aimed at accurately estimating the relevance of the candidate features, so that only the most relevant features may be extracted on all trees.

The above two examples demonstrate the need for a data acquisition procedure with the goal of accurate feature relevance estimation. Data acquisition has traditionally been studied in machine learning under the topic of *active learning*. Our formulation of the active learning problem differs from the traditional setting of active learning where the class labels of unlabeled examples are queried [3, 17, 19]. Differently from most previous work in active learning, we consider a situation where class-labeled instances or subjects are monitored. A set of potentially informative features are proposed by experts in order to learn a predictive model. It is conceivable that some of these can-

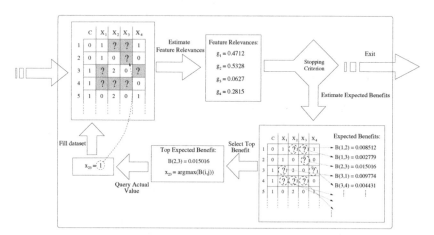

FIGURE 5.1: Active sampling process for feature relevance estimation. The entries shown in gray in the dataset in the top left corner indicate the ones that are missing at a particular instance. The bottom right-hand corner shows the process of computing the benefit of sampling at a particular missing value. The missing value with the highest benefit is chosen and actually sampled.

didate features are useless or redundant. Moreover, the measurement of the features on all the instances may be costly. It is therefore necessary to evaluate the efficacy (or relevance) of a new candidate feature by measuring it on a subsample of the instances to permit discarding useless features inexpensively.

As opposed to random subsampling, we propose to choose this subsample actively, that is, to choose the subsample that is likely to provide the best estimate of the relevance of the feature to the class. The active sampling strategy is to iteratively choose the most informative missing value to fill given all the previous data. When the budget for the data acquisition has been exhausted, the estimate of the feature relevances can then be used for feature selection. The selected features are then measured on all the instances and the resulting database is used to generate the model. The overall savings in cost comes from discarding irrelevant features without having measured them on all instances. The process is illustrated in Figure 5.1.

Although our final goal is feature selection, we do not review the extensive previous work in this area [1, 9] because we are mainly concerned with feature value *acquisition*. We would, however, like to mention that research in feature selection distinguishes between the so-called *wrapper* and *filter* approaches, depending upon whether or not the feature selection is explicitly based upon the accuracy of the final classifier [11]. Our active feature sampling algorithm that interleaves feature value acquisition and feature relevance estimation is independent of the final classifier that might be used and therefore is a *filter* method.

There has been previous work in active feature value acquisition for classifier induction [13, 22, 14] or testing [18] where the goal is to minimize the number of feature values acquired to learn and deploy a classifier on the *entire* feature set. This approach is inappropriate in some domains such as medical and agricultural, because the entire set of feature values is expensive to obtain not just on the *training* examples but also on the *test* instances (i.e., after the classifier is deployed). We have recently shown that active sampling heuristics developed for classifier induction perform poorly when the goal is feature relevance estimation [21].

Although the general theory of active learning derives from the theory of optimal experimentation [5, 16], its application to problems in learning raises practical issues such as finding good approximations to the theory, dealing with missing values, and learning with sampling bias (which is a side effect of active sampling).

In Section 5.2 we present the task of active sampling for feature relevance estimation and present the active sampling algorithm in abstract terms. In Section 5.3 we derive a sampling benefit function from a statistical formulation of the problem. Our specific implementation of the proposed method is presented in Section 5.4, where all necessary derivations and choices are described. In Section 5.5 we show the results of our experiments with discussion. Conclusions and future work are presented in Section 5.6.

5.2 Active Sampling for Feature Relevance Estimation

Let us consider a dataset $D = \{d_i\}_{i=1,\ldots,N}$ of a finite number of records, each corresponding to a pattern instance (or a subject). Let the random variable corresponding to the class label be denoted by $\mathbf{c}$ taking values in $\mathcal{C}$. The random vector $\mathbf{x} = (\mathbf{x}_1, \ldots, \mathbf{x}_F)$ corresponds to F features (or attributes) that can be extracted on any instance, taking on values in $\mathcal{X}_1 \times \ldots \times \mathcal{X}_F$. Each record comprises a class value and feature values: $d_i = (\mathbf{c}, \mathbf{x}_1, \ldots, \mathbf{x}_F)$. Initially the class labels are known for every instance in D. However, all of the feature values are missing for every instance.[1] The goal is to incrementally select a specified number of missing values to fill so as to estimate the relevance of all the features most accurately. For now we leave the definitions of the terms "feature relevance" and "most accurately" unspecified.

Our proposed incremental feature value acquisition process is illustrated in general terms in Figure 5.1. At any stage of the process the dataset has some feature values missing, as indicated in the figure. We estimate the vector $\hat{\mathbf{g}}$ of *feature relevances* according to which the features may be ranked. For each missing entry we can calculate the *benefit* of acquiring the value of that entry. We then choose the entry with the maximum benefit and actually acquire its value, which is entered into the dataset.[2] We perform this process iteratively until some stopping criterion has been met. The core step of the process, where one missing value is acquired, is described in Algorithm 5.2.1.

Algorithm 5.2.1: AcquireOneMissingValue(D_k)

$\hat{\mathbf{g}}(D_k) = EstimateRelevances(D_k)$
for each (i, f) such that record i has feature value f missing
$\quad B[i, f] \leftarrow 0$ **comment:** Initialize the value of the benefit to zero

$\quad$**for each** $x \in \mathcal{X}_f$
$\quad\quad D_{temp} = D_k.FillValue(\mathbf{x}_{if} = x)$
$\quad\quad \hat{\mathbf{g}}(D_{Temp}) = EstimateRelevances(D_{Temp})$
$\quad\quad B[i, f] = B[i, f] + ComputeBenefit(\hat{\mathbf{g}}(D_{temp}), \hat{\mathbf{g}}(D_k))$
$\quad$**end**
end
comment: Now find the missing entry with the highest benefit

$(i^*, f^*) = \underset{i,f}{\mathrm{argmax}} \, (B[i, f])$
comment: Now query the value for the missing entry

$x^* = SampleMissingValue(i^*, f^*)$
comment: Fill the missing value

$D_{k+1} = D_k.FillValue(\mathbf{x}_{i^*f^*} = x^*)$
return (D_{k+1})

The benefit of acquiring the value of a missing entry is the amount of information such an acquisition provides about the feature relevances. We will derive the benefit function in Section 5.3. The stopping criterion will be problem dependent and may involve factors such as the total cost of feature value acquisition, the change in the feature relevances, or confidence in the feature relevance estimates.

5.3 Derivation of the Sampling Benefit Function

We will now derive our active sampling algorithm in more abstract terms. Consider the dataset D as described above. Let the probability distribution over $\mathcal{C} \times \mathcal{X}_1 \times \ldots \times \mathcal{X}_F$ be parametrized by $\boldsymbol{\theta} \in \boldsymbol{\Theta}$. It is required to estimate a vector valued function of the distribution parameter vector $\boldsymbol{\theta}$ $g(\boldsymbol{\theta}) = (g_1(\boldsymbol{\theta}), \ldots)$ accurately under a sum-of-squared-error loss function by filling as few missing values of D as possible.

After k sampling steps, the dataset is partially filled and the remaining values are missing. The current dataset is denoted D_k. The Bayes minimum mean square error (MMSE) estimate of g given D_k is given by $\hat{g}(D_k) = E[\mathbf{g}|D_k]$. The current mean squared error is given by

$$\text{MSE}_k = \sum_{j=1}^{F} \int_{\mathcal{G}_j} (E[\mathbf{g}_j|D_k] - g_j)^2 p(g_j|D_k) dg_j \tag{5.1}$$

Currently for an instance i we have the class label c_i available and perhaps also some subset of the feature values $obs(i)$ (observed values for instance i). Let $mis(i)$ be the subset of features values currently missing for instance i, and $\mathbf{x}_f$ be a particular feature whose value is missing for instance i. If we assume that for the $(k+1)^{th}$ sampling step this missing feature value is measured and a value of x was obtained, then the new dataset, denoted $(D_k, \mathbf{x}_{if} = x)$, has the value x for the feature f for instance i. The new mean squared error would be

$$\sum_{j=1}^{F} \int_{\mathcal{G}_j} (E[\mathbf{g}_j|D_k, \mathbf{x}_{if} = x] - g_j)^2 p(g_i|D_k, \mathbf{x}_{if} = x) dg_j$$

Since we do not know in advance what value would be obtained if we did sample at $\mathbf{x}_{if}$, we need to average the above quantity over all the possible outcomes, so as to estimate the predicted mean square error (denoted $\hat{\text{MSE}}(i, f)_{k+1}$) if we sampled the missing value for the feature f, for instance i. That is,

$$\hat{\text{MSE}}(i,f)_{k+1} = \sum_{j=1}^{F} \int_{\mathcal{X}_f} \int_{\mathcal{G}_j} (E[\mathbf{g}_j|D_k, \mathbf{x}_{if} = x] - g_j)^2 p(g_j|D_k, \mathbf{x}_{if} = x)$$
$$p(\mathbf{x}_{if} = x|D_k)$$

$$= \sum_{j=1}^{F} \int_{\mathcal{X}_f} \int_{\mathcal{G}_j} (E[\mathbf{g}_j|D_k, \mathbf{x}_{if} = x] - g_j)^2 p(g_j, \mathbf{x}_{if} = x|D_k)$$

$$(5.2)$$

Now the best missing value (i, f) to measure is the one that yields the lowest predicted mean squared error $\hat{\text{MSE}}(i,f)_{k+1}$. This criterion is akin to *Bayesian A-optimality* in experiment design [2].

Adding and subtracting $E[\mathbf{g}_j|D_k]$ inside the squared term in Equation 5.2, we obtain

$$\hat{\text{MSE}}(i,f)_{k+1} = \sum_{j=1}^{F} \int_{\mathcal{X}_f} \int_{\mathcal{G}_j} (E[\mathbf{g}_j|D_k, \mathbf{x}_{if} = x] - E[\mathbf{g}_j|D_k])^2 p(g_j, \mathbf{x}_{if} = x|D_k)$$

$$+ 2\sum_{j=1}^{F} \int_{\mathcal{X}_f} \int_{\mathcal{G}_j} (E[\mathbf{g}_j|D_k, \mathbf{x}_{if} = x] - E[\mathbf{g}_j|D_k]) \ldots$$

$$\ldots (E[\mathbf{g}_j|D_k] - g_j) p(g_j, \mathbf{x}_{if} = x|D_k)$$

$$+ \sum_{j=1}^{F} \int_{\mathcal{X}_f} \int_{\mathcal{G}_j} (E[\mathbf{g}_j|D_k] - g_j)^2 p(g_j, \mathbf{x}_{if} = x|D_k)$$

Since $p(g_j, \mathbf{x}_{if} = x|D_k) = p(\mathbf{x}_{if} = x|D_k)p(g_j|D_k, \mathbf{x}_{if} = x)$ and both $E[\mathbf{g}_j|D_k]$ and $E[\mathbf{g}_j|D_k, \mathbf{x}_{if} = x]$ are functionally independent of g_j, it can be shown that the second summand is -2 times the first summand. Furthermore, the third summand is functionally independent of (i, f) since x integrates out (by interchanging the order of integration). Therefore, we have

$$\hat{\text{MSE}}(i,f)_{k+1} = A - \int_{\mathcal{X}_f} \sum_{j=1}^{F} (E[\mathbf{g}_j|D_k, \mathbf{x}_{if} = x] - E[\mathbf{g}_j|D_k])^2 p(\mathbf{x}_{if} = x|D_k)$$

where A is independent of i and f. That is, in order to minimize the predicted mean squared error if the missing value at (i, f) is measured, it is sufficient to maximize the sum of the squared differences between the Bayes estimates of $\mathbf{g}$ before and and after the value at (i, f) is measured, averaged over the possible outcomes.

Therefore, to minimize the predicted mean squared error, the objective function to be maximized is

$$B(i,f) = \int_{\mathcal{X}_f} \sum_{j=1}^{F} (E[\mathbf{g}_j|D_k, \mathbf{x}_{if} = x] - E[\mathbf{g}_j|D_k])^2 p(\mathbf{x}_{if} = x|D_k) \qquad (5.3)$$

Our active sampling method based on this benefit criterion is called the *Maximum Average Change (MAC)* sampling algorithm.

For the purposes of feature relevance estimations, the function $\mathbf{g}$ we wish to estimate is the vector of feature relevances, i.e., $\mathbf{g} = (\mathbf{g}_1, \ldots, \mathbf{g}_F)$, where $\mathbf{g}_j$ is the relevance of the jth feature. Since we need to know the prior on $\mathbf{g}$ to compute the Bayes MMSE estimate, we approximate the objective function in Equation 5.3 by

$$B(i,f) = \int_{\mathcal{X}_f} \sum_{j=1}^{F} (\hat{\mathbf{g}}_j(D_k, \mathbf{x}_{if} = x) - \hat{\mathbf{g}}_j(D_k))^2 p(\mathbf{x}_{if} = x | D_k) \qquad (5.4)$$

where $\hat{\mathbf{g}}_j(D_k)$ is any reasonable estimate of $\mathbf{g}_j$ from dataset D_k.

5.4 Implementation of the Active Sampling Algorithm

Algorithm 5.2.1 for active feature value acquisition is general and can be used with any measure for feature relevance for which the squared-error loss is reasonable. That is, the choice for the function $EstimateRelevances(D)$ in the pseudocode can be any estimate of feature relevance that can be estimated from a dataset with missing values.

In addition, the implementation of the benefit criteria introduced above also requires the computation of the conditional probabilities $p(\mathbf{x}_{if} = x | D_k)$.

Although our active sampling algorithm is quite general, we implemented it for a particular choice of the model for data generation (i.e., the joint class-and-feature distribution), which we present below. We then explain how the conditional probabilities and feature relevances can be computed given the joint distribution.

Our model is applicable for problems with categorical valued features. That is, we assume that every feature $\mathbf{x}_f$ takes on a discrete set of values $\mathcal{X}_f = \{1, \ldots, V_f\}$.

5.4.1 Data Generation Model: Class-Conditional Mixture of Product Distributions

We assume that each class-conditional feature distribution is a mixture of M product distributions over the features. (Although for our implementation it is not necessary that the number of components is constant across classes, we make this assumption for simplicity.) That is, the class-conditional feature distribution for class $c \in \mathcal{C}$ is

$$P(\mathbf{x}_1 = x_1, \ldots, \mathbf{x}_F = x_F | c) = \sum_{m=1}^{M} \alpha_{cm} \prod_{f=1}^{F} \prod_{x=1}^{V_f} \theta_{cmfx}^{\delta(x,x_f)} \qquad (5.5)$$

where α_{cm} is the mixture weight of component m for class c, θ_{cmfx} is the probability that the feature f takes on the value x for component m and class c, and $\delta(.)$ is the Kronecker delta function. Note that if $M = 1$, our model is equivalent to the *Naïve Bayes* model.

Therefore, the full class-and-feature joint distribution can be written as

$$P(\mathbf{c} = c, \mathbf{x}_1 = x_1, \ldots, \mathbf{x}_F = x) = \sum_{c \in C} p(\mathbf{c} = c) \sum_{m=1}^{M} \alpha_{cm} \prod_{f=1}^{F} \prod_{x=1}^{V_f} \theta_{cmfx}^{\delta(x, x_f)} \quad (5.6)$$

where $p(\mathbf{c} = c)$ is class probability. The class-and-feature joint distribution is completely specified by the parameters αs, θs, and the class probabilities.

Before we describe how the α and θ parameters can be estimated from a dataset with missing values, we will explain how feature relevances and the conditional probability $p(\mathbf{x}_{if} = x|D_k)$ are calculated if the parameters are known.

5.4.2 Calculation of Feature Relevances

We use the mutual information between a feature and the class variable as our measure of the relevance of that feature. That is,

$$\mathbf{g}_f = I(\mathbf{x}_f; \mathbf{c}) = H(\mathbf{x}_f) - H(\mathbf{x}_f|\mathbf{c}) \quad (5.7)$$

Although we are aware of the shortcomings of mutual information as a feature relevance measure, especially for problems where there are inter-feature correlations, we chose it because it is easy to interpret and to compute given the joint class-and-feature distribution. We did not use approaches such as Relief [10] and SIMBA [7], which provide feature weights (that can be interpreted as relevances), because they do not easily generalize to data with missing values.

The entropies in Equation 5.7 can be computed as follows:

$$H(\mathbf{x}_f) = - \sum_{c=1}^{C} \sum_{x=1}^{V_f} p(\mathbf{c}, \mathbf{x}_f = x) \log(p(\mathbf{c}, \mathbf{x}_f = x)) \quad (5.8)$$

$$H(\mathbf{x}_f|\mathbf{c}) = - \sum_{c=1}^{C} \sum_{x=1}^{V_f} p(\mathbf{x}_f = x|\mathbf{c}) \log(p(\mathbf{x}_f = x|\mathbf{c})) p(\mathbf{c}) \quad (5.9)$$

If the α and θ parameters and $p(c)$ of the model are known, the mutual information can be computed as follows:

$$H(\mathbf{x}_f) = - \sum_{x=1}^{V_f} \left(\sum_{c=1}^{C} p(c) \sum_{m=1}^{M} \alpha_{cm} \theta_{cmfx} \right) \log \left(\sum_{c=1}^{C} p(c) \sum_{m=1}^{M} \alpha_{cm} \theta_{cmfx} \right) \quad (5.10)$$

$$H(\mathbf{x}_f | \mathbf{c}) = - \sum_{c=1}^{C} p(c) \sum_{x=1}^{V_f} \left(\sum_{m=1}^{M} \alpha_{cm} \theta_{cmfx} \right) \log \left(\sum_{m=1}^{M} \alpha_{cm} \theta_{cmfx} \right) \quad (5.11)$$

5.4.3 Calculation of Conditional Probabilities

Since the instances in the dataset D are assumed to be drawn independently, we have

$$p(\mathbf{x}_{if} = x | D_k) = p(\mathbf{x}_{if} = x | \mathbf{x}_{obs(i)} = \boldsymbol{x}_{obs(i)}, c_i)$$
$$= \frac{p(\mathbf{x}_{if} = x, \mathbf{x}_{obs(i)} = \boldsymbol{x}_{obs(i)} | c_i)}{p(\mathbf{x}_{obs(i)} = \boldsymbol{x}_{obs(i)} | c_i)} \quad (5.12)$$

where, as before, $\mathbf{x}_{obs(i)}$ are features that are observed for instance i that take on values $\boldsymbol{x}_{obs(i)}$, and c_i is the class label for instance i.

Therefore, the conditional probability in Equation 5.12 can be written in terms of the parameters of the joint distribution as

$$p(\mathbf{x}_{if} = x | D_k) = \frac{\sum_m^M \alpha_{c_i m} \theta_{c_i m f x} \prod_{\phi \in obs(i)} \theta_{c_i m \phi \boldsymbol{x}_{i\phi}}}{\sum_m^M \alpha_{c_i m} \prod_{\phi \in obs(i)} \theta_{c_i m \phi \boldsymbol{x}_{i\phi}}} \quad (5.13)$$

5.4.4 Parameter Estimation

Since after each sampling step we only have a dataset with missing values and not the parameters αs, θs, and $p(c)$ that describe our model, they need to be estimated from the data. Once we have the estimates, the conditional probabilities and feature relevances can be computed by using the estimates in place of the parameters in Equations 5.13, 5.10, and 5.11. We will now describe how these parameters are estimated.

Estimation of $p(c)$: Since class labels of all the records in the dataset are available, the estimates of the class probabilities are obtained from the (Laplace smoothed) relative frequencies of the classes in the dataset.

Estimation of αs and θs: We need to estimate the parameters of the class-conditional mixture distribution for all classes. Since we have class labeled instances, we can perform the estimation separately for each class, considering only the data from that particular class. We therefore suppress the subscript c for the parameters corresponding to the class variable in the following equations.

Let D_c be the part of the dataset corresponding to class c. The data likelihood is given by

$$l(D_c; \boldsymbol{\theta}) = \sum_{i=1}^{N} \log \sum_{m=1}^{M} p(\boldsymbol{x}_i | \alpha_m, \boldsymbol{\theta}_m) p(\alpha_m) \tag{5.14}$$

The maximum-likelihood estimates of the parameters are the values that maximize the above likelihood function. One approach to perform the above maximization is the Expectation-Maximization (EM) algorithm [4] that iterates the following two steps :

$$\text{E-step:} \qquad Q(\boldsymbol{\theta}|\boldsymbol{\theta}^t) = E[l_c(D_c, Z, \boldsymbol{\theta})|D_c, \boldsymbol{\theta}^t]$$

$$\text{M-step:} \qquad \boldsymbol{\theta}^{t+1} = \underset{\boldsymbol{\theta}}{\text{argmax}} \ Q(\boldsymbol{\theta}|\boldsymbol{\theta}^t)$$

where l_c is the log-likelihood of an associated *complete* problem where each record in D_c is generated by a component of the mixture specified indicated by $Z = \{\mathbf{z}_i\}_{i=1}^{N}$, $\mathbf{z}_i = (z_1, \ldots, z_M)$ and $z_{ij} = 1$ iff instance i is generated by component j.

When the dataset D_c has no missing values the *EM* update equation for θs can be shown to be

$$\theta_{mfx}^{t+1} = \frac{\sum_{i=1}^{N} \delta(x, x_{if}) h_{im}}{\sum_{i=1}^{N} h_{im}} \tag{5.15}$$

$$\alpha_m^{t+1} = \frac{1}{N} \sum_{i=1}^{N} h_{im} \tag{5.16}$$

where

$$h_{im} = E[z_{im} = 1 | \boldsymbol{x}_i, \boldsymbol{\theta}^t] = \frac{\alpha_m \prod_{f=1}^{F} \theta_{mjx_{if}}^t}{\sum_{m=1}^{M} \alpha_m \prod_{f=1}^{F} \theta_{mjx_{if}}^t} \tag{5.17}$$

Since for our problem there are missing values, we can derive the *EM* update equation as described in [6] to obtain

$$\theta_{mfx}^{t+1} = \frac{\sum_{i=1}^{N} h_{im}^{obs} \left(\theta_{mfx}^t IsMissing(i, f) + \delta(x, x_{if})(1 - IsMissing(i, f)) \right)}{\sum_{i=1}^{N} h_{im}^{obs}} \tag{5.18}$$

$$\alpha_m^{t+1} = \frac{1}{N} \sum_{i=1}^{N} h_{im}^{obs} \tag{5.19}$$

where

$$h_{im}^{obs} = E[z_{im} | \boldsymbol{x}_{obs(i)}] = \frac{\alpha_m \prod_{j \in obs(i)} \theta_{mjx_{ij}}^t}{\sum_{m=1}^{M} \alpha_m \prod_{j \in obs(i)} \theta_{mjx_{ij}}^t} \tag{5.20}$$

and where $IsMissing(i, f)$ takes on the value one or zero depending upon whether or not the feature f for record i is missing.

Note that in the actual implementation of Equation 5.18 we perform Laplace smoothing to reduce estimation variance.

5.5   Experiments

We conducted experiments on synthetic data and on datasets from the UCI repository [15]. For a particular dataset, the experimental setup is as follows. We start with the assumption that the class labels for all the samples are initially known and all of the feature values are missing. At each sampling step a single missing entry in the dataset is selected by the sampling policy and the actual value in the dataset is disclosed. The experiment ends when all entries of the dataset are sampled and all the original feature values are fully disclosed. After each sample is disclosed, we estimate the feature relevances from all the data that are currently available, which are compared to the "true" feature relevance values (the feature relevances estimated from the entire dataset). The comparison measure is the average sum-of-squared errors, which is plotted as a function of the number of missing entries filled thus far. The average is computed over 100 sampling runs to reduce fluctuations introduced by the random selection of entries in the case of multiple equivalent choices occurring at certain steps. The plots show the comparison of our active sampling algorithm to the random sampling algorithm.[3]

Although the models we presented are general, we only experimented with mixture distributions (cf. Section 5.4.1) of only one component per class (i.e., a Naïve Bayes model). We did not perform experiments with a higher number of components because of estimation problems during the initial sampling steps and also because of computational issues. In the future we intend to develop methods to adjust the number of components depending on the amount of data available at any sampling step.

5.5.1   Synthetic Data

We now describe how the synthetic dataset was generated. We created a dataset of size $N = 200$ samples with binary class labels and three binary features with exactly 100 records per class (i.e., $p(c = 0) = p(c = 1) = 0.5$). The features are mutually class-conditionally independent and with different relevances to the class labels.

The feature values are generated randomly according to the following scheme. For feature F_i we generate the feature values according to the probability $p(F_i = 0|c = 0) = p(F_i = 1|c = 1) = p_i$. Clearly, if p_i is closer to 0 or 1, the feature is more relevant for classification than if p_i is closer to 0.5. For our three features we chose $p_1 = 0.9$, $p_2 = 0.7$, and $p_3 = 0.5$, meaning that the first feature is highly relevant and the third is completely irrelevant for classification. The true feature relevances (mutual information values) are $r_1 = 0.37$, $r_2 = 0.08$, and $r_3 = 0$, respectively.

Since by construction there is no inter-feature dependence given the class, we conducted experiments using a product distribution for each class (i.e., a mixture of just one component). The average squared distance between the estimated and the true feature relevances is plotted as function of the number feature values sampled in Figure 5.2 for both the random and our active sampling policies.[4]

The graph in Figure 5.2 shows that our proposed active scheme clearly outperforms the random acquisition policy. For example, note that in order to reduce the difference between the estimated and true relevances to a fourth of the initial value (when all feature values are missing), the random policy requires 45 samples instead of 30 by our active method.

In Figure 5.3 we show, separately, the estimates of each of the individual feature relevances. In Figure 5.4 we show the average number of times each feature is sampled as a function of the number of samples. We observe that the frequency with which a feature is sampled is correlated to its relevance. This is a desirable property because the least relevant features will eventually be discarded and therefore sampling them would be wasteful.

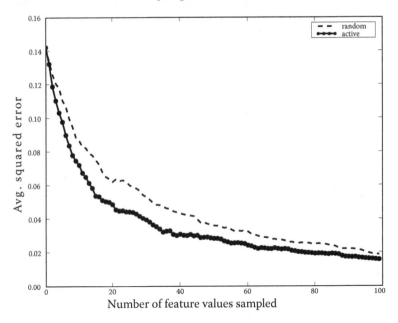

FIGURE 5.2: Squared sum of the differences between estimated and true relevances at each sampling step on artificial data for random and active policies.

5.5.2   UCI Datasets

We performed experiments on the Zoo, Solar Flares, Monks, and Cars datasets from the UCI repository. These datasets present larger class label spaces (from 2 to 6 classes) and an increased number of features (from 6 to 16). Also, some of the features take on more values (from 2 to 6 values) than our

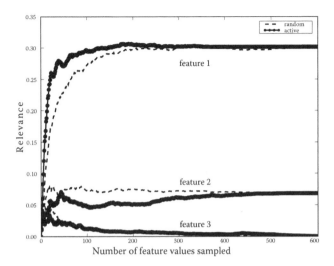

FIGURE 5.3: Estimated relevances at each sampling step for every single feature on artificial data. Random (dashed line) and active (solid-dotted line) policies are compared. Since there are three features and 200 instances, the x axis goes to 600.

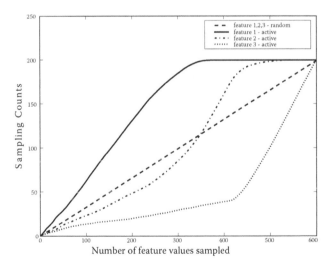

FIGURE 5.4: Average cumulative sampling counts at each sampling step for each feature on artificial data. The more relevant features are sampled more frequently than less relevant features in case of active policy. As a comparison, the random policy samples features independently of their relevance.

artificial datasets. Figure 5.5 shows the plots of the average sum-of-squared errors between the estimated and "true" feature relevances as a function of the number of samples acquired for both the active and the random sampling schemes. The error values are normalized such that at step 0 (i.e, when none of the missing entries has been filled) the error is 1.0.

Figure 5.5 illustrates the advantage of the active sampling policy over the random scheme in reducing the number of feature samples necessary for achieving comparable accuracy. We note that in order to reduce the estimation error of feature relevances to one fourth of the initial value, the number of samples required is 25% - 75% lower for the active policy than for the random policy. Again, we have observed in all datasets that most relevant features are sampled more frequently than less relevant features.

5.5.3   Computational Complexity Issues

The computational complexity of our active sampling algorithm due to the expensive EM estimation (which is repeated for every missing entry and every possible feature value) limits its applicability to large datasets. One way we reduced the computational expense was to memoize the calculation of the benefit function for equivalent entries (i.e., entries having the same non-missing feature values, thus having the same benefit value). Another strategy to reduce computation is to perform sub-optimal active sampling by considering only a random subset of the missing entries at each time step. This latter strategy can be used to trade off sampling cost versus computational cost.

In Figure 5.6 (upper panel) the active and random policies are shown together with the active policy that considers 0.1% and 1% of the missing entries (randomly selected) at each sampling step; results are based on the artificial dataset described in Section 5.5.1. We observe that the dominance of active policy compared to random increases monotonically with the subsample size, but in general this increase is not uniform. A similar experiment was performed on the Solar Flares dataset (see Figure 5.6, bottom panel) where active and random policies are plotted together with the active policy that considers 0.05%, 0.25%, and 1% of the missing entries (randomly selected). Again we observe that performing active policy considering a random subportion of the dataset (0.25% of the total number of missing entries at any instance) is an effective strategy to obtain a reduction in the number of samples acquired at a reduced computational cost.

5.6   Conclusions and Future Work

We have presented a general active feature sampling method for feature relevance estimation in domains where the feature values are expensive to

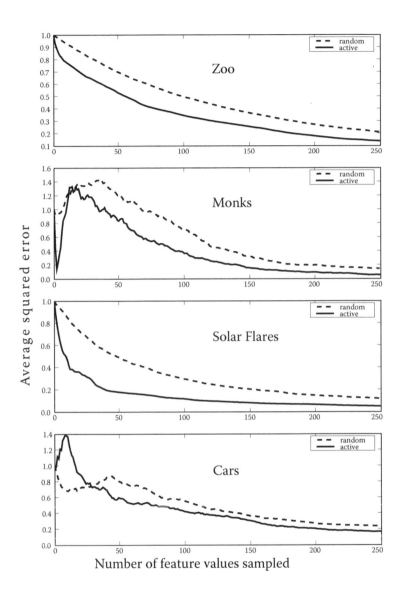

FIGURE 5.5: The normalized difference between final relevances and estimated relevances at each sampling step is plotted for random (dashed line) and active (solid line) policies on four UCI datasets (Zoo, Monks, Solar Flares, Cars). The value at step 0 (all feature values unknown) is normalized to 1.0 in all cases. For the Zoo dataset, after measuring 100 feature values using a random policy, the normalized difference in the estimated and true feature relevances is 0.5 as opposed to 0.3 for active sampling.

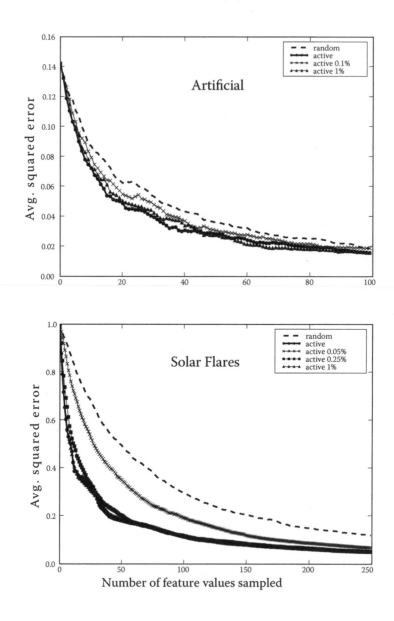

FIGURE 5.6: Average squared sum of the differences between estimated and true relevances at each sampling step on artificial and UCI Solar Flares datasets. Random and active policies are compared to the active policy that considers only a small random subset of the missing entries at every sampling step.

measure. At any stage, the feature sampling method evaluates the benefit of sampling the value of every individual missing entry in the dataset and selects the one with the highest benefit. The value of the selected entry is then queried. We have derived a benefit function that attempts to minimize the mean-squared error in the estimates of the feature relevances and showed that the minimum mean-squared error criterion is equivalent to the maximum average change criterion. Although we implemented the active sampling algorithm for a class-conditional mixture of product distribution model and mutual information, a measure of feature relevance, we argued that the active sampling algorithm can be applied with other models and measures for feature relevance. We experimentally demonstrated that the active sampling algorithm can be applied to perform feature relevance estimation at a reduced sampling cost over a random subsampling approach.

We intend to study the effect of an incorrect choice of number of components for the mixture distribution. Since the final goal is to obtain a classifier with high accuracy, we plan to investigate the advantage of active sampling over random sampling for the accuracy of the final classifier built after feature selection. Some other directions of future work are the online choice of number of components for the mixtures and extending the algorithm for continuous feature values (either by discretization or by developing models for continuous data).

Notes

1 Even if some of the feature values are initially available, the methods described below are still applicable.

2 If multiple entries have the same highest benefit, then the actual entry to measure is selected randomly from the equivalent alternatives.

3 At each sampling step, then, random policy chooses one of the missing entries uniformly at random.

4 In the case of the synthetic dateset, this difference never goes to zero since it is computed with respect to the *true* relevances (i.e., the relevances computed from the probabilities used to generate the data) rather than the estimated relevances from the completely filled dataset.

References

[1] A. Blum and P. Langley. Selection of relevant features and examples in machine learning. *Artificial Intelligence*, 97(1-2):245–271, 1997.

[2] K. Chaloner and I. Verdinelli. Bayesian experimental design: A review. *Statistical Science*, 10:273–304, 1995.

[3] D. A. Cohn, Z. Ghahramani, and M. I. Jordan. Active learning with statistical models. In G. Tesauro, D. Touretzky, and T. Leen, editors, *Advances in Neural Information Processing Systems*, volume 7, pages 705–712. MIT Press, Cambridge, MA, 1995.

[4] A. P. Dempster, N. M. Laird, and D. B. Rubin. Maximum likelihood estimation from incomplete data via the em algorithm (with discussion). *Journal of the Royal Statistical Society Series B*, 39:1–38, 1977.

[5] V. V. Fedorov. *Theory of Optimal Experiments*. Academic Press, New York, 1972.

[6] Z. Ghahramani and M. I. Jordan. Supervised learning from incomplete data via an EM approach. In J. D. Cowan, G. Tesauro, and J. Alspector, editors, *Advances in Neural Information Processing Systems*, volume 6, pages 120–127. Morgan Kaufmann Publishers, Inc., 1994.

[7] R. Gilad-Bachrach, A. Navot, and N. Tishby. Margin based feature selection - theory and algorithms. In *Proceedings of the Twenty-First International Conference on Machine learning (ICML-04)*, page 43, New York, NY, ACM Press, 2004.

[8] T. Golub, D. Slonim, P. Tamayo, C. Huard, M. Gaasenbeek, J. Mesirov, H. Coller, M. Loh, J. Downing, M. Caligiuri, C. Bloomfield, and E. Lander. Molecular classification of cancer: class discovery and class prediction by gene expression monitoring. *Science*, 286(5439):531–537, 1999.

[9] G. H. John, R. Kohavi, and K. Pfleger. Irrelevant features and the subset selection problem. In *Proceedings of the Eleventh International Conference on Machine Learning (ICML-94)*, pages 121–129, 1994.

[10] K. Kira and L. A. Rendell. A practical approach to feature selection. In *Proceedings of the Ninth International Workshop on Machine Learning (ML-92)*, pages 249–256, San Francisco, CA, USA, 1992. Morgan Kaufmann Publishers Inc.

[11] R. Kohavi and G. H. John. Wrappers for Feature Subset Selection. *Artificial Intelligence*, 97(1-2):273–324, 1997.

[12] J. Kononen, L. Bubendorf, A. Kallioniemi, M. Barlund, P. Schraml, S. Leighton, J. Torhorst, M. Mihatsch, G. Seuter, and O. P. Kallioniemi. Tissue microarrays for high-throughput molecular profiling of tumor specimens. *Nature Medicine*, 4(7):844–847, 1998.

[13] D. Lizotte, O. Madani, and R. Greiner. Budgeted learning of naive-bayes classifiers. In *Proceedings of the 19th Annual Conference on Uncertainty in Artificial Intelligence (UAI-03)*, pages 378–385, 2003.

[14] P. Melville, M. Saar-Tsechansky, F. Provost, and R. Mooney. Active feature-value acquisition for classifier induction. In *Proceedings of the Fourth IEEE International Conference on Data Mining (ICDM'04)*, pages 483–486, Washington, DC, IEEE Computer Society, New York, 2004.

[15] D. J. Newman, S. Hettich, C. L. Blake, and C. J. Merz. UCI repository of machine learning databases, 1998.

[16] P. Sebastiani and H. P. Wynn. Maximum entropy sampling and optimal Bayesian experimental design. *Journal of Royal Statistical Society*, pages 145–157, 2000.

[17] H. S. Seung, M. Opper, and H. Sompolinsky. Query by committee. In *Proceedings of the Fifth Annual Workshop on Computational Learning Theory (COLT-92)*, pages 287–294, 1992.

[18] V. S. Sheng and C. X. Ling. Feature value acquisition in testing: A sequential batch test algorithm. In *Proceedings of the 23rd International Conference on Machine Learning (ICML-06)*, pages 809–816, New York, NY, ACM Press, 2006.

[19] S. Tong and D. Koller. Support vector machine active learning with applications to text classification. In *Proceedings of the Seventeenth International Conference on Machine Learning (ICML-00)*, pages 999–1006, 2000.

[20] S. Veeramachaneni, F. Demichelis, E. Olivetti, and P. Avesani. Active sampling for knowledge discovery from biomedical data. In *Proceeding of the 9th European Conference on Principles and Practice of Knowledge Discovery in databeses (PKDD-05)*, pages 343–354, 2005.

[21] S. Veeramachaneni, E. Olivetti, and P. Avesani. Active sampling for detecting irrelevant features. In *Proceedings of the 23rd international conference on Machine learning (ICML-06)*, pages 961–968, New York, NY, ACM Press, 2006.

[22] Z. Zheng and B. Padmanabhan. On active learning for data acquisition. In *Proceedings of the International Conference on Datamining (ICDM-02)*, pages 562–570, 2002.

Chapter 6

A Study of Feature Extraction Techniques Based on Decision Border Estimate

Claudia Diamantini

Università Politecnica delle Marche

Domenico Potena

Università Politecnica delle Marche

6.1   Introduction

Feature extraction is the core of methodologies aimed at building new and more expressive features from the existing ones. This representation change typically allows one to enlighten characteristics of data that are not immediately evident in the original space. As a consequence, performance can be improved at the expenses of reduced interpretation capability by domain experts.

Feature extraction can be considered as a mapping from the original space to a lower dimensional feature space. The mapping can be carried out with respect to different criteria. They can be roughly divided into data representation and data discrimination criteria. In the former case, the goal is to find the set of reduced features that best approximate the original data, so the criteria are based on the minimization of a mean-squared error or distortion measure. One of the best-known methods based on this criterion is the *principal component analysis* (PCA) or Karhunen-Loeve expansion [7], which calculates eigenvalues and eigenvectors of the data covariance matrix, and defines the mapping as an orthonormal transformation based on the set of eigenvectors corresponding to the highest eigenvalues. The squared error of the transformation is simply the sum of the leftover eigenvalues. The PCA is an optimum method for data compression and signal representation, however, it presents several limitations for discriminating between data belonging to different classes. In particular, for data discrimination, criteria to evaluate the effectiveness of features should be a measure of the class separability. For this task, Bayes error probability is the best criterion to evaluate a feature set. Unfortunately, Bayes error is unknown in general. A family of methods that is frequently used in practice, but that is only indirectly related to Bayes error, is called *discriminant analysis* (DA), based on a family of functions of scatter matrices. In the simplest form, linear DA (LDA), also known as canonical analysis (CA), considers a within-class scatter matrix for each class, measuring the scatter of samples around the respective class mean, and the between-class scatter matrix, measuring the scatter of class means around the mixture mean, and finds a transformation that maximizes the between-class scatter and minimizes the within-class scatter, so that the class separability is maximized in the reduced dimensional space [7, 1]. Other approaches use upper bounds of Bayes error, like the Bhattacharyya distance [2]. In [12] Lee and Landgrebe introduced the principle that, in classification tasks, the relevance of features can be measured on the basis of properties of the decision border, the geometrical locus of points of the feature space separating one class from the others.

Following this approach, some authors proposed the use of artificial neural networks (ANNs) to estimate the unknown decision border. In early works [8, 13], authors suggested the use of multi-layer perceptron, which is the most widely used type of feedforward ANN, consisting of multiple layers of interconnected neurons. More recently, it was proposed to use ANNs targeted to the accurate estimate of the optimal decision border [17, 4]. In particular, [17] exploits support vector machines (SVMs), a class of powerful kernel-based learning algorithms [16]. In [4] a truly bayesian approach to feature extraction for classification is introduced that is based on an appropriately trained labeled vector quantizer (LVQ). We call the approach truly bayesian since the LVQ is trained with the Bayes risk weighted vector quantization (BVQ) learning algorithm, which is, to the best of our knowledge, the only learning algorithm based on the minimization of the misclassification risk [3]. Under this truly classification-based algorithm, an LVQ moves toward a locally optimal linear approximation of the bayesian decision border. In this chapter we present these approaches and compare them.

The rest of this section is devoted to the introduction of the basic notions about statistical pattern classification. Section 6.2 presents the decision boundary feature extraction (DBFE) principle in general and DBFE methods based on MLP and SVM in particular. Then, in Section 6.3, we introduce vector quantizers and the BVQ algorithm. The details of the BVQ-based feature extraction (BVQFE) method are given in Section 6.4. Comparative experiments are presented in Section 6.5. Finally, Section 6.6 ends the chapter.

6.1.1 Background on Statistical Pattern Classification

In statistical approaches to classification, data are described by a continuous random vector $\boldsymbol{X} \in \mathcal{R}^N$ (feature vector) and classes by a discrete random variable $Y \in \mathbf{Y} = \{y_1, y_2, \ldots, y_C\}$. For each class y_i, the distribution of data in the feature space is described by the conditional probability density function (cpdf) $p_{\boldsymbol{X}|Y}(\mathbf{x}|y_i)$. The cumulative probability density function of the random vector $\boldsymbol{X}$ is $p_{\boldsymbol{X}}(\mathbf{x}) = \sum_{i=1}^{C} P_Y(y_i) p_{\boldsymbol{X}|Y}(\mathbf{x}|y_i)$, where $P_Y(y_i)$ is the a-priori probability of class y_i.

A classification rule is a mapping $\Psi : \mathcal{R}^N \to \mathbf{Y}$, which assigns a class label to data on the basis of the observation of its feature vector. A classification rule partitions the feature space in C *decision regions* $D_1, \ldots, D_C$ such that $D_i = \{\mathbf{x} \in \mathcal{R}^N \mid \Psi(\mathbf{x}) = y_i\}$. The border separating decision regions is called the *decision border*. Figure 6.1 presents a set of data drawn from two gaussian classes (symbolized by $*$ and o); the straight line represents the decision border of a rule that assigns all points at the left of it to the $*$ class, and those at the right to the o class.

The predictive accuracy of a classification rule is evaluated by the *average error probability*, *err*. In the two-class case, *err* takes the form:

$$err = \int_{D_1} p_{Y|\boldsymbol{X}}(y_2|\mathbf{x}) p_{\boldsymbol{X}}(\mathbf{x}) dV_{\mathbf{x}} + \int_{D_2} p_{Y|\boldsymbol{X}}(y_1|\mathbf{x}) p_{\boldsymbol{X}}(\mathbf{x}) dV_{\mathbf{x}}, \qquad (6.1)$$

where $dV_{\mathbf{x}}$ denotes the differential volume in the $\mathbf{x}$ space, and $p_{Y|\boldsymbol{X}}(y_i|\mathbf{x})$ is the a-posteriori probability that can be derived from the cpdf by the Bayes Theorem.

The classification rule that minimizes the average error probability (6.1) is the Bayes rule: $\Psi_B(\mathbf{x}) = if \ p_{Y|\boldsymbol{X}}(y_1|\mathbf{x}) > p_{Y|\boldsymbol{X}}(y_2|\mathbf{x}) \ then \ y_1 \ else \ y_2$.

The decision border related to $\Psi_B(\mathbf{x})$ is the optimal decision border (or Bayes decision border). Indeed, it is defined by the geometrical locus of points such that $h_B(\mathbf{x}) = 0$, where $h_B(\mathbf{x}) = p_{Y|\boldsymbol{X}}(y_1|\mathbf{x}) - p_{Y|\boldsymbol{X}}(y_2|\mathbf{x})$. In general, for any decision rule Ψ, there always exists a function $h(\mathbf{x})$ such that $h(\mathbf{x}) = 0$ is the decision border, so the decision rule takes the form: $\Psi(\mathbf{x}) = if \ h(\mathbf{x}) > 0 \ then \ y_1 \ else \ y_2$. For this reason $h(\mathbf{x})$ is usually called the *decision function*.

Often, in practice, misclassifying y_1 and y_2 samples may have different consequences. Hence it is appropriate to assign a cost to each situation as: $b(y_i, y_j) \geq 0$ is the cost of deciding in favor of class y_j when the true class is y_i, with $b(y_i, y_i) = 0 \ \forall i$. In such a situation, the average error probability generalizes to the *average misclassification risk*:

$$R(\Psi) = b(y_2, y_1) \int_{D_1} p_{Y|\boldsymbol{X}}(y_2|\mathbf{x}) p_{\boldsymbol{X}}(\mathbf{x}) dV_{\mathbf{x}} + b(y_1, y_2) \int_{D_2} p_{Y|\boldsymbol{X}}(y_1|\mathbf{x}) p_{\boldsymbol{X}}(\mathbf{x}) dV_{\mathbf{x}}$$

$$(6.2)$$

and Bayes rule becomes: $\Psi_B(x) = if \ \dfrac{p_{Y|\boldsymbol{X}}(y_1|\mathbf{x})}{p_{Y|\boldsymbol{X}}(y_2|\mathbf{x})} > \dfrac{b(y_2, y_1)}{b(y_1, y_2)} \ then \ y_1 \ else \ y_2$.

These equations can be easily generalized to the case of C classes. We refer the interested readers to [7] for details.

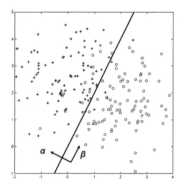

FIGURE 6.1: A two-class classification problem in a 2-dimensional space. α and β represent the informative direction and the redundant direction, respectively.

The development of most of the learning algorithms and non-parametric methods for classification try to overcome the limits of applicability of the Bayes rule, related to the the fact that cpdfs are in general unknown. Thus, one of the main efforts is that of obtaining cpdf estimates on the basis of a set of samples drawn from the C classes called training sets, and hereafter denoted by $\mathcal{TS}$. However, it is recognized that accurate cpdf estimation does not necessarily lead to good classification performance [6].

6.2 Feature Extraction Based on Decision Boundary

Decision boundary feature extraction (DBFE) is a discriminative approach proposed by Lee and his co-authors [12]. The approach is based on the geometry of the Bayes decision border in order to predict the minimum number of features needed to achieve the same classification accuracy as in the original space. The DBFE algorithm is based on the idea that moving along the direction of the decision border, the classification of each observation will remain unchanged (see Figure 6.1). Hence, the direction of the decision border is redundant. In contrast, a normal vector to the decision border at a point represents an informative direction and its effectiveness is proportional to the area of decision border that has the same normal vector.

In [12], starting from the normal vectors to the decision border, the authors define the effective decision boundary feature matrix (EDBFM) as

$$\Sigma_{EDBFM} = \frac{1}{\int_{S'} p(\mathbf{x})d\mathbf{x}} \int_{S'} \mathbf{N}^T(\mathbf{x})\mathbf{N}(\mathbf{x})p(\mathbf{x})d\mathbf{x}, \qquad (6.3)$$

where $\mathbf{N}(\mathbf{x})$ is the normal vector at a point $\mathbf{x}$, $\mathbf{N}^T(\mathbf{x})$ denotes the transposed normal vector, and S' is the portion of decision border containing most of the training data (the effective decision boundary). It is proved [12] that:

- The rank of the EDBFM represents the *intrinsic discriminant dimension*, that is, the minimum number of feature vectors needed to achieve the same Bayes error probability as in the original space.

- The eigenvectors of the EDBFM corresponding to nonzero eigenvalues are the necessary feature vectors.

The DBFE algorithm uses the knowledge of cpdf to define the Bayes decision border. However, true cpdfs are generally unknown in real classification problems. In order to overcome this limitation, [11] estimates the cpdf and the decision border by the Parzen method [7]. However, more effective techniques for the estimation of the decision border exist, which are based on neural networks. In the rest of this section, we will present approaches based on MLPs and SVMs. In the next section we will introduce a formal derivation from Equation (6.3) based on the LVQ.

6.2.1   MLP-Based Decision Boundary Feature Extraction

In [13], Lee and Landgrebe introduce the use of MLP to estimate the decision border. For this reason, we call this version of the method the MLP-feature extraction (MLPFE) method. Such an approach exploits an MLP with one hidden layer and C output neurons, with backpropagation used to train the network. Backpropagation is based on the minimization of the squared error, allowing a trained MLP to estimate class a-posteriori probability distributions.

Let us consider the case of a two-class problem, and let $h(\mathbf{x})$ be the decision function of an MLP (remember that if $h(\mathbf{x}) = 0$, then $\mathbf{x}$ is a point of the decision border). Given a point $\mathbf{p}$ on the decision border, the MLPFE algorithm estimates numerically the vectors normal to $\mathbf{p}$ as follows:

$$\mathbf{N}(\mathbf{p}) = \frac{\nabla h(\mathbf{p})}{\|\nabla h(\mathbf{p})\|} \approx \frac{1}{\|1/\xi\| \cdot \|\mathbf{p}\|} \left(\frac{1}{\xi_1}, \frac{1}{\xi_2}, \ldots, \frac{1}{\xi_N}\right),$$

where ξ_i, $i = 1, 2, \ldots, N$, are the smallest values such that $h(p_1, \ldots, p_i + \xi_i, \ldots, p_N) \neq 0$.

In order to find the point $\mathbf{p}$ for each training sample $\mathbf{x}_a$ correctly classified as class y_1, the algorithm finds the nearest sample $\mathbf{x}_b$ correctly classified as class y_2. The same procedure is repeated for the samples classified as class y_2. Then, a segment $\mathbf{s} = \alpha \cdot \mathbf{x}_a + (1 - \alpha) \cdot \mathbf{x}_b, 0 \leq \alpha \leq 1$, is built. Such a segment must pass through the decision border since the given points are classified differently. Then, the point $\mathbf{p}$ can be detected by moving along s stepwise, until the decision function is near to zero. The algorithm can be easily generalized to the case of a multi-class problem.

In the estimate of $\mathbf{N}(\mathbf{p})$, searching for ξ means evaluating a certain number of differences between the activation functions, resulting in an inaccurate estimation and a long computational time. So, in [8], the authors describe an improvement of the algorithm, called the analytical decision boundary feature extraction (ADBFE), where the normal vectors are calculated analytically from the equations of the decision border.

6.2.2   SVM Decision Boundary Analysis

In order to accurately reconstruct the Bayes decision border, in [17] the authors proposed SVM decision boundary analysis (SVMDBA), a method that combines the DBFE principle and the support vector machine algorithm. The maximum margin classifier principle underlying SVM [16] is developed for two-class problems. Exploiting the decision function $h(\mathbf{x})$ of an SVM adapted on the training set, the unit normal vector to the decision border at a point $\mathbf{x}$ can be analytically computed as follows:

$$\mathbf{N}(\mathbf{x}) = \frac{\nabla h(\mathbf{x})}{\|\nabla h(\mathbf{x})\|} \quad \text{where} \quad \nabla h(\mathbf{x}) = \frac{\partial h(\mathbf{x})}{\partial \mathbf{x}} = \sum_{i=1}^{l} \alpha_i y_i \frac{\partial K(\mathbf{x}, \mathbf{x}_i)}{\partial \mathbf{x}}$$

where $\mathbf{x}_i \in \mathcal{R}^N$, $i = \{1, 2, \ldots, l\}$ is the support vector, and $y_i \in \{\pm 1\}$ is its class label. $K(\mathbf{x}, \mathbf{x}_i)$ is the chosen kernel function and α_i are the parameters of the adapted SVM. Like in MLP-based approaches, the point $\mathbf{x}$ of the decision border is estimated by building a segment $\mathbf{s}$ connecting two differently classified points, and estimating the point of s such that $h(\mathbf{x})$ is less than a threshold ε. Unlike them, in SVMDBA only a part of the training set $\mathcal{TS}$ is used to evaluate the Σ_{EDBFM}. Such a subset consists of the $r \times \mid \mathcal{TS} \mid$ observations, $0 < r \leq 1$, such that the absolute decision function $\mid h(\mathbf{x}) \mid$ assumes the first $r \times \mid \mathcal{TS} \mid$ smallest values. In other words, such a subset consists of the observations nearest to the decision border.

The principle of maximum margin classifiers allows a very accurate reconstruction of the border. However, the literature discusses the high computational cost of the quadratic optimization underlying SVM [14, 9], which limits its application on huge amounts of data [15]. SVM is used in multi-class problem, by exploiting a one-against-all schema, which constructs C SVM classifiers with the ith one separating class y_i from all the remaining classes. Of course this leads to increased complexity.

In the following we introduce an alternative approach, based on the Bayes risk weighted vector quantization algorithm, which has shown performances at least comparable with SVM, with lower computational cost [3].

6.3   Generalities About Labeled Vector Quantizers

The goal of this section is to introduce the basic definitions about labeled vector quantizers and to present the Bayes risk weighted vector quantization algorithm.

DEFINITION 6.1   *A Euclidean nearest neighbor vector quantizer (VQ) of dimension N and order Q is a function $\Omega : \mathcal{R}^N \to \mathbf{M}$, $\mathbf{M} = \{\mathbf{m}_1, \mathbf{m}_2, \ldots, \mathbf{m}_Q\}$, $\mathbf{m}_i \in \mathcal{R}^N$, $\mathbf{m}_i \neq \mathbf{m}_j$, that defines a partition of $\mathcal{R}^N$ into Q regions $\mathcal{V}_1, \mathcal{V}_2, \ldots, \mathcal{V}_Q$, such that*

$$\mathcal{V}_i = \{\mathbf{x} \in \mathcal{R}^N : \| \mathbf{x} - \mathbf{m}_i \|^2 < \| \mathbf{x} - \mathbf{m}_j \|^2, \ j \neq i\}, \tag{6.4}$$

$\mathbf{M}$ is called the *code*. Elements of $\mathbf{M}$ are called *code vectors*. The region $\mathcal{V}_i$ defined by (6.4) is called the *Voronoi region* of the code vector $\mathbf{m}_i$. Note that the Voronoi region is completely defined by the code $\mathbf{M}$. In particular, the border of Voronoi region $\mathcal{V}_i$ is defined by the intersection of a finite set of hyperplanes $\mathcal{S}_{i,j}$ with equation $(\mathbf{m}_i - \mathbf{m}_j) \cdot (\mathbf{x} - \dfrac{\mathbf{m}_i + \mathbf{m}_j}{2}) = 0$, where $\mathbf{m}_j$ is a neighbor code vector to $\mathbf{m}_i$.

DEFINITION 6.2   *A labeled vector quantizer (LVQ) is a pair $LVQ =< \Omega, \mathcal{L} >$, where $\Omega : \mathcal{R}^N \to \mathbf{M}$ is a vector quantizer, and $\mathcal{L} : \mathbf{M} \to \mathbf{Y}$ is a labeling function, assigning to each code vector in $\mathbf{M}$ a class label.*

An LVQ defines a classification rule:

DEFINITION 6.3   *The classification rule associated with a labeled vector quantizer $LVQ =< \Omega, \mathcal{L} >$ is $\Psi_{LVQ} : \mathcal{R}^N \to \mathbf{Y}, \mathbf{x} \mapsto \mathcal{L}(\Omega(\mathbf{x}))$.*

Note that the nearest neighbor nature of this classification rule: each vector in $\mathcal{R}^N$ is assigned to the same class as its nearest code vector. Thus, decision regions are defined by the union of Voronoi regions of code vectors with the same label. Note also that decision borders are defined only by those hyperplanes $\mathcal{S}_{i,j}$ such that $\mathbf{m}_i$ and $\mathbf{m}_j$ have different labels.

An LVQ can be trained by the Bayes risk weighted vector quantization algorithm (BVQ) to find the best linear approximation to the true Bayes decision border. The BVQ formally derives from the minimization of the average misclassification risk. However, for the sake of simplicity, hereinafter we will refer to the version of the algorithm for the minimization of average error probability.

Let $\mathcal{LM} = \{(\mathbf{m}_1, l_1), \ldots, (\mathbf{m}_Q, l_Q)\}$ be a labeled code, where $l_i \in \mathbf{Y}$ denotes the class of the code vector $\mathbf{m}_i$, and let $\mathcal{TS} = \{(\mathbf{t}_1, u_1), \ldots, (\mathbf{t}_M, u_M)\}$ be the training set, where $\mathbf{t}_i \in \mathcal{R}^N$ denotes the feature vector and $u_i \in \mathbf{Y}$ is the class the sample belongs to. The BVQ algorithm is an interactive punishing-rewarding adaptation schema. At each iteration, the algorithm considers a training sample randomly picked from $\mathcal{TS}$. If the training sample turns out to fall "on" the decision border, then the position of the two code vectors determining the border is updated, moving the code vector with the same label of the sample toward the sample itself and moving away that with a different label. Since the decision border is a null measure subspace of the feature space, we have zero probability to get samples falling exactly on it. Thus, an approximation of the decision border is made, considering those samples falling close to it (at a maximum distance of $\Delta/2$). In the following, the BVQ algorithm at the k-th iteration is given:

> *BVQ Algorithm - k-th iteration*
> 1. randomly pick a training pair $(\mathbf{t}^{(k)}, u^{(k)})$ from $\mathcal{TS}$;
> 2. find the code vectors $\mathbf{m}_i^{(k)}$ and $\mathbf{m}_j^{(k)}$ nearest to $t^{(k)}$;
> 3. $\mathbf{m}_q^{(k+1)} = \mathbf{m}_q^{(k)}$ for $q \neq i, j$;
> 4. compute $\mathbf{t}_{i,j}^{(k)}$, the projection of $\mathbf{t}^{(k)}$ on $S_{i,j}^{(k)}$;
> 5. if $\mathbf{t}^{(k)}$ falls at a distance $d \leq \Delta/2$ from the border $S_{i,j}^{(k)}$, then

$$\mathbf{m}_i^{(k+1)} = \mathbf{m}_i^{(k)} - \gamma^{(k)} \frac{\delta(u^{(k)} = l_j^{(k)}) - \delta(u^{(k)} = l_i^{(k)})}{\parallel \mathbf{m}_i - \mathbf{m}_j \parallel} (\mathbf{m}_i^{(k)} - \mathbf{t}_{i,j}^{(k)})$$

$$\mathbf{m}_j^{(k+1)} = \mathbf{m}_j^{(k)} + \gamma^{(k)} \frac{\delta(u^{(k)} = l_j^{(k)}) - \delta(u^{(k)} = l_i^{(k)})}{\parallel \mathbf{m}_i - \mathbf{m}_j \parallel} (\mathbf{m}_j^{(k)} - \mathbf{t}_{i,j}^{(k)})$$

> else $\mathbf{m}_q^{(k+1)} = \mathbf{m}_q^{(k)}$ for $q = i, j$.

where $\delta(expr) = 1$ if $expr$ is true and 0 otherwise.

More details on the formal derivation of the algorithm and on the proper setting of parameters can be found in [3].

6.4   Feature Extraction Based on Vector Quantizers

Having a trained LVQ, the extraction of the most discriminating features is straightforward. As a matter of fact, according to the DBFE principle, the most informative directions are defined by the normal vectors to the decision borders. Such normal vectors are simply defined by $\mathbf{N}_{ij} = \mathbf{m}_i - \mathbf{m}_j$, where $l_i \neq l_j$ (see Figure 6.2).

The normal vectors $\mathbf{N}_{ij}$ can then be combined together to extract the informative features as in the Lee and Landgrebe approach.

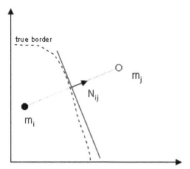

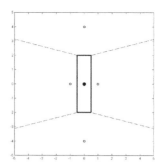

FIGURE 6.2: A piece of true decision border, its linear approximation, and the local discriminative direction $\mathbf{N}_{ij} = \mathbf{m}_i - \mathbf{m}_j$.

FIGURE 6.3: An example of the uneven contribution of normal vectors. White dots: class A code vectors. Black dot: class B code vector.

TABLE 6.1: Relevance of the features, of Pima Indians diabetes database. "Weight" represents the absolute value of eigenvector components associated with the feature. "Acc. W." is accumulation of weights.

#	Feature	Weight (%)	Acc. W.
2	Plasma glucose concentration at 2 hours in an oral glucose tolerance test	47.35	47.35
6	Body mass index	15.94	63.29
1	Number of times pregnant	15.07	78.36
3	Diastolic blood pressure	13.15	91.51
7	Diabetes pedigree function	2.94	94.45
8	Age	2.87	97.32
4	Triceps skin fold thickness	2.47	99.79
5	2-hour serum insulin	0.21	100

In order to illustrate the effectiveness of the method, we refer to the Pima Indians diabetes database from the UCI machine learning repository [5]. It presents data of female patients at least 21 years old from the Pima Indian tribe. The goal of this problem is to predict whether the patient will result positive for diabetes. The database consists of 8 features plus the class, and it contains 768 instances. In this experiment it can be shown that an LVQ with only 2 code vectors can reach an error of 0.24, while SVM can reach an error of 0.23 (with 400 SVs). Hence, we can assume a linear decision border to be a good approximation of the true one. In this special case, only one eigenvalue turns out to be different from zero. The corresponding eigenvector is the normal vector to the unique decision hyperplane. Table 6.1 lists features ordered with respect to the value of the related eigenvector component, hence with respect to their discriminative relevances.

Note that the most important feature agrees with the criterion of the World Health Organization for diagnosing diabetes (i.e., if the 2-hour post-load plasma glucose was at least 200 mg/dl at any survey examination or if found during routine medical care).

In more complex situations, where many pieces of the hyperplane form the decision border, the normal vectors should be appropriately weighted to take into account the extent of the portion of the hyperplane actually forming the decision border. To better explain this point, let us consider the example of the decision border shown in Figure 6.3.

For this example, we get four normal vectors to the piecewise linear decision border: $[1\ 0]$ and $[0\ 1]$, each repeated two times. Since the LVQ defines a piecewise linear decision border, the estimate of Σ_{EDBFM} turns out to be proportional to $\sum_i \mathbf{N}_i^T \cdot \mathbf{N}_i$, where $\mathbf{N}_i$ is the unit normal vector to a piece of the decision border. Eigenvalues and eigenvectors of such a Σ_{EDBFM} matrix turn out to be $\lambda_1 = \lambda_2 = 0.5$, $\mathbf{u}_1 = [1\ 0]$, $\mathbf{u}_2 = [0\ 1]$, suggesting that the two dimensions have the same discriminative power, while it is clear that projecting on the first dimension results in a minor accuracy loss than projecting on the second dimension. Exploiting only the normal vectors, we don't fully consider the geometry of the decision border, which greatly depends on the statistics of the classification problem. Indeed, if in this example we consider a square instead of a rectangle, we obtain the same Σ_{EDBFM} matrix. By defining the Σ_{EDBFM} matrix as a weighted sum of normal vectors, where each normal vector is weighted by the length of the related segment of decision border over the total length of the decision border, we get $\lambda_1 = 0.8$, $\lambda_2 = 0.2$, $\mathbf{u}_1 = [1\ 0]$, $\mathbf{u}_2 = [0\ 1]$; hence the first dimension correctly results four times more important than the second one.

In order to take into account the statistics of the problem, normal vectors should be appropriately weighted. Then we give the following general form of the BVQ-based feature extraction (BVQFE) algorithm:

BVQ-based Feature Extraction

1. Train the LVQ $\{(\mathbf{m}_1, l_1), \ldots, (\mathbf{m}_Q, l_Q)\}, \mathbf{m}_i \in \mathcal{R}^N$, $l_i \in \mathbf{Y}$ on a training set TS by using the BVQ algorithm;

2. set the elements of the matrix Σ_{BVQFM} to 0;

3. set w_{tot} to 0;

4. for each pair y_i, $y_j \in \mathbf{Y}$, where $i \neq j$ do

    1. set the elements of the matrix $\Sigma_{BVQFM_{ij}}$ to 0;

    2. for each pair $\mathbf{m}_k$, $\mathbf{m}_z \in \mathbf{M}$ defining a piece of decision border, where $l_k = y_i$ and $l_z = y_j$ do

        1. calculate the unit normal vector to the decision border as: $\mathbf{N}_{kz} = \frac{(\mathbf{m}_k - \mathbf{m}_z)}{\|\mathbf{m}_k - \mathbf{m}_z\|}$;

        2. calculate the weight w_{kz} of the unit normal vector $\mathbf{N}_{kz}$;

        3. $w_{tot} = w_{tot} + w_{kz}$;

        4. $\Sigma_{BVQFM_{ij}} = \Sigma_{BVQFM_{ij}} + w_{kz}\mathbf{N}_{kz}^T\mathbf{N}_{kz}$;

    3. $\Sigma_{BVQFM} = \Sigma_{BVQFM} + P(y_i)P(y_j)\Sigma_{BVQFM_{ij}}$;

5. $\Sigma_{BVQFM} = \frac{\Sigma_{BVQFM}}{w_{tot}}$.

The eigenvectors $\mathbf{u}_i$ of the $\boldsymbol{\Sigma}_{BVQFM}$ define the matrix $\mathbf{U} = [\mathbf{u}_1, \mathbf{u}_2, \ldots, \mathbf{u}_N]$, which is exploited to transform the original space into a new space such that $\mathbf{x}' = \mathbf{U} \cdot \mathbf{x}$. The eigenvectors corresponding to the largest eigenvalues represent the most discriminant features. So, the matrix $\mathbf{U}'$ built with only the first $\mathbf{n}'$ most discriminant features defines the transformation of the original space $\mathcal{R}^N$ in the reduced space $\mathcal{R}^{N'}$.

Algorithms to calculate the weight of the normal vectors $\mathbf{N}_{kz}$ are discussed in the following subsections.

6.4.1   Weighting of Normal Vectors

Since the LVQ decision border is piecewise linear, the EDBFM equation (6.3) becomes

$$\boldsymbol{\Sigma}_{BVQFM} = \frac{1}{\int_{S'} p(\mathbf{x})d\mathbf{x}} \sum_{\lambda=1}^{\Lambda} \mathbf{N}_\lambda^T \cdot \mathbf{N}_\lambda \int_{S_\lambda} p(\mathbf{x})d\mathbf{x}, \tag{6.5}$$

where $S' = \sum_{\lambda=1}^{\Lambda} S_\lambda$ is the piecewise effective decision border and $\mathbf{N}_\lambda$ is the unit normal vector to the piece of border S_λ, $\lambda = 1, 2, \ldots, \Lambda$. Hence, weight w_λ is represented by the probability distribution of data on S_λ: $w_\lambda = \int_{S_\lambda} p(\mathbf{x})d\mathbf{x}$. In order to estimate w_λ, one can resort to nonparametric density estimation methods, and in particular to the Parzen method [7]:

$$\widehat{p}(\mathbf{x}) = \frac{1}{M} \sum_{i=1}^{M} k(\mathbf{x} - \mathbf{x}_i),$$

where $k(.)$ is the kernel function. Different forms of the kernel can be chosen. In the following, we consider the uniform hypercubic window, that is, $k(\mathbf{x} - \mathbf{x}_i) = \Delta^{-N}$ over a N-dimensional hypercube of side Δ centered on the training sample $\mathbf{x}_i$ $(i = 1, 2, \ldots, M)$ and $k(\mathbf{x} - \mathbf{x}_i) = 0$ elsewhere. With this choice, after some manipulations, we get

$$\widehat{w}_\lambda(\Delta) = \sum_{i=1}^{M} \delta(d(\mathbf{x}_i, S_\lambda) \le \frac{\Delta}{2}), \tag{6.6}$$

where $d(\mathbf{x}_i, S_\lambda)$ is the Euclidean distance between $\mathbf{x}_i$ and the piece of decision border S_λ, that is, we can approximate the true weights by counting how many training samples fall "on" (i.e., at a distance less than $\Delta/2$ from) each piece of decision border S_λ. In [10, 4] it is proposed to weight the normal vectors by the volumes of the decision border. It is simple to see that this method is a special case of the previous one. In fact, when $p(\mathbf{x}) = p$ is constant along each piece of decision border, Equation (6.5) becomes

$$\widehat{\boldsymbol{\Sigma}}_{BVQFM} = \frac{1}{p \sum_{\lambda=1}^{\Lambda} \int_{S_\lambda} d\mathbf{x}} \sum_{\lambda=1}^{\Lambda} \mathbf{N}_\lambda^T \cdot \mathbf{N}_\lambda \cdot p \int_{S_\lambda} d\mathbf{x} = \frac{1}{w_{tot}} \sum_{\lambda=1}^{\Lambda} \mathbf{N}_\lambda^T \cdot \mathbf{N}_\lambda \cdot w_\lambda$$

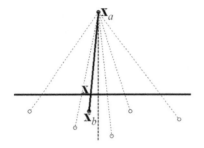 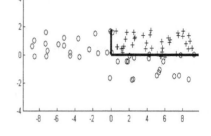

FIGURE 6.4: Two differently classified nearest samples. The horizontal line is the decision border.

FIGURE 6.5: A two-class problem with uneven contribution of normal vectors. The bold line is the Bayes decision border.

$w_\lambda = \int_{S_\lambda} d\mathbf{x}$ is now simply the volume of the piece of decision border S_λ. Volumes can be estimated by resorting to the concept of numerical integration of an N-dimensional function, by using a grid of equally spaced points [4].

The DBFE feature extraction techniques discussed in Sections 6.2.1 and 6.2.2 evaluate unit normal vectors $\mathbf{N}(\mathbf{x})$ by considering for each sample $\mathbf{x}_a$ the nearest sample $\mathbf{x}_b$ differently classified. $\mathbf{x}$ turns out to be the point belonging to the segment $\mathbf{s} = \alpha \cdot \mathbf{x}_a + (1 - \alpha) \cdot \mathbf{x}_b$, $0 \leq \alpha \leq 1$, and such that the decision function $h(\mathbf{x})$ is set to zero within a threshold.

To choose $\mathbf{x}_b$ such that it is the nearest sample to $\mathbf{x}_a$ means that $\mathbf{s}$ is almost normal to the decision border (see Figure 6.4). Hence, $\mathbf{x}$ is close to the projection of $\mathbf{x}_a$ on the decision border, and the distance between $\mathbf{x}_a$ and the decision border is approximated by the distance between $\mathbf{x}$ and $\mathbf{x}_a$. This observation allows us to recast both MLP-based and SVM-based approaches in the theoretical framework discussed above. In order to grasp the intuition behind this statement, let us consider a piecewise linear decision border: On each piece S_λ, SVMDBA, and MLPFE, find as many $\mathbf{N}_\lambda$ as the number of samples falling at a certain distance from S_λ. Hence, they implicitly perform a Parzen estimate of the probability density function along the decision border. In the MLP-based approach, the size of Δ is fixed and set to $2 * \max\{d(\mathbf{x}_a, S)\}$, so each sample contributes to the Parzen estimate, while in the SVM-based approach, Δ is implicitly set such that only $r \times \mid \mathcal{TS} \mid$ samples are considered. If $r = 1$, the SVM-based and MLP-based approaches work in the same way. In order to understand how these parameters influence performance, let us consider the classification problem depicted in Figure 6.5: Class 1, represented by "+", is described by a uniform distribution over the rectangle $[-0.2, 9.8] \times [-0.2, 1.8]$. Class 2, represented by "o", is distributed according to two uniform distributions over the rectangles $[-9.8, -0.2] \times [-0.2, 1.8]$ and $[-0.2, 9.8] \times [-1.8, 0.2]$. The classes are equiprobable. The bayesian decision border is given by $S : \{x_1 = 0, x_2 \in [0, 1.8] \wedge x_2 = 0, x_1 \in [0, 9.8]\}$. Then, from

Equation (6.3), it turns out that the eigenvalues and related eigenvectors of the Σ_{EDBFM} matrix are $\lambda_1 \cong 0.1591, \mathbf{u}_1 = [1,0]; \; \lambda_2 \cong 0.841, \mathbf{u}_2 = [0,1]$.

For each class, 1000 samples are generated. These are used to estimate the normal vectors by the SVMDBA, MLPFE, and BVQFE approaches. In order to eliminate the influence of the classifier characteristics (MLP, SVM, and BVQ), we use as the decision function $h(\mathbf{x})$ the equation of the bayesian decision border S. In Table 6.2 we show the eigenvalues and related eigenvectors of the Σ_{EDBFM} averaged over 100 different datasets, obtained with the best setting of $\Delta_{BVQFE} = 0.5$ and $r = 0.2$. With such a setting of r it turns out that the most far sample is at a distance of 0.248 from the decision border. Hence we can assume that SVMDBA implicitly defines a Parzen window of size 0.496. Note the similarity with Δ_{BVQFE}. As a matter of fact, eigenvalues estimated by SVMDBA and BVQFE are similar and close to the real values. They both perform much better than MLPFE.

TABLE 6.2: Comparison of eigenvalues estimated by MLPFE, SVMDBA, and BVQFE.

		MLPFE	SVMDBA	BVQFE
$\mathbf{u}_1 = [1,0]$	λ_1	0.3068	0.1521	0.1608
$\mathbf{u}_2 = [0,1]$	λ_2	0.6932	0.8478	0.8392

Table 6.3 shows how the eigenvalues calculated by SVMDBA depend on the parameter r. In the table, Σ_{EDBFM} is averaged over 100 different datasets. Moving from $r = 0.2$ to $r = 1$, the estimate of the eigenvalues is worse, and when the size of the Parzen window is such that all samples contribute to the estimate of Σ_{EDBFM} (that is, r is set to 1), the eigenvalues tend to the values obtained by MLPFE in the previous experiment.

TABLE 6.3: SVMDBA eigenvalues estimate vs. the value of r.

r	0.2	0.4	0.6	0.8	1
λ_1	0.152	0.143	0.134	0.127	0.296
λ_2	0.848	0.857	0.866	0.873	0.704

Hence, in general, contrary to what the authors state in [17], the parameter r may greatly affect the performance of the SVMDBA approach.

As observed by Fukunaga [7, p.328], the optimal value of the Parzen window is not easy to obtain, and it has to be searched experimentally. We note that BVQ returns the minimum error probability when the parameter Δ_{BVQFE} is set to the side of the optimal Parzen window. So, this value is given as a byproduct of the BVQ training [3], while SVM training does not give any suggestion on how to set r.

6.5 Experiments

In the present section we experimentally compare the performance of DBFE-based methods. We first examine their accuracy and robustness on the synthetic experiment proposed in [13, 8]. Then we show the performances on real-world datasets from the UCI Machine Learning Repository [5], and in particular we exploit the Waveform dataset in order to examine the complexity of the methods.

6.5.1 Experiment with Synthetic Data

The dataset consists of three equiprobable classes y_1, y_2, y_3 distributed according to the following statistics:

$$\mu_1 = \begin{bmatrix} 0 \\ 0 \\ 0 \end{bmatrix}, \Sigma_1 = \begin{bmatrix} 4 & 0 & 0 \\ 0 & 4 & 0 \\ 0 & 0 & 9 \end{bmatrix}$$

$$\mu_{21} = \begin{bmatrix} 5 \\ 0 \\ 0 \end{bmatrix}, \Sigma_{21} = \begin{bmatrix} 2 & 0 & 0 \\ 0 & 2 & 0 \\ 0 & 0 & 9 \end{bmatrix} \text{ and } \mu_{22} = \begin{bmatrix} -5 \\ 0 \\ 0 \end{bmatrix}, \Sigma_{22} = \begin{bmatrix} 2 & 0 & 0 \\ 0 & 2 & 0 \\ 0 & 0 & 9 \end{bmatrix}$$

$$\mu_{31} = \begin{bmatrix} 0 \\ 5 \\ 0 \end{bmatrix}, \Sigma_{31} = \begin{bmatrix} 9 & 0 & 0 \\ 0 & 2 & 0 \\ 0 & 0 & 9 \end{bmatrix} \text{ and } \mu_{32} = \begin{bmatrix} 0 \\ -5 \\ 0 \end{bmatrix}, \Sigma_{32} = \begin{bmatrix} 9 & 0 & 0 \\ 0 & 2 & 0 \\ 0 & 0 & 9 \end{bmatrix}$$

The intrinsic discriminant dimension (Section 6.2) of the problem is 2, and the pairs of eigenvalues and related eigenvectors are ($\lambda_1 = 0.56$, $\mathbf{u}_1 = [0\ 1\ 0]$); ($\lambda_2 = 0.44$, $\mathbf{u}_2 = [1\ 0\ 0]$); and ($\lambda_3 = 0$, $\mathbf{u}_3 = [0\ 0\ 1]$).

Similarly to [8], 2000 samples from each class were generated, of which 500 were used for the training and the remaining for the test. We initialized an LVQ of order 20 with the first 20 training vectors, and we set $\Delta = 0.5$, $\gamma^{(0)} = 0.5$. We did not stress the setting of the parameters deliberately, in order not to take advantage of either the knowledge of the class statistics or of the results in [13, 8]. As a result, the net shows an average error probability on the test set of 0.1694, which is slightly worse than that in [13] (0.143) and [8] (0.152). Nevertheless, the feature extraction algorithm produces comparable eigenvalues and eigenvectors:

$$\lambda_1 = 0.5237, \lambda_2 = 0.4689, \lambda_3 = 0.0074,$$

$$\mathbf{u}_1 = \begin{bmatrix} -0.29 \\ 0.96 \\ -0.03 \end{bmatrix}, \mathbf{u}_2 = \begin{bmatrix} 0.96 \\ 0.29 \\ -0.01 \end{bmatrix}, \mathbf{u}_3 = \begin{bmatrix} 0.00 \\ -0.03 \\ -1.00 \end{bmatrix}.$$

As a weighting method we used the one based on the training samples, while in [4] experiments with the volume calculus are presented. We employed a

TABLE 6.4:   Average nearest neighbor error probabilities vs. dimensions of the transformed spaces for the PCA, CA, MLPFE, ADBFE, SVMDBA, and BVQFE approaches.

feature No.	Error Probability (Variance)					
	PCA	CA	MLPFE	ADBFE	SVMDBA	BVQFE
1	0.489	0.576	0.467	0.483	0.424	0.469
	$(8.1 \cdot 10^{-5})$	$(4.7 \cdot 10^{-3})$	$(2.1 \cdot 10^{-3})$	$(2.6 \cdot 10^{-3})$	$(1.9 \cdot 10^{-4})$	$(1.5 \cdot 10^{-3})$
2	0.229	0.408	0.212	0.220	0.211	0.208
	$(9.4 \cdot 10^{-4})$	$(5.6 \cdot 10^{-3})$	$(4.9 \cdot 10^{-5})$	$(7.0 \cdot 10^{-4})$	$(7.8 \cdot 10^{-5})$	$(5.0 \cdot 10^{-5})$
3	0.219	0.218	0.219	0.219	0.218	0.219
	$(3.9 \cdot 10^{-5})$	$(4.4 \cdot 10^{-5})$	$(3.9 \cdot 10^{-5})$	$(3.9 \cdot 10^{-5})$	$(3.9 \cdot 10^{-5})$	$(3.9 \cdot 10^{-5})$

gaussian radial basis kernel to train the SVM, and we set r to 0.2, which gave the best results for this experiment. Average error probability on the test set is 0.1666. Table 6.4 compares the accuracy of the proposed method with that of competing methods, namely of MLPFE, ADBFE, SVMBDA, PCA, and CA, by showing the error performed by a nearest neighbor (NN) classifier on the data transformed according to the above approaches. In particular, we present the average error probability when the most important feature is considered when the first two features are considered and on the whole transformed space. By using the same classifier for each approach, we eliminate the influence of the classifier characteristics (in particular, MLP vs. LVQ vs. SVM) and we can better appreciate the performance of the feature extraction methods. The error probabilities in Table 6.4 are averaged over 10 different datasets, and the related variances are also shown in brackets.

We can see that the accuracies obtained by using the methods based on the DBFE principle are substantially the same, and they are definitely better than those of the methods that do not exploit information about the decision border.

It was noted that MLPFE and ADBFE indirectly define the decision border from the estimation of the a-posteriori class probabilities, while BVQ and SVM are devoted to directly finding the Bayes decision border. The use of direct information about the decision border is an advantage in many cases since it is well known that an accurate estimation of the a-posteriori class probabilities leads to an accurate estimation of the decision border; however, if a-posteriori class probabilities are not well estimated, nothing can be said about the accuracy of the estimated decision border. This advantage can be experimentally observed if, for the same experiment, we consider a training set of reduced size. Table 6.5 reports the average error probabilities and variances of the error performed by the DBFE-based methods when only 50 training vectors and 150 test vectors are used for each class. The results are averaged over 10 different datasets.

Note that BVQFE and SVMDBA give comparable results. They both find the best features and are more robust: The variance of the error in the case of the best pair of features is an order of magnitude lower than that of both MLPFE and ADBFE. Nevertheless, the MLPs used in this experiment have on average the same mean squared error as the MLPs used in the previous

TABLE 6.5:   Average nearest neighbor error probabilities vs.
dimensions of the transformed spaces for the BVQFE, ADBFE,
SVMDBA, and MLPFE methods. Reduced dataset.

feature No.	Error Probability (Variance)			
	MLPFE	ADBFE	SVMDBA	BVQFE
1	$0.495\ (5.4 \cdot 10^{-3})$	$0.460\ (3.9 \cdot 10^{-3})$	$0.459\ (1.2 \cdot 10^{-3})$	$0.475\ (2.3 \cdot 10^{-3})$
2	$0.227\ (1.3 \cdot 10^{-3})$	$0.236\ (2.0 \cdot 10^{-3})$	$0.221\ (6.5 \cdot 10^{-4})$	$0.219\ (6.9 \cdot 10^{-4})$
3	$0.246\ (3.2 \cdot 10^{-4})$	$0.246\ (3.2 \cdot 10^{-4})$	$0.246\ (3.2 \cdot 10^{-4})$	$0.246\ (3.2 \cdot 10^{-4})$

experiment.

6.5.2   Experiment with Real Data

We evaluate the performance of DBFE methods on four real-world datasets
drawn from the UCI repository: Waveform, Pima Indians Diabetes, Liver
Disorders, and Letter. Waveform is a three-class problem in a 40-dimensional
space, with known statistics. The first 21 features of each class represent a
wave generated from a combination of two of three shifted triangular wave-
forms, plus Gaussian noise with mean 0 and variance 1. The latter 19 features
are all Gaussian noise. It can be proven that the intrinsic discriminant dimension for this problem is 2. The Waveform dataset contains 5000 instances.
Pima Indians Diabetes is described in Section 6.4. The Liver Disorders dataset
consists of 345 observations of males, each with 6 features, classified on the basis of their sensitivity to liver disorders. Finally, samples of the Letter dataset
are 20000 images of 26 English capital letters, described by 16 features.

TABLE 6.6:   Parameter settings for
experiments with UCI datasets.

UCI Datasets	SVMDBA		BVQFE	
	Kernel	r	VQ size	Δ
Waveform	3-polynomial	1.0	10	0.5
Pima Indians	2-polynomial	0.2	2	0.3
Liver	rbf	0.2	8	0.15
Letter	rbf	0.2	52	0.05

For Waveform, Pima Indians, and Liver we use 10-fold, 12-fold, and 5-fold
cross validation, respectively, while we split the Letter dataset into 15000 samples for the training and 5000 for the test. For the Pima Indians, Waveform,
and Letter experiments we used the same setup as in [17]. Table 6.6 shows
the parameter settings of SVMDBA and BVQFE for the experiments.

Figures 6.6(a - d) show the error performed by a nearest neighbor classifier
on the data transformed according to DBFE-based methods. In particular,
we plot error probability vs. the first N' most discriminative features.

In Pima Indians, Liver, and Letter, we see that from a certain number of
features (the intrinsic discriminant dimension) on, the error becomes nearly
constant. The trend of the error in the Waveform experiment is due to the

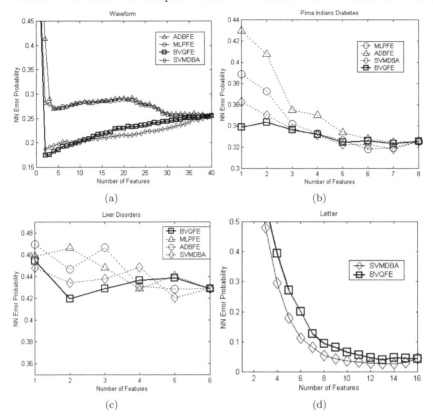

FIGURE 6.6: Comparison of DBFE-based methods over the first discriminative features. Average error computed by using NN.

fact that each dimension added to the intrinsic discriminant ones introduces noise, which greatly affects the nearest neighbor classifier. In any case, it is clear that the performance given by the methods on smaller numbers of features better illustrates the capability to find good discriminative features.

We can observe that BVQFE clearly outperforms MLP-based methods on the Waveform and Pima Indians datasets, and it is slightly better on Liver. It is at least comparable with SVMDBA everywhere. We do not show the performance of MLP-based methods on the Letter dataset since we were not able to obtain results in a reasonable time on an Intel Pentium M 1.73GHz-1.0GB RAM. This raises the issue of computational complexity.

In order to compute Σ_{EDBFM}, BVQFE calculates $Q \cdot |\mathcal{TS}|$ distances, where Q is the number of code vectors. Since the optimal number of code vectors depends on the classification problem, but it does not depend on the size M of the training set [3], the complexity of BVQFE is linear in the size of the training set. On the other hand, the MLP-based methods require, for each correctly classified training vector, the calculus of about $|\mathcal{TS}|$ distances and a certain number of MLP activation functions in order to find the point on the decision border. Hence these methods show a complexity that is quadratic in the training set size. As regards SVMDBA, it calculates $|\mathcal{TS}|$ decision functions $h(.)$ in order to extract the subset of cardinality r, plus $(r \cdot |\mathcal{TS}|)^2$ distances and a certain number of $h(.)$ in order to find the point on the decision border and the normal vector. Thus the complexity of SVMDBA is quadratic in $|\mathcal{TS}|$. Since each calculus of $h(.)$ implies the calculus of a kernel function for each support vector, SVMDBA is more time-consuming than MLP-based methods. This analysis is confirmed by the experiment in

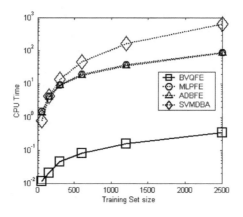

FIGURE 6.7: Comparison of DBFE-based methods over the number of training samples. Waveform dataset. Average cpu-time computed over 10 different datasets. Logarithmic scale.

Figure 6.7, which draws graphically the differences in cpu-time of the DBFE-based methods vs. training set size on the Waveform dataset. The parameter Δ_{BVQFE} is the optimal one, while we chose $r = 0.2$ in order to evaluate SVMDBA computational time on a favorable situation where only a subset of training samples is used. In such an experiment, the learning time of the nets is not considered. We observe that the BVQFE cpu-time is around two orders of magnitude lower than that of the other methods. If we also consider the learning time, the performance of SVMDBA becomes better than the MLP-based ones, since SVM is quicker than MLP on high-dimensional datasets. Unlike SVM, the BVQ learning time is not dependent on the size of the training set, then BVQFE keeps on being faster than SVMDBA. This fact is emphasized in general C-class problems, where we have to train C different SVMs and compute the same number of Σ_{EDBFM}. Finally, note that considering scalability w.r.t. the training set is significant for data mining problems, typically characterized by huge amounts of data.

6.6 Conclusions

This chapter presented a study on feature extraction techniques for classification based on the decision boundary feature extraction principle. We detailed the techniques that apply the DBFE principle on approximations of the Bayes decision border extracted by neural network classifiers. In particular, multi-layer perceptron and support vector machines are considered. We also introduced a formal derivation of the DBFE principle for labeled vector quantizers. It is shown that LVQ allows for a cheap representation of decision borders, from which the most discriminative features can be easily extracted. Furthermore, the use of the Bayes risk weighted vector quantization algorithm to train the LVQ allows one to define a robust and effective procedure for the estimation of a true decision border. Experimentally we observed that BVQFE and SVMDBA give comparable results, and they perform better than MLP-based methods. Furthermore, the BVQFE method shows a lower computational cost than the others.

In the development of the theory and in the experiments, we focused on the average error probability as a performance measure. However, we note that BVQ is a general algorithm for the minimization of the average misclassification risk. Hence, BVQFE is the only DBFE-based method that can directly manage real problems where the misclassification costs differ from one class to another.

References

[1] C. H. Park, H. Park and P. Pardalos. A Comparative Study of Linear and Nonlinear Feature Extraction Methods. In *Proceedings of the Fourth IEEE International Conference on Data Mining*. IEEE, Piscataway, NJ, 2004.

[2] E. Choi and C. Lee. Feature Extraction Based on the Bhattacharyya Distance. *Pattern Recognition*, 36(8):1703–1709, 2003.

[3] C. Diamantini and M. Panti. An Efficient and Scalable Data Compression Approach to Classification. *ACM SIGKDD Explorations*, 2(2):54–60, 2000.

[4] C. Diamantini and D. Potena. Feature Extraction for Classification: an

LVQ-based Approach. In *Proc. of Int. Workshop on Feature Selection for Data Mining: Interfacing Machine Learning and Statistics*, pages 2–9. SIAM, Philadelphia, PA, 2006.

[5] D.J. Newman, S. Hettich, C.L. Blake and C.J. Merz. UCI Repository of Machine Learning Databases. http://kdd.ics.uci.edu/, 1998.

[6] J. H. Friedman. On Bias, Variance, 0/1-Loss, and the Curse-of-Dimensionality. *Data Mining and Knowledge Discovery*, 1(1):55–77, 1997.

[7] K. Fukunaga. *Introduction to Statistical Pattern Recognition (2nd Edition)*. Academic Press, San Diego, 1990.

[8] J. Go and C. Lee. Analytical Decision Boundary Feature Extraction for Neural Networks. In *Proc. IEEE Int. Symposium on Geoscience and Remote Sensing*, volume 7, pages 3072–3074. IEEE, Piscataway, NJ, 2000.

[9] T. Joachims. Making Large-Scale SVM Learning Practical. In B. Scholkopf, C. J. Burges, A. J. Smola, editors, *Advances in Kernel Methods - Support Vector Learning*. MIT Press, Cambridge, MA, 1999.

[10] C. Lee and D. Landgrebe. Feature Selection Based on Decision Boundaries. In *Int. Geoscience and Remote Sensing Symposium*, volume 3, pages 1471–1474. IEEE, Piscataway, NJ, June 1991.

[11] C. Lee and D. Landgrebe. Decision Boundary Feature Extraction for Nonparametric Classification. *IEEE Trans. on Man and Cybernetics*, 23(2):433–444, March/April 1993.

[12] C. Lee and D. Landgrebe. Feature Extraction Based on Decision Boundaries. *IEEE Trans. on Pattern Analysis and Machine Intelligence*, 15(4):388–400, Apr. 1993.

[13] C. Lee and D. Landgrebe. Decision Boundary Feature Extraction for Neural Networks. *IEEE Trans. on Neural Networks*, 8(1):75–83, Jan. 1997.

[14] Osuna, E. and Girosi, F. Reducing the Run-time Complexity of Support Vector Machines. In *Proc. International Conference on Pattern Recognition (ICPR98)*, 1998.

[15] B. Schoelkopf, C. Burges, and V. Vapnik. Extracting Support Data for a Given Task. In U. Fayyad and R. Uthurusamy, editors, *Proc. 1st Int. Conf. on Knowledge Discovery and Data Mining*, Menlo Park, CA, AAAI Press, 1995.

[16] V. Vapnik. *Statistical Learning Theory*. J. Wiley and Sons, New York, 1998.

[17] J. Zhang and Y. Liu. SVM Decision Boundary Based Discriminative

Subspace Induction. *Pattern Recognition*, 1(38):1746–1758, 2005.

Chapter 7

Ensemble-Based Variable Selection Using Independent Probes

Eugene Tuv

Analysis & Control Technology Department, Intel

Alexander Borisov

Analysis & Control Technology Department, Intel

Kari Torkkola

Intelligent Systems Lab, Motorola

7.1 Introduction

Traditional multivariate statistics has approached variable selection using stepwise selection and best subset selection within linear-regression models. More recent trends are nonlinear models and addressing the question of instability (a small change in the data might result in a drastic change in the inferred results). This chapter discusses an approach that covers both of these concerns. ¡Nonlinearity is addressed using decision trees as the underlying regressors or classifiers, and instability is addressed by employing *ensembles* of decision trees.

Assuming now that we have a possibly nonlinear and stable system that ranks variables in the order of importance, the last missing component is finding a threshold to include only truly important variables. This is the main topic of the current chapter.

Variable selection generally contains two components. There needs to be a criterion that, given a set of variables, evaluates the joint relevance of the set. The second component is a search mechanism that adds or removes variables to the current set. It may also be that the criterion only evaluates the relevance of a single variable, or a small number of variables at a time. This search is iterated until a desired number of variables is reached.

A general problem is, however, where to terminate the search. Given some kind of a ranking of variables, or sets of variables, it is not clear how to threshold the ranking in order to select only truly important variables and to exclude noise. If the number of true variables is known, this is naturally not a problem, but in real-world cases this information is seldom available.

We present a principled approach to doing this for datasets of any type and complexity by means of independent probe variables. We describe first how ensembles of trees can produce measures for relevant variable ranking with multiple interacting variables. The main idea of independent probes is then presented together with an algorithm incorporating these ideas. Experimentation with artificial as well as real datasets demonstrates the performance of the method.

7.2 Tree Ensemble Methods in Feature Ranking

In this chapter we try to address a problem of feature selection in very general supervised settings: The target variable could be numeric or categorical, the input space could have variables of mixed type with non-randomly missing values, the underlying $X - Y$ relationship could be very complex and multivariate, and the data could be massive in both dimensions (tens of thousands of variables and millions of observations). Ensembles of unstable but very fast and flexible base learners such as trees (with embedded feature weighting) can address all of the listed challenges. They have proved to be very effective in variable ranking in problems with up to 100,000 predictors [2, 11]. A more comprehensive overview of feature selection with ensembles is given in [13].

A decision tree partitions the input space into a set of disjoint regions and assigns a response value to each corresponding region. It uses a greedy, top-down recursive partitioning strategy. At every step a decision tree uses an exhaustive search by trying all combinations of variables and split points to achieve the maximum reduction in impurity of the node. Therefore, the tree constructing process itself can be considered as a type of variable selection, and the impurity reduction due to a split on a specific variable could indicate the relative importance of that variable to the tree model. Note that this relative importance is based on a multivariate model, and it is different from the relevance measured by standard, univariate filter methods. For a single decision tree, a measure of variable importance has been defined in [4]:

$$VI(x_i, T) = \sum_{t \in T} \Delta I(x_i, t) \qquad (7.1)$$

where $\Delta I(x_i, t) = I(t) - p_L I(t_L) - p_R I(t_R)$ is the decrease in impurity due to an actual (or potential) split on variable x_i at a node t of the optimally pruned tree T. The sum in (7.1) is taken over all internal tree nodes where x_i is a primary splitter, as proposed in [4]. Node impurity $I(t)$ for regression is defined as $\frac{1}{N(t)} \sum_{s \in t} (y_s - \bar{y})^2$, where the sum and mean are taken over all observations s in node t, and $N(t)$ is the number of observations in node t. For classification $I(t) = Gini(t)$, where $Gini(t)$ is the Gini index of node t:

$$Gini(t) = \sum_{i \neq j} p_i^t p_j^t \tag{7.2}$$

and p_i^t is the proportion of observations in t whose response label equals i ($y = i$) and i and j run through all response class numbers. The Gini index is in the same family of functions as *cross-entropy*, $-\sum_i p_i^t log(p_i^t)$, and measures node impurity. It is zero when t has observations only from one class and reaches its maximum when the classes are perfectly mixed.

Random Forest [3] is a representative of tree ensembles that extends the "random subspace" method [8]. The randomness originates from sampling both the data and the variables. It grows a forest of random trees on bagged samples showing excellent results comparable with the best-known classifiers. Random Forest (RF) does not overfit, and can be summarized as follows:

1. A number n is specified much smaller than the total number of variables N (typically $n \sim \sqrt{N}$).

2. For each tree to be constructed, a different sample of training data is drawn with replacement (bootstrap sample). The size of the sample is the same as that of the original dataset. This bootstrap sampling typically leaves 30 percent of the data out-of-bag. These data help provide an unbiased estimate of the tree's performance later. Each tree is constructed up to a maximum pre-specified depth.

3. At each node, n out of the N variables are selected at random.

4. The best split among these n variables is chosen for the current node, in contrast to typical decision tree construction, which selects the best split among all variables.

The computational complexity for each tree in the RF is $O(\sqrt{N} M \log M)$, where M is the number of the training cases. Therefore, it can handle very large numbers of variables with a moderate number of observations. Note that for every tree grown in RF, about one-third of the cases are out-of-bag (out of the bootstrap sample). The out-of-bag (OOB) samples can serve as a test set for the tree grown on the non-OOB data.

Averaging how often different variables were used in the splits of the trees (and from the quality of those splits) gives a measure of variable importance as a byproduct of the construction. For a stochastic tree ensemble of S trees the importance measure (7.1) is thus

$$I(x_i) = \frac{1}{S} \sum_{s=1}^{S} VI(x_i, T_s) \qquad (7.3)$$

The regularization effect of averaging makes this measure much more reliable than a single tree.

Relative feature ranking (7.3) provided by such ensembles, however, does not separate relevant features from irrelevants. Only a list of importance values is produced without a clear indication of which variables to include and which to discard. Also, trees tend to split on variables with more distinct values. This effect is more pronounced for categorical predictors with many levels. Trees often make a less relevant (or completely irrelevant) input variable more "attractive" to split on only because they have high cardinality.

The main idea of this work relies on the following fact. We add a number of randomly generated features, all independent of the target variable Y, to the set of original features. A stable feature ranking method, such as an ensemble of trees, which measures the relative relevance of an input to a target variable Y, would assign a significantly (in statistical sense) higher rank to a legitimate variable X_i than to an independent probe variable. These independent probe variables thus act as a baseline that determines the ranking cutoff point. If the sample size is small, the process of variable generation and ranking must be repeated several times in order to gain statistical significance. We present now an algorithm for variable ranking or selection based on this idea.

7.3 The Algorithm: Ensemble-Based Ranking Against Independent Probes

Our method is a combination of three ideas: **A)** Estimating variable importance using an RF ensemble of trees of a fixed depth (3-6 levels) with the split weight re-estimation using OOB samples (gives a more accurate and unbiased estimate of variable importance in each tree and filters out noise variables), **B)** comparing variable importance against artificially constructed noise variables using a formal statistical test, and **C)** iteratively removing the effect of identified important variables to allow the detection of less important variables (trees and parallel ensemble of trees are not well suited for additive models). All the advantages of ensembles of trees listed in Section 7.2, such as the capability to handle missing variables, are now inherited by the algorithm.

A) **Split weight re-estimation**

We propose a modified scheme for calculating split weight and selecting the best split in each node of a tree. The idea is to use training samples to find the best split point on each variable, and then use samples that were not used for building the tree (out-of-bag) to select the best split variable in a node. Split weight used for variable importance estimation is also calculated using out-of-bag samples.

This helps to prevent irrelevant features from entering the model, or at least makes their weight close to zero, because the irrelevant variable weight calculated on out-of-bag samples will not depend on the variable type (continuous, discrete) or on the number of distinct variable values.

B) **Selecting important features**
In order to determine a cutoff point for the importance scores, there needs to be a *contrast* variable that is known to be truly independent of the target. By comparing variable importance to this contrast (or several), one can then use a statistical test to determine which variables are truly important. We propose to obtain these independent probe variables by randomly permuting values of the original N variables across the M examples. Generating contrasts using unrelated distributions, such as Gaussian or uniform, is not sufficient, because the values of the original variables may exhibit some special structure.

For each of the T permutations a short ensemble of $L = 10 - 50$ trees is constructed. For each ensemble, variable importance is then computed for all variables, including the independent probes for each series. Using a series of ensembles is important when the number of variables is large or tree depth is small, because some (even important) features can be absent in a particular tree. To gain statistical significance, the importance score of all variables is compared to a percentile (we used 75th) of importance scores of the N contrasts. A statistical test (Student's t-test) is performed to compare the scores over the T series. Variables that are scored significantly higher than contrasts are selected.

C) **Removing effects of identified important variables**
To allow detection of less important variables, after a subset of relevant variables is discovered by step B, we remove their effects on the response. To accomplish this, the target is predicted using only these important variables, and a residual of the target is computed. Then we return to step A, until no variables remain with scores significantly higher than those of the contrasts. The last step is identical to stage-wise regression, but applied to a nonlinear model. It is important that the step (A) uses all variables to build the ensemble, and does not exclude identified important ones.

To accommodate step C in classification problems we adopted the multiclass logistic regression approach described in [5]. We predict log-odds of class probabilities for each class with an ensemble, and then take pseudo-residuals as summarized in Algorithm 7.3.2. The main difference in the regression

TABLE 7.1: Notation in the Algorithms

C	Number of classes (if classification problem)
X	Set of original variables
Y	Target variable
Z	Permuted versions of X
F	Current working set of variables
Φ	Set of important variables
$\mathbf{V}_{i.}$	ith row of variable importance matrix $\mathbf{V}$
$\mathbf{V}_{.j}$	jth column of matrix $\mathbf{V}$
$g_I(F, Y)$	Function that trains an ensemble of L trees based on variables F and target Y, and returns a row vector of importance for each variable in F
$g_Y(F, Y)$	Function that trains an ensemble based on variables F and target Y, and returns a prediction of Y (the number of trees is typically larger than L)
$G_k(F)$	Current predictions for log-odds of k-th class
$I(Y_i = k)$	Indicator variable, equals one if $(Y_i = k)$
x	Data vector in the original variable space

case is that variable selection and removal of the influence of the discovered variables to the target are done separately for each class in each iteration (loop 2). The important feature set is then grouped from all classes. The stopping criteria is the absence of important features for all C classes.

The algorithms are now presented using the notation in Table 7.1. Note that the computational complexity of our method is of the same order as of an RF model, but it could be significantly faster for datasets with large numbers of cases since trees in RF are built to the maximum depth.

Algorithm 7.3.1 Independent Probes with Ensembles (IPE), Regression

---

1.   set $\Phi \leftarrow \{\}$;
2.   for $i = 1, ..., T$ do
3. $\{Z_1, ..., Z_N\} \leftarrow$ permute$\{X_1, ..., X_N\}$
4.   set $F \leftarrow X \cup \{Z_1, ..., Z_N\}$
5. $\mathbf{V}_{i.} = g_I(F, Y)$;
     endfor
6. $\mathbf{v}_n =_{(1-\alpha)} percentile_{j \in \{Z_1,...,Z_N\}} \mathbf{V}_{.j}$
7.   Set $\hat{\Phi}$ to those $\{X_k\}$ for which $\mathbf{V}_{.k} > max(\mathbf{v}_n, \mathbf{V}_{.Z_k})$ with t-test significance 0.05
8.   If $\hat{\Phi}$ is empty, then quit.
9. $\Phi \leftarrow \Phi \cup \hat{\Phi}$;
10. $Y = Y - g_Y(\hat{\Phi}, Y)$
11. Go to 2.

Algorithm 7.3.2 Independent Probes with Ensembles (IPE), Classification

1.   set $\Phi \leftarrow \{\}$; $G_k(F) = 0$;

2.   for $k = 1, ..., C$ do

    a.   Compute class prob. $p_k(x) = exp(G_k(x))/\sum_{l=1}^{K} exp(G_l(x))$

    b.   Compute pseudo-residuals $Y_i^k = I(Y_i = k) - p_k(x_i)$

    c.   Set $\mathbf{V} = 0$

    d.   for $i = 1, ..., T$ do

        $\{Z_1, ..., Z_N\} \leftarrow$ permute$\{X_1, ..., X_N\}$

        set $F \leftarrow X \cup \{Z_1, ..., Z_N\}$

        $\mathbf{V}_{i.} = \mathbf{V}_{i.} + g_I(F, Y^k)$;

        endfor

    e. $\mathbf{v}_n =_{(1-\alpha)} percentile_{j \in \{Z_1,...,Z_N\}} \mathbf{V}_{.j}$

    f.   Set $\hat{\Phi}_k$ to those $\{X_s\}$ for which $\mathbf{V}_{.s} > max(\mathbf{v}_n, \mathbf{V}_{.Z_s})$

        with t-test significance 0.05

    g. $\Phi \leftarrow \Phi \cup \hat{\Phi}_k$;

    h. $G_k(F) = G_k(F) + g_Y(\hat{\Phi}, Y^k)$

    endfor

3.   If $\hat{\Phi}_k$ is empty for all $k = 1, ..., C$, then quit.

4.   Go to 2.

As the function $g(., .)$ we have used ensembles of trees. Any classifier/regressor function can be used, from which variable importance from all variable interactions can be derived. To our knowledge, only ensembles of trees can provide this conveniently.

7.4   Experiments

As advocated by [9], an experimental study must have relevance and must produce insight. The former is achieved by using real datasets. However, such studies often lack the latter component, failing to show exactly why and under which conditions one method excels over another. This can be achieved by using synthetic datasets because they let one vary systematically the domain characteristics of interest, such as the number of relevant and irrelevant attributes, the amount of noise, and the complexity of the target concept. We describe first experiments with the proposed method using synthetic datasets followed by a real example.

7.4.1 Benchmark Methods

7.4.1.1 CFS

As one benchmark method, we use correlation-based feature selection (CFS), [7] as implemented in the Weka machine learning package [15]. CFS assumes that useful feature subsets contain features that are predictive of the target variable but uncorrelated with one another. CFS computes a heuristic measure of the "merit" of a feature subset from pairwise feature correlations. A heuristic search is used to traverse the space of feature subsets in reasonable time; the subset with the highest merit found during the search is reported. CFS thus also determines the number of returned features. CFS discretizes internally every continuous feature and can thus work with mixed-type input variables. In the experiments, CFS using forward search is labeled as "CFS," and using genetic search as "CFS-Gen."

Computational complexity is light, linear in the number of samples, but quadratic in the number of variables [7].

7.4.1.2 RFE

Another benchmark method is recursive feature elimination (RFE) [6], as implemented in the Spider machine learning library [14]. The idea is to compute the change in the cost function of a classifier $\partial J(i)$ caused by removing a given feature i or, equivalently, by bringing its weight to zero. RFE trains a support vector machine (SVM) as the classifier optimizing the weights w_i with respect to criterion J. A ranking criterion w_i^2 (or $\partial J(I)$) is computed for all features. The feature with the smallest ranking criterion is removed and the iteration is repeated. RFE thus considers the current feature set as a whole rather than ranking features individually. However, RFE has no intrinsic threshold. The desired number of features has to be determined by other means. In the experiments with artificial data we give an unfair advantage to RFE by retaining the top N features returned by RFE, where N is the known number of relevant features in the generated dataset. Another unfair advantage we gave for RFE is kernel selection for the intrinsc SVM classifier or regressor. We used an RBF kernel for the data generated by Friedman's generator where the target is a sum of multivariate Gaussians, and a linear kernel for the dataset where the target is a linear combination of variables.

As the method trains an SVM for each feature removed, the computational complexity is linear in the number of variables and retains the complexity of the SVM in the number of data samples, which is quadratic.

7.4.1.3 Breiman's RF Error Sensitivity Method

As the third benchmark we use the *sensitivity*-based measure of variable relevance evaluated by a Random Forest as proposed by Breiman [3]. For each tree, the prediction accuracy on the out-of-bag portion of the data is recorded. Then the same is done after permuting each predictor variable.

The differences between the two accuracies are then averaged over all trees and normalized by the standard error. For regression, the MSE is computed on the out-of-bag data for each tree, and then the same is computed after permuting a variable. The differences are averaged and normalized by the standard error to get a z-score, and assign a significance level to the z-score assuming normality. The null hypothesis tested is that the mean score is zero, against the one-sided alternative that the mean score is positive. It is clear that for a large number of variables this method would be extremely computationally challenging.

Since IPE uses a learner internally, it can be considered an embedded feature selection method. RFE, which uses internally an SVM, and RF error sensitivity methods can also similarly be called embedded methods. However, unlike RFE and RF, which just rank all the variables, IPE also determines the number of important variables. In this sense, the IPE operates in a similar fashion to CFS.

7.4.2 Data and Experiments

7.4.2.1 Synthetic Complex Nonlinear Data - Friedman's Generator

A very useful data generator is described in [5]. This generator produces datasets with multiple non-linear interactions between input variables. Any greedy method that evaluates the importance of a single variable one at a time is bound to fail with these datasets.

For each data set, 20 $N(0,1)$ distributed input variables were generated. The target is a multivariate function of ten of those, thus ten are pure noise. The target function is generated as a weighted sum (weights are random) of 20 multidimensional Gaussians, each Gaussian involving about four randomly drawn input variables at a time. Thus all of the important 10 input variables are involved in the target, to a varying degree. We derive the "true importance" of a variable as the sum of absolute values of the weights of those Gaussians that the variable was involved in. Mixed-type data were generated by discretizing a randomly chosen half of the variables, each into a randomly generated number of levels, which varied between 2 and 32.

Four different experiments are illustrated in Figures 7.1 and 7.2. Each experiment involves averaging results from 50 generated datasets, 400 samples each: 1) continuous variables - regression, 2) continuous variables - classification, 3) mixed-type variables - regression, 4) mixed-type variables - classification. Note that the smaller the dataset is, the harder it will be to detect true variables among the noise variables. Even though we claim that IPE handles massive datasets (because of its low computational complexity), here we are really evaluating the sensitivity of the method.

Each of the pairs of panels in Figures 7.1 and 7.2 illustrates two things. 1) How well the true important variables can be detected as a function of their "true" importance, and 2) What is the rate of erroneously detecting a noise variable as an important variable?

Figures 7.1 and 7.2 show that for all four scenarios of complex nonlinear dependencies (regression and classification with numeric and mixed-type predictors) IPE and RF methods are notably superior to CFS, CFS-Gen, and RFE. IPE and RF have similar detection rates, but RF consistently produced twice as many false alarms.

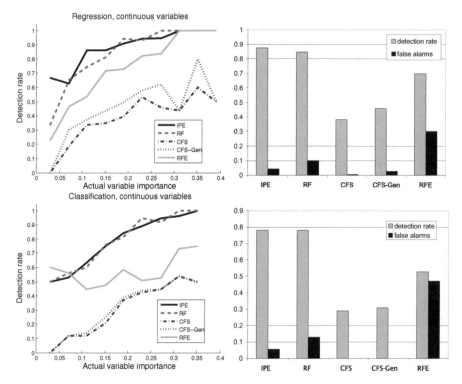

FIGURE 7.1: Data with nonlinear dependencies. Top: continuous variables, regression. Bottom: continuous variables, classification. Legend: IPE - Independent Probes with Ensembles, RF - Random Forest, CFS - Correlation-Based Feature Selection with forward search, CFS-Gen - same but with genetic search, RFE - Recursive Feature Elimination. Graph panels (left), Horizontal axis: True importance of an input variable. Vertical axis: The fraction of times such a variable was detected as an important variable. Bar graphs display the detection rates as well as the false alarm rates for each of the methods averaged over 50 datasets.

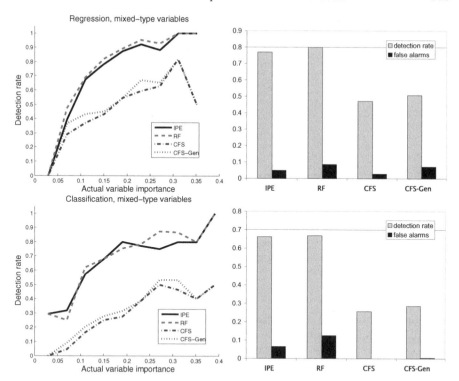

FIGURE 7.2: Data with nonlinear dependencies. Top: mixed-type variables, regression. Bottom: mixed-type variables, classification. Legend: IPE - Independent Probes with Ensembles, RF - Random Forest, CFS - Correlation-Based Feature Selection with forward search, CFS-Gen - same but with genetic search, RFE - Recursive Feature Elimination. Graph panels (left), Horizontal axis: True importance of an input variable. Vertical axis: The fraction of times such a variable was detected as an important variable. Bar graphs display the detection rates as well as the false alarm rates for each of the methods averaged over 50 datasets.

7.4.2.2   Linear Models Challenging for Trees

We also experimented using data with linear relationships, where the target is a simple linear combination of a number of input variables plus noise as follows:

$$Y = -0.25x(1) + 0.1x(2) + 0.05x(3) + 0.025x(4) +$$
$$0.015x(5) + 0.01N(0, 1) \tag{7.4}$$

where each $x(i)$ is drawn from $N(0, 1)$. Fifteen independent noise variables drawn from $N(0, 1)$ were joined to the data columns. This would be a simple problem for stage-wise linear regression, but typically linear problems are harder for trees. These results are illustrated in Figure 7.3.

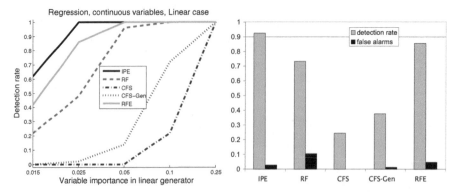

FIGURE 7.3: Data with linear dependencies. Continuous variables, regression. Left panel, Horizontal axis: the importance of the variable. Vertical axis: the fraction of times such a variable was detected as an important variable. Right panel displays the detection rates as well as the false alarm rates for each of the methods averaged over 50 datasets.

With these data sets of 200 samples, IPE detected 100% of the important variables, as long as their variance was larger than 2.5 times the variance of additive noise. The false acceptance rate remained at 1.5% with an overall detection rate of 93%. RFE was the second best, detecting 86% of relevant features with 4.8% false detects. RF recovered 72% of relevant features with 11% false detects. CFS and CFS-Gen performed poorly.

7.4.2.3 Real Data from Semiconductor Manufacturing with Large Number of Multilevel Categorical Predictors

Semiconductor fabrication is becoming increasingly complex, with routes stretching to several hundred process steps. Even with highly advanced process control systems in place, there is inevitable variation in yield and performance between and within manufacturing lots. A common practice is to associate this variation with equipment differences by performing analysis of variance at every process step where multiple pieces of equipment are used. Looking for operations where there are significant differences between process tools can reveal the sources of process variation. The one-at-a-time approach to equipment commonality studies has many shortcomings: Most target variables of interest are affected by multiple process steps. For example, yield can be reduced by high particle counts at nearly any manufacturing step. The maximum frequency (Fmax) at which a part can run can be affected by a variety of lithography, etch, implant, and diffusion operations.

We used one such dataset with known signals to test the feature selection methods discussed in this chapter to detect tools that had non-random effects on the maximum frequency. The data had 223 categorical predictors (manufacturing steps) with the number of levels (different pieces of equipment) ranging from 3 to 31, and a numeric response (Fmax). There were 6 out of 223 variables that had an actual effect. Operation 43 had the strongest effect, and had nearly complete confounding with operations 7, 95, and 106 (100% redundancy). Operations 46, 52, 53, and 54 had progressively weaker effects on a single tool only. Operation 223 had small differences between a large number of tools.

The proposed method (IPE) showed significantly superior performance on this task, identified all relevant variables, and 5 irrelevant steps. That translates to 100% detection and 2.2% false alarm rates. The RF sensitivity method identified only 4 relevant redundant variables with the strongest effect and 15 irrelevant steps. That translates to 44% detection and 6.7% false alarm rates. The CFS method identified all relevant variables except step 54, and 28 irrelevant steps. That translates to a decent detection rate of 89% and a relatively high percent of false alarms - 13%. The CFS-Gen method performed poorly, identified 107 variables with 45% false alarm rate, and missed the weakest contributor-step 223.

In all the experiments we had the "ground truth," the knowledge of which variables were truly the important ones. Thus we did not have to have another indirect layer in the evaluation process, but we could directly assess the variable selection performance of IPE and the benchmark methods.

7.5   Discussion

This chapter presented an efficient approach to feature selection using independent probe variables. The result is a truly autonomous variable selection method that considers all variable interactions and does not require a pre-set number of important variables. It showed excellent results on a variety of simulated and real-life complex datasets. It performed favorably when tested against several different and reportedly powerful feature selection methods.

In earlier work, the idea of adding random "probe variables" to the data for feature selection purposes has been used in [1]. Adding permuted original variables as random "probes" has been used in [12] in the context of comparing gene expression differences across two conditions. A univariate filter method based on the permutation test is considered in [10]. However, flexible tree ensembles with robust split estimation and variable scoring mechanisms in a combination with formal statistical tests have not been used to compare *ranks* of artificial probes to real variables in the context of variable selection.

The presented method retains all the good features of ensembles of trees: mixed-type data can be used, missing variables can be tolerated, and variables are not considered in isolation. The method does not require any preprocessing, and it is applicable to both classification and regression. It will report redundant features if at least one of them is relevant to a response. However, if the best-subset problem is of interest, a penalization strategy can be easily added to the ensemble construction. Redundant variables are prevented from entering the model by penalizing the algorithm for adding new variables. The computational complexity of IPE is the same as that of the Random Forest, $O(\sqrt{N}M \log M)$, where N is the number of variables, and M is the number of observations. This is in fact lighter than that of any of the benchmark methods.

References

[1] J. Bi, K. Bennett, M. Embrects, C. Breneman, and M. Song. Dimensionality reduction via sparse support vector machines. *Journal of Machine Learning Research*, 3:1229–1243, March 2003.

[2] A. Borisov, V. Eruhimov, and E. Tuv. Dynamic soft feature selection for tree-based ensembles. In I. Guyon, S. Gunn, M. Nikravesh, and L. Zadeh, editors, *Feature Extraction, Foundations and Applications*, volume 207 of *Studies in Fuzziness and Soft Computing*, pages 363–378. Springer, New York, 2006.

[3] L. Breiman. Random forests. *Machine Learning*, 45(1):5–32, 2001.

[4] L. Breiman, J. Friedman, R. Olshen, and C. Stone. *Classification and Regression Trees*. CRC Press, Boca Raton, FL, 1984.

[5] J. Friedman. Greedy function approximation: a gradient boosting machine. Technical report, Dept. of Statistics, Stanford University, 1999.

[6] I. Guyon, J. Weston, S. Barnhill, and V. Vapnik. Gene selection for cancer classification using support vector machines. *Machine Learning*, 46:389422, 2002.

[7] M. Hall. *Correlation-based Feature Selection for Machine Learning*. PhD thesis, Waikato University, Department of Computer Science, Hamilton, NZ, 1998.

[8] T. K. Ho. The random subspace method for constructing decision forests. *IEEE Transactions on Pattern Analysis and Machine Intelligence*, 20(8):832–844, 1998.

[9] P. Langley. Relevance and insight in experimental studies. *IEEE Expert*, 11:11–12, October 1996.

[10] P. Radivojac, Z. Obradovic, A. Dunker, and S. Vucetic. Feture selection filters based on the permutation test. In *Proc. European Conference on*

Machine Learning, pages 334–346, 2004.

[11] K. Torkkola and E. Tuv. Ensembles of regularized least squares classifiers for high-dimensional problems. In I. Guyon, S. Gunn, M. Nikravesh, and L. Zadeh, editors, *Feature Extraction, Foundations and Applications*, volume 207 of *Studies in Fuzziness and Soft Computing*, pages 301–317. Springer, New York, 2006.

[12] V. Tusher, R. Tibshirani, and G. Chu. Significance analysis of microarrays applied to the ionizing radiation response. *PNAS*, 98(9):5116–5121, April 24 2001.

[13] E. Tuv. Feature selection and ensemble learning. In I. Guyon, S. Gunn, M. Nikravesh, and L. Zadeh, editors, *Feature Extraction, Foundations and Applications*, volume 207 of *Studies in Fuzziness and Soft Computing*, pages 189–207. Springer, New York, 2006.

[14] J. Weston, A. Elisseeff, G. Bakir, and F. Sinz. The Spider. http://www.kyb.tuebingen.mpg.de/bs/people/spider/, 2004.

[15] I. H. Witten and E. Frank. *Data Mining: Practical machine learning tools and techniques*. Morgan Kaufmann, San Francisco, second edition, 2005.

Chapter 8

Efficient Incremental-Ranked Feature Selection in Massive Data

Roberto Ruiz

Pablo de Olavide University

Jesús S. Aguilar-Ruiz

Pablo de Olavide University

José C. Riquelme

University of Seville

8.1   Introduction

In recent years, there has been an explosion in the growth of databases in all areas of human endeavor. Progress in digital data acquisition and storage technology has resulted in the growth of huge databases. In this work, we address the feature selection issue under a classification framework. The aim is to build a classifier that accurately predicts the classes of new unlabeled instances. Theoretically, having more features and instances should give us more discriminating power. However, this can cause several problems: increased computational complexity and cost; too many redundant or irrelevant features; and estimation degradation in the classification error.

The problem of feature selection received a thorough treatment in pattern recognition and machine learning. Most of the feature selection algorithms approach the task as a search problem, where each state in the search specifies a distinct subset of the possible attributes [2]. The search procedure is combined with a criterion in order to evaluate the merit of each candidate subset of attributes. There are a lot of possible combinations between each procedure search and each attribute measure [17, 4, 16]. However, search methods can be prohibitively expensive in massive datasets, especially when a data mining algorithm is applied as an evaluation function.

There are various ways in which feature selection algorithms can be grouped according to the attribute evaluation measure, depending on the type (filter or wrapper technique) or on the way that features are evaluated (individual or subset evaluation). The filter model relies on general characteristics of the data to evaluate and select feature subsets without involving any mining algorithm. The wrapper model requires one predetermined mining algorithm and uses its performance as the evaluation criterion. It searches for features better suited to the mining algorithm, aiming to improve mining performance, but it also is more computationally expensive [15, 13] than filter models. Feature ranking (FR), also called feature weighting [2, 8], assesses individual features and assigns them weights according to their degrees of relevance, while the feature subset selection (FSS) evaluates the goodness of each found feature subset. (Unusually, some search strategies in combination with subset evaluation can provide a ranked list.)

In order to compare the effectiveness of feature selection, feature sets chosen by each technique are tested with three well-known learning algorithms: a probabilistic learner (naïve Bayes), an instance-based learner (IB1), and a decision tree learner (C4.5). These three algorithms have been chosen because they represent three quite different approaches to learning, and their long-standing tradition in classification studies.

The chapter is organized as follows. In the next two sections, we will review previous work, and notions of feature relevance and redundancy, respectively. In Section 8.4, we will present our proposed measures of feature relevance and redundancy using a wrapper or filter approach, and describe our algorithm. Experimental results are shown in Section 8.5, and the most interesting conclusions are summarized in Section 8.6.

8.2   Related Work

Traditional feature selection methods in some specific domain often select the top-ranked features according to their individual discriminative powers [7]. This approach is efficient for high-dimensional data due to its linear time complexity in terms of dimensionality. They can only capture the relevance of features to the target concept, but cannot discover redundancy and basic interactions among features. In the FSS algorithms category, candidate feature subsets are generated based on a certain search strategy. Different algorithms address these issues distinctively. In [17], a great number of selection methods are categorized. We found different search strategies, namely *exhaustive*, *heuristic*, and *random* searches, combined with several types of measures to form different algorithms. The time complexity is exponential in terms of data dimensionality for exhaustive searches and quadratic for heuristic searches. The complexity can be linear to the number of iterations in a random search, but experiments show that in order to find the best feature subset, the number of iterations required is usually at least quadratic to the number of features [5]. The most popular search methods in pattern recognition and machine learning cannot be applied to massive datasets due to the large number of features and instances (sometimes tens of thousands). One of the few used search techniques in these domains is sequential forward (SF, also called hill-climbing or greedy search). Different subset evaluation measures in combination with an SF search engine can be found. We are specially interested in the wrapper approach.

A key issue of wrapper methods is how to search into the space of subsets of features. Although several heuristic search strategies exist such as greedy sequential search, best-first search, and genetic algorithm, most of them are still computationally expensive $O(N^2)$ (with N the number of features of the original dataset), which prevents them from scaling well to datasets containing thousands of features. A rough estimate of the time required by most of these techniques is in the order of thousands of hours, assuming that the method does not get caught in a local minima first and stops prematurely. For example, if we have chosen 50 features from 20,000 (0.0025% of the whole set) through a greedy search, the subset evaluator would be run approximately one million times (N times to find the best single feature, then it tries each of the remaining features in conjunction with the best to find the most suited pair of features $N - 1$ times, and so on, more or less $20,000 \times 50$ times). Assuming 4 seconds on average by each evaluation, the results would take more than 1,000 hours.

The limitations of both approaches, FR and FSS, clearly suggest that we should pursue a hybrid model. Recently, a new framework of feature selection has been used, where several of the above-mentioned approaches are combined. [21] proposed a fast correlation-based filter algorithm (FCBF) that uses correlation measure to obtain relevant features and to remove redundancy. There are other methods based on relevance and redundancy concepts. Recursive feature elimination (RFE) is a proposed feature selection algorithm described in [10]. The method, given that one wishes to find only r dimensions in the final subset, works by trying to choose the r features that lead to the largest margin of class separation, using an SVM classifier. This combinatorial problem is solved in a greedy fashion at each iteration of training by removing the input dimension that decreases the margin the least until only r input dimensions remain (this is known as backward selection). The authors in [6] have used mutual information for gene selection that has maximum relevance with minimal redundancy by solving a simple two-objective optimization, and [20] proposes a hybrid of filter and wrapper approaches to feature selection.

In [12], the authors propose a rank search method to compare feature selection algorithms. Rank search techniques rank all features, and subsets of increasing size are evaluated from the ranked list (i.e., the first attribute, the two first ones, etc.). The best attribute set is reported. The authors apply the wrapper approach to datasets up to 300 attributes and state that for the ADS dataset (1,500 attributes) the estimated time to only generate the ranking in a machine with a 1.4GHz processor would be about 140 days and to evaluate the ranked list of attributes would take about 40 days. In contrast, our method can be tested on datasets with 20,000 features on a similar machine in a few hours.

This chapter presents a feature selection method, named *BIRS* (Best Incremental Ranked Subset), based on the hybrid model, and attempts to take advantage of all of the different approaches by exploiting their best performances in two steps: First, a filter or wrapper approach provides a ranked list of features, and, second, ordered features are added using a wrapper or filter subset evaluation ensuring good performance (the search algorithm is valid for any feature ranked list). This approach provides the possibility of efficiently applying any subset evaluator, wrapper model included, in large and high-dimensional domains, obtaining good results. The final subset is obviously not the optimum, but it is unfeasible to search for every possible subset of features through the search space. The main goal of our research is to obtain a few features with high predictive power. The wrapper version of this algorithm has been proved to be efficient and effective in microarray domains [18].

8.3   Preliminary Concepts

8.3.1   Relevance

The purpose of a feature subset algorithm is to identify relevant features according to a definition of relevance. However, the notion of relevance in machine learning has not yet been rigorously defined in common agreement [1]. Reference [13] includes three disjointed categories of feature relevance: strong relevance, weak relevance, and irrelevance. These groups are important to decide what features should be conserved and which ones can be eliminated. The strongly relevant features are, in theory, important to maintain a structure in the domain, and they should be conserved by any feature selection algorithm in order to avoid the addition of ambiguity to the sample. Weakly relevant features could be important or not, depending on the other features already selected and on the evaluation measure that has been chosen (accuracy, simplicity, consistency, etc.). Irrelevant attributes are not necessary at all. Reference [1] makes use of information theory concepts to define the entropic or variable relevance of a feature with respect to the class. Reference [2] collects several relevance definitions. The above notions of relevance are independent of the specific learning algorithm being used. There is no guarantee that just because a feature is relevant, it will necessarily be useful to an algorithm (or vice versa). The definition of incremental relevance in [3] makes it explicit, since it is considered especially suited to obtain a predictive feature subset.

DEFINITION 8.1 Incremental usefulness *Given a sample of data* $\mathbf{X_L}$, *a learning algorithm* L, *a feature space* $\mathbf{F}$, *and a feature subset* $\mathbf{S}$ *(*$\mathbf{S} \subseteq \mathbf{F}$*)*, *the feature* F_i *is incrementally useful to* L *with respect to* $\mathbf{S}$ *if the accuracy of the hypothesis that* L *produces using the group of features* $\{F_i\} \cup \mathbf{S}$ *is better than the accuracy achieved using just the subset of features* $\mathbf{S}$.

We consider this definition to be especially suited to obtain a predictive feature subset. In the next section, concepts can be applied to avoid a subset that contains attributes with the same information.

8.3.2 Redundancy

Notions of feature redundancy are normally in terms of feature correlation. It is widely accepted that two features are redundant to each other if their values are completely correlated. There are two widely used types of measures for the correlation between two variables: linear and non-linear. In the first case, the Pearson correlation coefficient is used, and in the second one, many measures are based on the concept of entropy, or the measure of the uncertainty of a random variable. Symmetrical uncertainty is frequently used, defined as

$$SU(X,Y) = 2 \left[\frac{IG(X|Y)}{H(X) + H(Y)} \right]$$

where $H(X) = -\sum_i P(x_i) log_2(P(x_i))$ is the entropy of a variable X and $IG(X|Y) = H(X) - H(X|Y)$ is the information gain from X provided by Y.

The above-mentioned definitions are between pairs of variables. However, it may not be as straightforward in determining feature redundancy when one is correlated with a set of features. Reference [14] applies a technique based on cross-entropy, named Markov blanket filtering, to eliminate redundant features. This idea is formalized in the following definition.

DEFINITION 8.2 Markov blanket *Given a feature* $F_i \in \mathbf{S}$ *(a set of attributes) and the class* $\mathbf{Y}$, *the subset* $\mathbf{M} \subseteq \mathbf{S}$ *(*$F_i \notin \mathbf{M}$*) is a Markov blanket of* F_i *if, given* $\mathbf{M}$, F_i *is conditionally independent of* $\mathbf{S} - \mathbf{M} - \{F_i\}$ *and* $\mathbf{Y}$.

Two attributes (or sets of attributes) X, Y are said to be conditionally independent given a third attribute Z (or set) if, the given Z makes X and Y independent, i.e., the distribution of X, knowing Y and Z, is equal to the distribution X knowing Z; therefore, Y does not have influence on X ($P(X|Y, Z) = P(X|Z)$).

Theoretically, it can be shown that once we find a Markov blanket $\mathbf{M}$ of feature F_i in a feature set $\mathbf{S}$, we can safely remove F_i from $\mathbf{S}$ without increasing the divergence from the original distribution. Furthermore, in a sequential filtering process, in which unnecessary features are removed one by one, a feature tagged as unnecessary based on the existence of a Markov blanket $\mathbf{M}$ remains unnecessary in later stages when more features have been removed. The Markov blanket condition requires that $\mathbf{M}$ assumes not only the information that F_i has about $\mathbf{Y}$, but also about all the other features. In [14] it is stated that the cardinality of set $\mathbf{M}$ must be small and fixed.

References [20] and [21] are among the most cited works at present following the above-mentioned framework (FR+FSS). Both are based on this concept of Markov blanket. In the first one, the number of attributes of $\mathbf{M}$ is not provided, but it is a fixed number among the highly correlated features. In the second one, a fast correlation-based filter is implemented (FCBF), where $\mathbf{M}$ is formed by only one attribute, and gradually eliminates redundant attributes with respect to $\mathbf{M}$ from the first to the final attributes of an ordered list. Other methods based on relevance and redundancy concepts can be found in [10, 6].

8.4   Incremental Performance over Ranking

In this section, we introduce first our ideas of relevance and redundancy taking into account the aim of applying a wrapper model to massive datasets; second, changes introduced by the filter model; and then our approach is described.

As previously indicated, the wrapper model makes use of the algorithm that will build the final classifier to select a feature subset. Thus, given a classifier L, and given a set of features $\mathbf{S}$, a wrapper method searches in the space of $\mathbf{S}$, using cross-validation to compare the performance of the trained classifier L on each tested subset. While the wrapper model is more computationally expensive than the filter model, it also tends to find feature sets better suited to the inductive biases of the learning algorithm and therefore provides superior performance.

In this work, we propose a fast search over a minimal part of the feature space. Beginning with the first feature from the list ordered by some evaluation criterion, features are added one by one to the subset of selected features only if such inclusion improves the classifier accuracy. Then, the learning algorithm of the wrapper approach is always run N (number of features) times, usually with a few features. A feature ranking algorithm makes use of a scoring function computed from the values of each feature and the class label. By convention, we assume that a high score is indicative of a valuable feature and that we sort features in decreasing order of this score. We consider ranking criteria defined for individual features, independently of the context of others.

When a ranking of features is provided from a high dimensional data set, a large number of features with similar scores is generated, and a common criticism is that it leads to the selection of redundant subsets. However, according to [8], noise reduction and consequently better class separation may be obtained by adding variables that are presumably redundant. Moreover, a very high attribute correlation (in absolute value) does not mean the absence of attribute complementarity. Therefore, our idea of redundancy is not based only on correlation measures, but also on the learning algorithm target (wrapper or filter approach), in the sense that a feature is chosen if additional information is gained by adding it to the selected subset of features.

8.4.1 Incremental Ranked Usefulness

In feature subset selection, it is a fact that two types of features are generally perceived as being unnecessary: features that are irrelevant to the target concept, and features that are redundant given other features. Our approach is based on the concept of a Markov blanket, which is described in [14]. This idea was formalized using the notion of conditionally independent attributes, which can be defined by several approaches [20, 21]. We set this concept by a wrapper model, defining incremental ranked usefulness in order to devise an approach to explicitly identify relevant features and do not take into account redundant features.

Let $\mathbf{X_L}$ be a sample of labeled data, $\mathbf{S}$ be a subset of features of $\mathbf{X_L}$, and L be a learning algorithm; the *correct rate* (or accuracy) $\Gamma(\mathbf{X_L}/\mathbf{S}, L)$ is named to the ratio between the number of instances correctly classified by L and the total number of evaluated instances considering only the subset $\mathbf{S}$. In the training process, this accuracy will be an estimate of error by cross-validation.

Let $R = \{F_i\}$, $i = 1 \dots N$ be a ranking of all the features in $\mathbf{X_L}$ sorted in descending order, and $\mathbf{S}$ be named the subset of the i first features of R.

DEFINITION 8.3 Incremental ranked usefulness *The feature F_{i+1} in R is incrementally useful to L if it is not conditionally independent of the class $\mathbf{Y}$ given $\mathbf{S}$; therefore, the correct rate of the hypothesis that L produces using the group of features $\{F_{i+1}\} \cup \mathbf{S}$ is significantly better (denoted by $>$) than the correct rate achieved using just the subset of features $\mathbf{S}$.*

Therefore, if $\Gamma(\mathbf{X_L}/\mathbf{S} \cup \{F_{i+1}\}, L) \not> \Gamma(\mathbf{X_L}/\mathbf{S}, L)$, then F_{i+1} is conditionally independent of class $\mathbf{Y}$ given the subset $\mathbf{S}$, and then we should be able to omit

Input: $\mathbf{X_L}$ training U-measure, L-subset evaluator
Output: BestSubset

1	list R = {}
2	for each feature $F_i \in \mathbf{X_L}$
3	$Score$ = compute(F_i, U, $\mathbf{X_L}$)
4	append F_i to R according to $Score$
5	BestEvaluation = 0
6	BestSubset = $\emptyset$
7	for $i = 1$ to N
8	TempSubset = BestSubset $\cup$ $\{F_i\}$ ($F_i \in$ R)
9	TempEvaluation = WrapperOrFilter(TempSubset, L)
10	if (TempEvaluation > BestEvaluation)
11	BestSubset = TempSubset
12	BestEvaluation = TempEvaluation

FIGURE 8.1: *BIRS* algorithm.

F_{i+1} without compromising the accuracy of class prediction.

A fundamental question in the previous definition is how the *significant* improvement is analyzed in this wrapper model. A five-fold cross-validation is used to estimate if the accuracy of the learning scheme for a set of features is significantly better ($>$) than the accuracy obtained for another set. We conducted a Student's paired two-tailed t-test in order to evaluate the statistical significance (at 0.1 level) of the difference between the previous best subset and the candidate subset. This last definition allows us to select features from the ranking, but only those that increase the classification rate significantly. Although the size of the sample is small (five folds), our search method uses a t-test. We want to obtain a heuristic, not to do an accurate population study. However, on the one hand, it must be noticed that it is a heuristic based on an objective criterion, to determine the statistical significance degree of difference between the accuracies of each subset. On the other hand, the confidence level has been relaxed from 0.05 to 0.1 due to the small size of the sample. Statistically significant differences at the $p < 0.05$ significance level would not allow us to add more features, because it would be difficult for the test to obtain significant differences between the accuracy of each subset. Obviously, if the confidence level is increased, more features can be selected, and vice versa.

Following a filter model in the subset evaluation, we need a different way to find out if the value of measurement of a set is significantly better ($>$) than another one when adding an attribute. Simply, it is verified if the improvement surpasses a threshold (for example, 0.005), one resulted from the best previous subset and the other resulted from the joint candidate.

TABLE 8.1:   Example of feature selection
process by *BIRS*.

Rank	F_5	F_7	F_4	F_3	F_1	F_8	F_6	F_2	F_9

Subset Eval.		Acc	P-Val	Acc	Best Sub
1	F_5	80		80	$\mathbf{F_5}$
2	$\mathbf{F_5}, F_7$	82			
3	$\mathbf{F_5}, F_4$	81			
4	$\mathbf{F_5}, F_3$	83			
5	$\mathbf{F_5}, F_1$	84	< **0.1**	84	$\mathbf{F_5}, \mathbf{F_1}$
6	$\mathbf{F_5}, \mathbf{F_1}, F_8$	84			
7	$\mathbf{F_5}, \mathbf{F_1}, F_6$	86			
8	$\mathbf{F_5}, \mathbf{F_1}, F_2$	89	< **0.1**	89	$\mathbf{F_5}, \mathbf{F_1}, \mathbf{F_2}$
9	$\mathbf{F_5}, \mathbf{F_1}, \mathbf{F_2}, F_9$	87			

8.4.2   Algorithm

There are two phases in the algorithm, named *BIRS* (Best Incremental Ranked Subset), shown in Figure 8.1: Firstly, the features are ranked according to some evaluation measure (lines 1–4). In the second phase, we deal with the list of features once, crossing the ranking from the beginning to the last ranked feature (lines 5-12). We obtain the classification accuracy with the first feature in the list (line 9) and it is marked as selected (lines 10-12). We obtain the classification rate again with the first and second features. The second will be marked as selected depending on whether the accuracy obtained is significantly better (line 10). We repeat the process until the last feature on the ranked list is reached. Finally, the algorithm returns the best subset found, and we can state that it will not contain irrelevant or redundant features.

The first part of the above algorithm is efficient since it requires only the computation of N scores and to sort them, while in the second part, time complexity depends on the learning algorithm chosen. It is worth noting that the learning algorithm is run N (number of features) times with a small number of features, only the selected ones. Therefore, the running time of the ranking procedure can be considered to be negligible regarding the global process of selection. In fact, the results obtained from a random order of features (without previous ranking) showed the following drawbacks: 1) The solution was not deterministic; 2) a greater number of features were selected; 3) the computational cost was higher because the classifier used in the evaluation contained more features since the first iterations.

Consider the situation depicted in Table 8.1: an example of the feature selection process done by *BIRS*. The first line shows the features ranked according to some evaluation measure. We obtain the classification accuracy with the first feature in the list (F_5:80%). In the second step, we run the classifier with the first two features of the ranking (F_5,F_7:82%), and a paired t-test is performed to determine the statistical significance degree of the differences. Since it is greater than 0.1, F_7 is not selected. The same happens with the next two subsets (F_5,F_4:81%, F_5,F_3:83%). Later, the feature F_1 is added, because the accuracy obtained is significantly better than that with only F_5 (F_5,F_1:84%), and so on. In short, the classifier is run nine times to select, or not, the ranked features (F_5,F_1,F_2:89%): once with only one feature, four times with two features, three with three features, once with four, and once with four, features. Most of the time, the learning algorithm is run with few features. In short, this wrapper-based approach needs much less time than others with a broad search engine.

As we can see in the algorithm, the first feature is always selected. This does not mean a great shortcoming in high-dimensional databases, because usually several different sets of features share similar information. The main disadvantage of *sequential forward generation* is that it is not possible to consider certain basic interactions among features, i.e., features that are useless by themselves can be useful together. *Backward generation* remedies some problems, although there still will be many hidden interactions (in the sense of being unobtainable), but it demands more computational resources than the forward approach. The computer-load necessities of the forward search might become very inefficient in high-dimensional domains, as it starts with the original set of attributes and removes features increasingly.

8.5 Experimental Results

The aim of this section is to evaluate our approach in terms of classification accuracy, degree of dimensionality, and speed in selecting features, in order to see how good *BIRS* is in situations where there is a large number of features and instances.

The comparison was performed with two representative groups of datasets: Twelve datasets were selected from the UCI Repository (Table 8.2) and five from the NIPS 2003 feature selection benchmark [9]. In this group (Table 8.3), the datasets were chosen to span a variety of domains (cancer prediction from mass-spectrometry data, handwritten digit recognition, text classification, and prediction of molecular activity). One dataset is artificial. The input variables are continuous or binary, sparse or dense. All problems are two-class classification problems. The full characteristics of all the datasets are summarized in Tables 8.2 and 8.3. We chose three different learning algorithms: C4.5, IB1, and Naïve Bayes, to evaluate the accuracy on selected features for each feature selection algorithm.

Figure 8.2 can be considered to illustrate both blocks that always com-

TABLE 8.2: UCI Repository of Machine Learning
Databases. For each dataset we show the acronym
used in this text, the number of features, the number
of examples, and the number of possible classes.

Data	Acron.	#Feat.	#Inst.	#Classes
ads	ADS	1558	3279	2
arrhythmia	ARR	279	452	16
hypothyroid	HYP	29	3772	4
isolet	ISO	617	1559	26
kr vs kp	KRV	36	3196	2
letter	LET	16	20000	26
multi feat.	MUL	649	2000	10
mushroom	MUS	22	8124	2
musk	MUK	166	6598	2
sick	SIC	29	3772	2
splice	SPL	60	3190	3
waveform	WAV	40	5000	3

TABLE 8.3: NIPS 2003 challenge data sets. For each dataset we
show the acronym used in this text, the domain it was taken from, its type
(dense, sparse, or sparse binary), the number of features, the number of
examples, and the percentage of random features. All problems are
two-class classification problems.

Data	Acron.	Domain	Type	#Feat.	#Inst.	%Ran.
Arcene	ARC	Mass Spectro.	Dense	10000	100	30
Dexter	DEX	Text classif.	Sparse	20000	300	50
Dorothea	DOR	Drug discove.	S. bin	100000	800	50
Gisette	GIS	Digit recogn.	Dense	5000	6000	30
Madelon	MAD	Artificial	Dense	500	2000	96

pose algorithm *BIRS* (originally introduced in [21]). Therefore, this feature
selection algorithm needs measures to evaluate individual and subsets of attributes. Numerous versions of selection algorithms *BIRS* could be formed
combining the criteria of each group of measures (individual and subset). In
order to simplify, we will use the same evaluation measure in the two phases
(individual and subset). In the experiments, we used two criteria: one belongs
to the wrapper model, and one to the filter model. 1) In the wrapper approach
(denoted by BI_{NB}, BI_{C4}, or BI_{IB}) we order features according to their individual predictive power, using as criterion the performance of the target
classifier built with a single feature. The same classifier is used in the second
phase to evaluate subsets. 2) In the filter approach, a ranking is provided
using a non-linear correlation measure. We chose symmetrical uncertainty
(denoted by BI_{CF}), based on entropy and information gain concepts [11] in

FIGURE 8.2: Type of feature evaluation in *BIRS*.

both phases. Note the similarity among the results obtained in previous works with several ranking measure approaches [18]. Accuracy differences are not statistically significant, although wrapper ranking is a little bit better.

Also in these experiments, to find out if the value of measurement of a set is significantly better ($\gg$) than another one when adding an attribute, it is distinguished between filter and wrapper models in the subset evaluation. In the first case, it is simply verified if the improvement surpasses a threshold established in 0.005; nevertheless, in the second case, we conduct Student's paired two-tailed t-test in order to evaluate the statistical significance (at level 0,1) of the difference between two averaged accuracy values: one resulted from the joint candidate and the other resulted from the best previous subset.

Due to the high dimensionality of data, we limited our comparison to sequential forward (SF) techniques and a fast correlation-based filter (FCBF) algorithm [21] applied to the first group of datasets, and only FCBF with the NIPS datasets. We chose two representative subset evaluation measures in combination with the SF search engine. One, denoted by SF_{WR}, uses a target learning algorithm to estimate the worth of feature subsets; the other, denoted by SF_{CF}, is a subset search algorithm that exploits sequential forward search and uses the correlation measures (variation of the CFS correlation-based feature selection algorithm [11]) to guide the search.

The experiments were conducted using the WEKA's implementation of all these existing algorithms, and our algorithm is also implemented in the WEKA environment [19]. We must take into account that the proper way to conduct a cross-validation for feature selection is to avoid using a fixed set of features selected with the whole training dataset, because this induces a bias in the results. Instead, one should withhold a pattern, select features, and assess the performance of the classifier with the selected features using the leftout examples. The results reported in this section were obtained with a 5×2-fold cross-validation over each dataset, i.e., a feature subset was selected using the 50% of the instances; then, the accuracy of this subset was estimated over the unseen 50% of the data. In this way, estimated accuracies, selected attribute numbers, and time needed were the result of a mean over five executions of two cross-validation samples. We use two instead of ten cross-validations because of the time cost consuming with massive datasets. Standard methods have been used for the experimental section (sequential forward; Naïve Bayes, IB1, and C4.5 classifiers; and the t-Student statistical test). There exist other methods following the wrapper approach to extract

TABLE 8.4: Accuracy of NB on selected features for UCI data. The symbols $^+$ and $^-$ respectively identify statistically significant (at 0.05 level) wins or losses over BI_{NB}.

Data	Wrapper		Filter			Original
	BI_{NB}	SF_{NB}	BI_{CF}	SF_{CF}	FCBF	
ADS	95.42	95.83	95.38	95.81	95.64	96.38
ARR	66.99	67.70	66.50	68.05	63.98	60.13
HYP	95.10	95.32	94.15^-	94.15^-	94.90	95.32
ISO	83.30	82.28	77.61	80.79	74.62^-	80.42
KRV	94.27	94.32	90.43^-	90.43^-	92.50	87.50
LET	65.67	65.67	64.28^-	64.28^-	65.06	63.97
MUL	97.21	96.87	97.04	96.72	96.19	94.37
MUS	98.78	99.01	98.52	98.52	98.52	95.10
MUK	84.59	84.59	79.94	69.78^-	72.29	83.56
SIC	94.55	93.88	93.89	93.89	96.25	92.41
SPL	94.85	94.91	93.63^-	93.60^-	95.49	95.26
WAV	81.01	81.55	81.01	80.12	78.42^-	80.02
time(s)	6111	49620	49	133	68	

relevant features, which involve the selection process into the learning process (neural networks, Bayesian networks, support vector machines), although the source code of these methods is not freely available and therefore the experiments cannot be reproduced. In fact, some of them are designed for specific tasks, so the parameter settings are quite different for the learning algorithm.

Tables 8.4, 8.5, and 8.6 report accuracy by Naïve Bayes, IB1, and C4.5, respectively, by each feature selection algorithm and the original set. From the last row of each table, we can observe for each algorithm the running time. We conducted a Student's paired two-tailed t-test in order to evaluate the statistical significance of the difference between two averaged accuracy values: one resulted from the wrapper approach of $BIRS$ (BI_{NB}, BI_{C4} or BI_{IB}) and the other resulted from one of the wrapper version of SF (SF_{NB}, SF_{C4} or SF_{IB}), BI_{CF}, SF_{CF}, $FCBF$, and the original set. The symbols $+$ and $-$ respectively identify statistic significance, at 0.05 level, wins or losses over BI_{WR}.

We studied the behavior of BI_{WR} in three ways in Tables 8.4, 8.5, and 8.6: with respect to a whole set of features (last row, original); with respect to another wrapper approach (SF_{WR}); and with respect to three filter approaches (BI_{CF}, SF_{CF}, and $FCBF$).

As it is possible to be observed in the last column of Tables 8.4, 8.5, and 8.6, classification accuracies obtained with the wrapper approach of $BIRS$ (BI_{WR}) with respect to results obtained with the total set of attributes are statistically better in 4 and 3 occasions for classifiers NB and IB, respectively, and worse in 2 applying C4. Note that the number of selected attributes is drastically less than the original set, retaining on average 15% (NB, Ta-

TABLE 8.5: Accuracy of C4 on selected features for UCI data. The symbols $^+$ and $^-$ respectively identify statistically significant (at 0.05 level) wins or losses over BI_{C4}.

	Wrapper		Filter			Original
Data	BI_{C4}	SF_{C4}	BI_{CF}	SF_{CF}	FCBF	
ADS	96.55	96.85	96.43	96.39	95.85	96.46
ARR	68.01	67.39	66.42	67.04	64.87	64.29
HYP	99.07	99.30	96.56^-	96.56^-	98.03	99.36
ISO	69.43	N/D	72.68	71.94	66.63	73.38
KRV	95.11	94.26	90.43^-	90.43^-	94.07	99.07^+
LET	84.99	85.17	84.21^-	84.21^-	84.84	84.45
MUL	92.42	93.11	93.17	93.12	92.29	92.74
MUS	99.91	100.00^+	98.52^-	98.52^-	98.84^-	100.00^+
MUK	95.43	N/D	94.06	94.60	91.19^-	95.12
SIC	98.28	98.19	96.33^-	96.33^-	97.50	98.42
SPL	93.05	93.04	92.54	92.61	93.17	92.92
WAV	76.20	75.44	76.46	76.56	74.52	74.75
time(s)	17914	40098	49	133	68	

ble 8.4), 16.3% (C4, Table 8.5), and 13.1% (IB, Table 8.6) of the attributes. As we can see, BI_{WR} chooses less than 10% of the attributes in more than half of all the cases studied in these tables.

BI_{WR} versus SF_{WR}: No significant statistical differences are shown between the accuracy of our wrapper approach and the accuracy of the sequential forward wrapper procedure (SF_{WR}), except for the MUS dataset and C4 classifier (Table 8.5).

Notice that in two cases with C4 classifiers (ISO and MUK) and two with IBs (ADS and MUL), SF_{WR} did not report any results after three weeks running; therefore, there are no selected attributes or success rates. Without considering this lack of results with SF_{WR}, the chosen subset by $BIRS$ is considerably smaller with the IB classifiers, 13.1% versus 20%, and less difference with NB and C4, although it is supposed that the lack of results would favor $BIRS$, since SF has not finished because of the inclusion of many attributes.

On the other hand, the advantage of $BIRS$ with respect to the SF for NB, IB1, and C4.5 is clear having to take into account the running time needed. $BIRS$ takes 6,112, 5,384, and 21,863 seconds applying NB, C4, and IB, respectively, whereas SF takes 49,620, 40,098, and 210,642 seconds. We can observe that $BIRS$ is consistently faster than SF_W, because the wrapper subset evaluation is run less times. For example, for the ADS dataset and C4.5 classifier, $BIRS$ and SF retain 8.5 and 12.4 features, respectively, on average. To obtain these subsets, the first one evaluated 1,558 features individually (to generate the ranking) and 1,558 subsets, while the second one evaluated 18,630 subsets (1,558 features + 1557 pairs of features + ... + 1,547 sets of

TABLE 8.6: Accuracy of IB on selected features for UCI data. The symbols $^+$ and $^-$ respectively identify statistically significant (at 0.05 level) wins or losses over BI_{IB}.

Data	Wrapper		Filter			Original
	BI_{IB}	SF_{IB}	BI_{CF}	SF_{CF}	FCBF	
ADS	95.28	N/D	95.93	96.07	95.75	95.95
ARR	62.74	57.12	61.37	61.06	58.67	54.12
HYP	83.66	83.57	85.75	85.75	94.88	90.85
ISO	80.64	78.61	79.37	80.28	72.57^-	77.58
KRV	92.27	94.24	90.43	90.43	93.85	89.21
LET	95.52	95.58	93.62^-	93.62^-	94.81	94.23^-
MUL	96.72	N/D	97.54	97.70	97.53	97.52
MUS	98.36	99.99	98.52	98.52	98.88	100.00
MUK	93.34	94.72	92.59	93.17	89.04^-	95.14
SIC	96.55	97.05	94.73	94.73	95.82	95.58
SPL	86.35	85.62	86.40	86.34	79.21^-	73.74^-
WAV	76.39	77.18	78.89^+	78.72	71.76^-	73.42^-
time(s)	40253	210642	49	133	68	

TABLE 8.7: Number of features selected by each feature selection algorithm on UCI data. Last row shows number of features retained on average. N - number of features of the original set, N' - number of features selected.

Data	Wrapper						Filter		
	BI_{NB}	SF_{NB}	BI_{C4}	SF_{C4}	BI_{IB}	SF_{IB}	BI_{CF}	SF_{CF}	FCBF
ADS	10.5	16.4	8.5	12.4	5.2	N/A	6.7	9.2	83.1
ARR	5.8	8.4	6.7	8.6	14.1	12.7	11.4	17.2	8.0
HYP	4.6	8.5	4.2	5.9	1.0	1.0	1.0	1.0	5.3
ISO	68.5	29.0	22.5	N/A	35.5	29.4	68.8	95.2	22.9
KRV	5.0	5.2	6.2	4.9	6.5	10.0	3.0	3.0	6.5
LET	11.0	11.6	11.0	10.1	10.9	11.0	9.0	9.0	10.3
MUL	22.2	15.3	20.6	13.6	11.3	N/A	28.0	90.3	121.3
MUS	2.1	3.0	4.1	4.9	1.6	4.7	1.0	1.0	3.6
MUK	1.0	1.0	9.7	N/A	4.7	12.0	6.5	16.3	2.9
SIC	2.4	1.0	5.9	5.5	2.8	6.7	1.0	1.0	4.8
SPL	13.1	14.8	9.8	11.0	5.9	6.6	6.0	6.1	21.8
WAV	9.4	12.9	9.6	7.9	10.0	12.4	12.4	14.8	6.1
$\frac{N'}{N} * 100$	15.0	16.8	16.3	18.2	13.1	20.3	11.7	14.1	18.1

twelve features). The time savings of $BIRS$ became more obvious when the computer-load necessities of the mining algorithm increased. In many cases, the time savings were 10 times less, and we must take into account that SF did not report any results on several datasets.

These results verify the computational efficiency of incremental searches applied by $BIRS$ over greedy sequential searches used by SF, with a lower number of features selected and without significant statistical differences on accuracy.

$BIRS$ wrappers versus filters: We noticed that the computer-load necessities of filter procedures can be considered as negligible regarding wrapper models. Nevertheless, wrapper approaches of $BIRS$ (BI_{WR}) obtained better accuracies: They showed significant gains to the filter version of $BIRS$, $_{CF}BI_{CF}$, in 4, 5, and 1 cases for NB, C4, and IB respectively, and they only lost in one with IB; with respect to the sequential version SF_{CF}, $BIRS$ won in 5, 5, and 1 occasions for NB, C4, and IB, respectively; and with respect to $FCBF$, BI_{WR} was better in 2, 2, and 4 cases with each respective classifier.

Table 8.7 reports the number of features selected by each feature selection algorithm on UCI data, showing three different results for each wrapper approach, depending on the learning algorithm chosen. Obviously, there is one value for filter approaches because filters do not depend on the classifier used. From the last row, we can observe for each algorithm the number of features retained on average. The filter approach of $BIRS$ retains less attributes than the rest of the algorithms. BI_{CF} retains 11.7% of the attributes on average for the 12 databases, SF_{CF} retains 14.1% of the attributes on average for all datasets, whereas $FCBF$ retains 18.1%.

We used the WEKA implementation of the $FCBF$ algorithm with default values. However, if the threshold by which features can be discarded is modified, the results obtained might vary. Note that if this threshold is set to the upper value, the number of selected features diminishes considerably, together with a notable reduction of prediction.

Another comparison can be between the versions filters, that is to say, as the approach behaves filter of $BIRS$ (BI_{CF}) with respect to the sequential search SF_{CF} and to the $FCBF$ algorithm. About accuracies, results obtained with both $(BIRS$ and $SF)$ first are similar and a little less than those obtained with $FCBF$. Nevertheless, the most reduced datasets are obtained with the filter model of $BIRS$. In addition, the time needed to reduce each dataset with BI_{CF} was faster than the others.

NIPS datasets: Table 8.8 shows the results obtained by the three classifiers, Naïve Bayes (NB), C4.5 (C4), and IB1 (IB), from the NIPS 2003-*Neural Information Processing Systems* (Table 8.3) feature selection benchmark data. The table gives the accuracy and number of features selected by each feature selection algorithm and the original set. We conducted a Student's paired

TABLE 8.8: *BIRS* accuracy of Naïve Bayes (NB), C4.5 (C4), and IB1 (IB) on selected features for NIPS data: Acc records 5×2CV classification rate (%) and #Att records the number of features selected by each algorithm. The symbols $^+$ and $^-$ respectively identify statistically significant (at 0.05 level) wins or losses over BI_{WR}.

Data		BI_{WR}		BI_{CF}		FCBF		Original
		Acc	#Att	Acc	#Att	Acc	#Att	
NB	ARC	64.60	15.3	63.20	39.2	61.20	35.2	65.40
	DEX	81.33	30.2	82.47	11.3	85.07	25.1	86.47
	DOR	93.23	10.5	93.80	11.9	92.38	75.3	90.68^-
	GIS	92.66	35.3	90.83	11.6	87.58^-	31.2	91.88
	MAD	59.00	11.8	60.56	5.8	58.20	4.7	58.24
C4	ARC	65.80	7.9	59.00	39.2	58.80	35.2	57.00
	DEX	80.27	18.9	81.47	11.3	79.00	25.1	73.80
	DOR	92.13	7.2	91.63	11.9	90.33	75.3	88.73
	GIS	93.29	26.9	90.92	11.6	90.99^-	31.2	92.68
	MAD	73.02	17.0	69.77	5.8	61.11^-	4.7	57.73^-
IB	ARC	69.00	15.1	68.60	39.2	62.00	35.2	78.00
	DEX	81.00	34.1	81.73	11.3	79.20	25.1	56.67^-
	DOR	92.18	3.5	90.98	11.9	90.35	75.3	90.25
	GIS	82.25	2.3	90.07	11.6	90.06	31.2	95.21
	MAD	74.92	14.4	71.59	5.8	56.90	4.7	54.39

two-tailed t-test in order to evaluate the statistical significance of the difference between two averaged accuracy values: one resulted from BI_{WR} (BI_{NB}, BI_{C4}, or BI_{IB}) and the other resulted from one of BI_{CF}, $FCBF$, and the original set. The symbols $+$ and $-$ respectively identify statistic significance, at 0.05 level, wins or losses over BI_{WR}. Results obtained with *SF* algorithms are not shown. The wrapper approach is too expensive in time, and its filter approach selects so many attributes that the program ran out of memory after a long period of time due to its quadratic space complexity. On the other hand, the *CFS* algorithm has been modified to be able to obtain results with *BIRS* for the DEX and DOR databases. From Table 8.8 we can conclude the following:

- *BIRS* is a good method to select attributes, because with a very reduced set of attributes one can obtain similar results, even better, than with the whole set of features in a massive database. About accuracies obtained by the wrapper model of *BIRS*, it excels specially when the C4 classifier is applied, winning in four of the five datasets; with the NB classifier, *BIRS* obtains good results on the DEX dataset; and applying IB, it loses in ARC and GIS, but nevertheless wins by approximately 20 points in the DEX and MAD datasets. In all the cases, the reduction obtained with respect to the original data is drastic, emphasizing that obtained with the DOR dataset, where approximately 0.01% of the attributes (10 of 100,000) is always retained.

- The behavior of the filter approach of *BIRS* is excellent. It produces rates of successes similar to the wrapper approach, with the number of attributes equal or even lower. Note that the number of attributes in filter approaches does not depend on the classifier applied.

- If we study the comparison between *BIRS* approaches and the *FCBF* algorithm, it can be verified that, except for the DEX dataset with an NB classifier, the accuracies obtained applying *FCBF* are normally below those obtained applying *BIRS*, emphasizing the existing differences for MAD dataset with a C4 classifier, and for ARC and MAD datasets with IB. The subsets selected by *FCBF* are greater than those chosen by *BIRS* on average, however, the time cost is approximately six times less.

8.6   Conclusions

The success of many learning schemes, in their attempts to construct data models, hinges on the reliable identification of a small set of highly predictive attributes. Traditional feature selection methods often select the top-ranked features according to their individual discriminative powers. However, the inclusion of irrelevant, redundant, and noisy features in the model building process phase can result in poor predictive performance and increased computation. The most popular search methods in machine learning cannot be applied to massive datasets, especially when a wrapper approach is used as an evaluation function. We use the incremental ranked usefulness definition to decide at the same time whether or not a feature is relevant and non-redundant. The technique extracts the best non-consecutive features from the ranking, trying to avoid the influence of unnecessary features in further classifications.

Our approach, named *BIRS*, uses a very fast search through the attribute space, and any subset evaluation measure, the classifier approach included, can be embedded into it as an evaluator. Massive datasets take a lot of computational resources when wrappers are chosen. *BIRS* reduces the search space complexity as it works directly on the ranking, transforming the combinatorial search of a sequential forward search into a quadratic search. However, the evaluation is much less expensive as only a few features are selected, and therefore the subset evaluation is computationally inexpensive in comparison to other approaches involving wrapper methodologies.

In short, our technique *BIRS* chooses a small subset of features from the original set with similar predictive performance to others. For massive datasets, wrapper-based methods might be computationally unfeasible, so *BIRS* turns out to be a fast technique that provides good performance in predicting accuracy.

Acknowledgment

The research was supported by the Spanish Research Agency CICYT under grant TIN2004-00159 and TIN2004-06689-C03-03.

References

[1] D. Bell and H. Wang. A formalism for relevance and its application in feature subset selection. *Machine Learning*, 41(2):175–195, 2000.

[2] A. L. Blum and P. Langley. Selection of relevant features and examples in machine learning. *Artificial Intelligence*, 97(1-2):245–271, 1997.

[3] R. A. Caruana and D. Freitag. How useful is relevance? In *Working Notes of the AAAI Fall Symp. on Relevance*, pages 25–29, 1994.

[4] M. Dash and H. Liu. Feature selection for classification. *Intelligent Data Analisys*, 1(3):131–56, 1997.

[5] M. Dash, H. Liu, and H. Motoda. Consistency based feature selection. In *Pacific-Asia Conf. on Knowledge Discovery and Data Mining*, pages 98–109, 2000.

[6] C. Ding and H. Peng. Minimum redundancy feature selection from microarray gene expression data. In *IEEE Computer Society Bioinformatics*, pages 523–529, IEEE PRess, Poscataway, NJ, 2003.

[7] T. Golub, D. Slonim, P. Tamayo, C. Huard, M. Gaasenbeek, J. Mesirov, H. Coller, M. Loh, J. Downing, M. Caligiuri, C. Bloomfield, and E. Lander. Molecular classification of cancer: Class discovery and class prediction by gene expression monitoring. *Science*, 286:531–37, 1999.

[8] I. Guyon and A. Elisseeff. An introduction to variable and feature selection. *Journal of Machine Learning Research*, 3:1157–1182, 2003.

[9] I. Guyon, S. Gunn, A. Ben-Hur, and G. Dror. Result analysis of the nips 2003 feature selection challenge. In *Advances in Neural Information Processing Systems*, pages 545–552. MIT Press, Cambridge, MA, 2005.

[10] I. Guyon, J. Weston, S. Barnhill, and V. Vapnik. Gene selection for cancer classification using support vector machine. *Machine Learning*, 46(1-3):389–422, 2002.

[11] M. A. Hall. Correlation-based feature selection for discrete and numeric class machine learning. In *17th Int. Conf. on Machine Learning*, pages 359–366. Morgan Kaufmann, San Francisco, CA, 2000.

[12] M. A. Hall and G. Holmes. Benchmarking attribute selection techniques for discrete class data mining. *IEEE Transactions on Knowledge and Data Eng.*, 15(3), 2003.

[13] R. Kohavi and G. H. John. Wrappers for feature subset selection. *Artificial Intelligence*, 1-2:273–324, 1997.

[14] D. Koller and M. Sahami. Toward optimal feature selection. In *13th Int. Conf. on Machine Learning*, pages 284–292, 1996.

[15] P. Langley. Selection of relevant features in machine learning. In *Proceedings of the AAAI Fall Symposium on Relevance*, pages 140–144, 1994.

[16] H. Liu and H. Motoda. *Feature Selection for Knowlegde Discovery and Data Mining*. Kluwer Academic Publishers, London, UK, 1998.

[17] H. Liu and L. Yu. Toward integrating feature selection algorithms for classification and clustering. *IEEE Trans. on Knowledge and Data Eng.*, 17(3):1–12, 2005.

[18] R. Ruiz, J. C. Riquelme, and J. S. Aguilar-Ruiz. Incremental wrapper-based gene selection from microarray expression data for cancer classification. *Pattern Recognition*, 39:2383–2392, 2006.

[19] I. H. Witten and E. Frank. *Data Mining: Practical Machine Learning Tools and Techniques*. Morgan Kaufmann, San Francisco, CA, 2005.

[20] E. P. Xing, M. I. Jordan, and R. M. Karp. Feature selection for high-dimensional genomic microarray data. In *Proc. 18th Int. Conf. on Machine Learning*, pages 601–608. Morgan Kaufmann, San Francisco, CA, 2001.

[21] L. Yu and H. Liu. Efficient feature selection via analysis of relevance and redundancy. *Journal of Machine Learning Research*, 5:1205–24, 2004.

Part III

Weighting and Local Methods

Chapter 9

Non-Myopic Feature Quality Evaluation with (R)ReliefF

Igor Kononenko

University of Ljubljana

Marko Robnik Šikonja

University of Ljubljana

9.1 Introduction

Researchers in machine learning, data mining, and statistics have developed a number of methods that estimate the usefulness of a feature for predicting the target variable. The majority of these measures are myopic in a sense that they estimate the quality of one feature independently of the context of other features. Our aim is to show the idea, advantages, and applications of non-myopic measures, based on the Relief algorithm, which is context sensitive, robust, and can deal with datasets with highly interdependent features. For a more thorough overview of feature quality measures, see [15].

The next section briefly overviews myopic impurity based measures for feature evaluation and defines the basic algorithm Relief for non-myopic feature evaluation. The succeeding section develops a more realistic variant ReliefF that is able to evaluate the features in multi-class problems, can deal with missing feature values, and is robust with respect to noise. Afterwards, the basic idea is extended also to regressional problems, and we describe the Regressional ReliefF (RReliefF). Section 9.4 describes various extensions of the (R)ReliefF family of algorithms: evaluation of literals in inductive logic programming, cost-sensitive feature evaluation with ReliefF, and the ordEval algorithm for the evaluation of features with ordered values. In Section 9.5 we define two approaches to comprehensively interpret ReliefF's estimates. Section 9.6 discusses implementation issues, such as time complexity, the importance of sampling, and parallelization. Finally, in Section 9.7 we describe several modes of applications of the (R)ReliefF family of algorithms.

9.2   From Impurity to Relief

The majority of feature evaluation measures are impurity based, meaning that they measure the impurity of the class value distribution. These measures evaluate each feature separately by measuring the impurity of the splits resulting from the partition of the learning instances according to the values of the evaluated feature. Figure 9.1 illustrates the idea. The geometrical objects are learning instances, described with features: size (big, small), shape (circle, triangle, square, star, ellipse), and contains_circle (yes, no). The color (white, black) represents the class value.

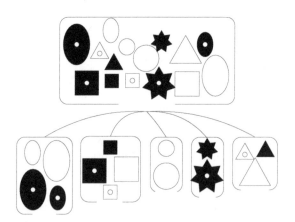

FIGURE 9.1: Illustration of the impurity based feature evaluation.

The impurity based measures assume the conditional independence of the features upon the class, evaluate each feature separately, and do not take the context of other features into account. In problems that possibly involve much feature interactions, these measures are not appropriate. The general form of all impurity based measures is

$$W(F_i) = \text{imp}(y) - \sum_{j=1}^{n_i} p(F_i = j)\, \text{imp}(y|F_i = j)$$

where $\text{imp}(y)$ is the impurity of class values before the split, $\text{imp}(y|F_i = j)$ is the impurity after the split on $F_i = j$, n_i is the number of values of feature F_i, and $p(F_i = j)$ is the (prior) probability of the feature value j. By subtracting the expected impurity of the splits from the impurity of unpartitioned instances we measure gain in the purity of class values resulting from the split. Larger values of $W(F_i)$ imply purer splits and therefore better features. We cannot directly apply these measures to numerical features, but we can use any discretization technique and then evaluate the discretized features.

9.2.1 Impurity Measures in Classification

There are several impurity based measures for classification problems, e.g., Two well-known impurity measures are entropy and Gini-index. With entropy we get the information gain measure, also referred to as mutual information due to its symmetry:

$$\text{Gain}(F_i) = H_Y - H_{Y|F_i} = H_Y + H_{F_i} - H_{YF_i} = I(F_i; Y) = I(Y; F_i) \quad (9.1)$$

where H_Y is the class entropy, and $H_{Y|F_i}$ is the conditional class entropy given the value of feature F_i. Gini-index gain [1] is obtained by the difference between the prior and the expected posterior Gini-indices:

$$\text{Gini}(F_i) = \sum_{j=1}^{n_i} p(F_i = j) \sum_{c=1}^{C} p(y = c|F_i = j)^2 - \sum_{c=1}^{C} p(y = c)^2 \quad (9.2)$$

where $p(y = c)$ is the (prior) probability of the class value c and $p(y = c|F_i = j)$ is the conditional probability of the class c given the feature value j.

Both measures, $\text{Gain}(F_i)$ and $\text{Gini}(F_i)$, are nonnegative and they tend to overestimate features with more values. Therefore, either all features have to be binary or we have to use a normalization. For information gain there are two frequently used normalizations. The first is gain-ratio, where information gain is normalized with the feature entropy [21]. This normalization eliminates the problem of overestimating the multi-valued features, however, the gain-ratio overestimates features with small feature entropy H_{F_i}. A better normalization is with the joint entropy H_{YF_i} [15].

Another possibility is to generalize entropy in terms of the minimum description length (MDL) principle. The impurity can be defined as the number of bits needed to code the classes. We need to code the class probability distribution and the class for each (training) instance. The MDL measure [14] is the most appropriate among impurity measures for estimating the quality of multi-valued features. Its advantage is also in the detection of useless (non-compressive) features. Since the optimal coding for both parts of the code uses binomial coefficients, one has to be careful with the implementation (due to incorrect implementations, some authors considered the MDL measure useless). The best way to avoid overly large numbers is to use the log of gamma function.

9.2.2   Relief for Classification

All measures described so far evaluate the quality of a feature independently of the context of other features, i.e., they assume the independence of features with respect to the class. The term "myopic" characterizes their inability to detect the information content of a feature that stems from a broader context and dependencies between features.

The context of other features can be efficiently taken into account with the algorithm ReliefF. Let us first describe a simpler variant, called Relief [12], which is designed for two-class problems without missing values. The basic idea of the algorithm, when analyzing learning instances, is to take into account not only the difference in feature values and the difference in classes, but also the *distance* between the instances. Distance is calculated in the feature space, therefore similar instances are close to each other and dissimilar are far apart. By taking the similarity of instances into account, the context of all the features is implicitly considered.

The basic algorithm Relief [12] (see Algorithm 9.2.1), for each instance from a random subset of m ($m \leq M$) learning instances, calculates the nearest instance from the same class (nearest hit $\mathbf{x}_H$) and the nearest instance from the opposite class (nearest miss $\mathbf{x}_M$). Then it updates the quality of each feature with respect to whether the feature differentiates two instances from the same class (undesired property of the feature) and whether it differentiates two instances from opposite classes (desired property). By doing so, the quality estimate takes into account the local ability of the feature to differentiate between the classes. The locality implicitly takes into account the context of other features.

Quality estimations W can also be negative, however, $W[F_i] \leq 0$ means that feature F_i is irrelevant.

Figure 9.2 illustrates the problem of conditionally dependent features and the way Relief deals with it. On the left-hand side we see why impurity based measures fail: Split on values of each feature (size or shape) does not reduce class (color) impurity. On the right-hand side we illustrate Relief: It randomly selects an instance and finds its nearest hit (small black square) and one of the nearest misses (small white ellipse or big white square, both containing circle). The values of both important features (size and shape) separate the selected instance and its miss and do not separate the instance and its hit, so they both get a positive update. The feature *contains_circle*, which is irrelevant to the class, does the opposite and gets a negative update.

Algorithm 9.2.1 Basic algorithm Relief.

Input: M learning instances $\mathbf{x}_k$ described by N features; sampling parameter m
Output: for each feature F_i a quality weight $-1 \leq W[i] \leq 1$

for i = 1 to N do W[i] = 0.0; **end for**;
for l = 1 to m do
  randomly pick an instance $\mathbf{x}_k$;
  find its nearest hit $\mathbf{x}_H$ and nearest miss $\mathbf{x}_M$;
  for i = 1 to N **do**
    W[i] = W[i] – diff(i,$\mathbf{x}_k$,$\mathbf{x}_H$)/m + diff(i,$\mathbf{x}_k$,$\mathbf{x}_M$)/m;
  end for;
end for;
return(W);

For (each) feature F_i the function $\text{diff}(i, \mathbf{x}_j, \mathbf{x}_k)$ in Algorithm 9.2.1 returns the difference of feature values of two instances:

$$\text{diff}(i, \mathbf{x}_j, \mathbf{x}_k) = \begin{cases} \frac{|x_{j,i} - x_{k,i}|}{\max(F_i) - \min(F_i)} & F_i \text{ is numerical} \\ 0 & x_{j,i} = x_{k,i} \wedge F_i \text{ is nominal} \\ 1 & x_{j,i} \neq x_{k,i} \wedge F_i \text{ is nominal} \end{cases} \quad (9.3)$$

If we have a dataset with mixed numerical and nominal features, the use of (9.3) would underestimate the numerical features. Let us illustrate this by taking two instances with 2 and 5 being values of feature F_i, respectively, where the possible values of F_i are integers from [1..8]. If F_i is nominal, the value of $\text{diff}(F_i, 2, 5) = 1$, since the two nominal values are different. If F_i is numerical, $\text{diff}(F_i, 2, 5) = \frac{|2-5|}{7} \approx 0.43$. The Relief algorithm uses results of the diff function to update their qualities; therefore, with (9.3) numerical features are underestimated. We can overcome this problem with the ramp

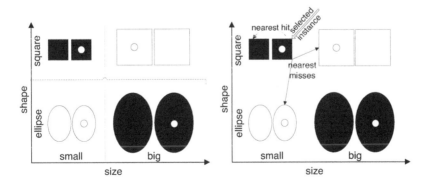

FIGURE 9.2: Problem of conditionally dependent features (left) and the idea of the Relief algorithm (right).

function as proposed by [8]. It can be defined as a generalization of the diff function for the numerical features:

$$\text{diff}(i, \mathbf{x}_j, \mathbf{x}_k) = \begin{cases} 0 & |x_{j,i} - x_{k,i}| \leq t_{eq} \\ 1 & |x_{j,i} - x_{k,i}| > t_{diff} \\ \frac{|x_{j,i} - x_{k,i}| - t_{eq}}{t_{diff} - t_{eq}} & t_{eq} < |x_{j,i} - x_{k,i}| \leq t_{diff} \end{cases} \quad (9.4)$$

where t_{eq} and t_{diff} are two user definable threshold values, t_{eq} is the maximum distance between two feature values to still consider them equal, and t_{diff} is the minimum distance between feature values to still consider them different. If we set $t_{eq} = 0$ and $t_{diff} = \max(F_i) - \min(F_i)$ we obtain (9.3). Default values are $t_{eq} = 0.05(\max(F_i) - \min(F_i)), t_{diff} = 0.10(\max(F_i) - \min(F_i))$.

Relief estimates the following difference of probabilities:

$$W(F_i) = P(\text{different value of } F_i | \text{near instance with different prediction})$$
$$- P(\text{different value of } F_i | \text{near instance with same prediction}) \quad (9.5)$$
$$= P(\text{same value of } F_i | \text{near instance with same prediction})$$
$$- P(\text{same value of } F_i | \text{near instance with different prediction}) \quad (9.6)$$

If we omit the nearness condition, we get a function that is closely related to Gini-index[13]:

$$Wm(F_i) = constant \times \sum_{j=1}^{n_i} p(F_i = j)^2 \times \text{Ginim}(F_i) \quad (9.7)$$

where $\text{Ginim}(F_i)$ is strongly related with $\text{Gini}(F_i)$ from Equation (9.2):

$$\text{Ginim}(F_i) = \sum_{j=1}^{n_i} \left(\frac{p(F_i = j)^2}{\sum_j p(F_i = j)^2} \times \sum_{c=1}^{C} p(y = c | F_i = j)^2 \right) - \sum_{c=1}^{C} p(y = c)^2$$

$$(9.8)$$

The only difference between $\text{Ginim}(F_i)$ and $\text{Gini}(F_i)$ is that instead of the factor $\frac{p(F_i=j)^2}{\sum_j p(F_i=j)^2}$ in Equation (9.2) we have $\frac{p(F_i=j)}{\sum_j p(F_i=j)} = p(F_i = j)$. However, the crucial difference between the myopic Relief, defined by Equation (9.7), and $\text{Gini}(F_i)$ is in the factor in front of Ginim in Equation (9.7): $\sum_j p(F_i = j)^2$. This factor represents the prior probability that two randomly selected instances have the same value of the given feature. The factor implicitly normalizes the Relief's quality estimates with respect to the number of feature values. While $\text{Gini}(F_i)$ overestimates multi-valued features, Relief and its myopic variant (9.7) have no such undesired bias.

Basic Relief is able to evaluate the quality of numerical and discrete features, which are highly interdependent. For example, for very hard parity problems of arbitrary order, where the learning instances are described with an additional number of irrelevant features, Relief is able to detect a subset of relevant features.

Algorithm 9.3.1 ReliefF.

Input: M learning instances $\mathbf{x}_k$ (N features and C classes);
Probabilities of classes p_y; Sampling parameter m;
Number n of nearest instances from each class;
Output: for each feature F_i a quality weight $-1 \leq W[i] \leq 1$;

1 **for** i $= 1$ **to** N **do** W[i] $= 0.0$; **end for**;
2 **for** $l = 1$ **to** m **do**
3   randomly pick an instance $\mathbf{x}_k$ (with class y_k);
4   **for** $y = 1$ **to** C **do**
5   find n nearest instances x$[j, y]$ from class y, $j = 1..n$;
6   **for** i $= 1$ **to** N **do** **for** $j = 1$ **to** n **do**
7   **if** $y = y_k$ { nearest hit? }
8   **then** W[i] $=$ W[i] $-$ diff(i,$\mathbf{x}_k$,x$[j, y]$)$/(m * n)$;
9   **else** W[i] $=$ W[i] $+ p_y/(1 - p_{y_k})*$ diff(i,$\mathbf{x}_k$,x$[j, y]$)$/(m * n)$;
10   **end if**;
11   **end for**; { j } **end for**; { i }
12   **end for**; { y }
13 **end for**; { l }
14 return(W);

9.3   ReliefF for Classification and RReliefF for Regression

A more realistic variant of Relief is its extension, called ReliefF [13] (see Algorithm 9.3.1). The original Relief was designed for two-class problems without missing values and is quite sensitive to noise. ReliefF is able to deal with incomplete and noisy data and can be used for evaluating the feature quality in multi-class problems:

Missing feature values: ReliefF can also use incomplete data. For that purpose we generalize the function diff to calculate the probability that two instances have different values of the given feature. We have two possibilities. One of instances ($\mathbf{x}_l$) has an unknown value of feature F_i:

$$\text{diff}(F_i, \mathbf{x}_l, \mathbf{x}_k) = 1 - p(F_i = x_{k,i} | y = y_l)$$

Both instances have unknown feature values:

$$\text{diff}(F_i, \mathbf{x}_l, \mathbf{x}_k) = 1 - \sum_{j=1}^{n_i} \left(p(F_i = j | y = y_l) \times p(F_i = j | y = y_k) \right)$$

Noisy data: The most important part of algorithm Relief is searching for the nearest hit and miss. Noise (mistake) in a class and/or feature value significantly affects the selection of nearest hits and misses. In order to make this process more reliable in the presence of noise, ReliefF uses n nearest hits and n nearest misses and averages their contributions to the features' quality estimates. n is a user defined parameter with typical values $n \in [5...10]$. This simple extension significantly improves the reliability of quality estimates.

Multi-class problems: Instead of n nearest hits and misses, ReliefF searches for n nearest instances from each class. The contributions of different classes are weighted with their prior probabilities. In Algorithm 9.3.1, the weighting factor is $p_y/(1 - p_{y_k})$. The class of an instance is y_k, while y is the class of its nearest miss. The factor is therefore proportional to the probability of class y, normalized with the sum of probabilities of all classes, different from y_k.

In regression, as an impurity measure, the *variance* of the numeric target variable is used. It is defined as the mean squared error:

$$s^2 = \frac{1}{M} \sum_{k=1}^{M} (y_k - \bar{y})^2$$

where $\bar{y}$ is the mean of the target variable over all M learning instances. Variance is closely related to Gini-index, which is an impurity measure. If in the binary classification problem one class is transformed into value 0 and the other into value 1 of the regression variable (the discrete class variable is transformed into a numerical form), we get the following equality [1]:

$$\text{Gini\_prior} = 2s^2$$

For evaluating the quality of a feature, *the expected change of variance* is used. It behaves similarly to the expected change of impurity in classification - it tends to overestimate the features with large numbers of values.

Like most of the feature quality measures defined for classification problems, the expected change of variance is also a myopic measure. When estimating the quality of a feature, it does not take into account the context of other features. In the following, we develop a non-myopic measure for regression by appropriately adapting algorithm ReliefF.

In regression problems the target variable is numerical, therefore nearest hits and misses cannot be used in a strict sense as in algorithm ReliefF. RReliefF (Regressional ReliefF) uses a kind of "probability" that two instances belong to two "different" classes [23]. This "probability" is modeled with the distance between the values of the target variable of two learning instances.

By omitting the nearness condition in Equation (9.6) we get

$$Wm(F_i) = P(\text{diff}(i, \mathbf{x}_j, \mathbf{x}_k) = 0 | y_j = y_k) - P(\text{diff}(i, \mathbf{x}_j, \mathbf{x}_k) = 0 | y_j \neq y_k)$$

Algorithm 9.3.2 RReliefF – Regressional ReliefF.

Input: M learning instances $\mathbf{x}_k$ described with N features;
Sampling parameter m; Number n of nearest instances;
Output: for each feature F_i a quality weight $-1 \leq W[i] \leq 1$;

> set all N_{dY}, $N_{dF}[i]$, $N_{dY \wedge dF}[i]$, $W[i]$ to 0;
> **for** $l = 1$ **to** m **do**
>> randomly pick an instance $\mathbf{x}_k$;
>> find indices k_j of n nearest instances, $j \in [1..n]$;
>> **for** $j = 1$ **to** n **do**
>>> { index 0 in diff corresponds to target (regression) variable }
>>> $N_{dY} = N_{dY} + \text{diff}(0, \mathbf{x}_{k_j}, \mathbf{x}_k)/n$;
>>> **for** $i = 1$ **to** N **do**
>>>> $N_{dF}[i] = N_{dF}[i] + \text{diff}(i, \mathbf{x}_{k_j}, \mathbf{x}_k)/n$;
>>>> $N_{dY \wedge dF}[i] = N_{dY \wedge dF}[i] + \text{diff}(0, \mathbf{x}_{k_j}, \mathbf{x}_k) \cdot \text{diff}(i, \mathbf{x}_{k_j}, \mathbf{x}_k)/n$;
>>> **end for**; { i }
>> **end for**; { j }
> **end for**; { l }
> { for each feature calculate the value of (9.10) }
> **for** $i = 1$ **to** N **do**
> $W[i] = N_{dY \wedge dF}[i]/N_{dY} - (N_{dF}[i] - N_{dY \wedge dF}[i])/(m - N_{dY})$;
> **end for**;
> return(W);

where y_l stands for the class of learning instance $\mathbf{x}_l$. Further, let

$$P_{eq\_val} = P(\text{diff}(i, \mathbf{x}_j, \mathbf{x}_k) = 0), \quad P_{samecl} = P(y_j = y_k) \ \text{ and}$$

$$P_{samecl|eq\_val} = P(y_j = y_k | \text{diff}(i, \mathbf{x}_j, \mathbf{x}_k) = 0)$$

By using the Bayesian rule we get

$$W(F_i) = \frac{P_{samecl|eq\_val} P_{eq\_val}}{P_{samecl}} - \frac{(1 - P_{samecl|eq\_val}) P_{eq\_val}}{1 - P_{samecl}} \tag{9.9}$$

The trick now is to bring back the nearness condition. For estimating the quality in Equation (9.9) we need the (posterior) probability $P_{samecl|eq\_val}$ that two (nearest) instances belong to the same class provided they have the same feature value, and the prior probability P_{samecl} that two instances belong to the same class. We can transform the equation, so that it contains the probability that two instances belong to different classes provided they have different feature values:

$$W(F_i) = \frac{P_{\text{diffcl}|\text{diff}} P_{\text{diff}}}{P_{\text{diffcl}}} - \frac{(1 - P_{\text{diffcl}|\text{diff}}) P_{\text{diff}}}{1 - P_{\text{diffcl}}} \tag{9.10}$$

Here P_{diff} denotes the prior probability that two instances have different feature values, and P_{diffcl} denotes the prior probability that two instances belong to different classes.

Algorithm RReliefF has to approximate the probabilities in Equation (9.10). The details are provided in Algorithm 9.3.2. The algorithm calculates the "frequencies":

- N_{dY} – sum of "probabilities" that two nearest instances belong to different classes;

- $N_{dF}[i]$, – sum of "probabilities" that two nearest instances have different feature values;

- $N_{dY \wedge dF}[i]$ – sum of "probabilities" that two nearest instances belong to different classes and have different feature values.

Finally, from the above "frequencies," it calculates the feature qualities $W[i]$ using Equation (9.10).

Both algorithms, ReliefF and RReliefF, calculate the quality of features according to Equation (9.10), which represents a unified view of the feature quality estimation – in classification and regression.

When computing the diff function, it also makes sense to take distance into account. The rationale is that closer instances should have greater influence, so we exponentially decrease the influence of the near instances with the distance from the selected instance. Details of the implementation are in [26].

9.4 Extensions

9.4.1 ReliefF for Inductive Logic Programming

When dealing with the classification problems, inductive logic programming (ILP) systems often lag behind the state-of-the-art attributional learners. Part of the blame can be ascribed to a much larger hypothesis space that, therefore, cannot be so thoroughly explored. ReliefF is suitable for the propositional representation of training instances. A slightly different approach is needed when estimating the quality of literals when inducing the first order theories with an ILP system.

The main difference stems from the fact that, while learning in the propositional language, we are only interested in the boundaries between different classes. On the other hand, when learning in the first order language, we are not searching for boundaries but for a theory that explains positive learning instances and does not cover negative ones. A crucial part of ReliefF is the function that measures the difference (distance) between the training instances.

Algorithm 9.4.1 Literal quality assessment with ReliefF. Note that Diff is used for nearest hits and Diff$_A$ for nearest misses.

Input:: Literal space LS; Current training set $T = T^+ \cup T^-$;
T^+, T^-: positive and negative instances respectively; Sampling parameter m;
Output: Weight vector W where $W[L]$ estimates the quality of literal L;

```
set all weights W[L] := 0.0;
for l := 1 to m do
    randomly select an instance x_k ∈ T⁺;
    find n nearest hits x_H[i] and n nearest misses x_M[i];
    for L := 1 to #literals do
        for i := 1 to n do
            W[L] := W[L] + (Diff_A(L, x_k, x_M[i]) − Diff(L, x_k, x_H[i]))/(n × m);
        end for; { i }
    end for; { L }
end for; { l }
```

The key idea of using ReliefF within ILP is to estimate literals according to how well they distinguish between the instances that are logically similar [20]. Algorithm 9.4.1 searches for n nearest hits/misses. The search for the nearest hits and misses is guided by the *total distance* between the two instances Diff$_T$, computed as

$$\text{Diff}_T(\mathbf{x}_k, \mathbf{x}_l) = \frac{1}{|LS|} \sum_{L \in LS} \text{Diff}(L, \mathbf{x}_k, \mathbf{x}_l) \tag{9.11}$$

It is simply a normalized sum of differences over the literal space LS. It estimates the logical similarity of two instances relative to the background knowledge.

Both the total distance Diff$_T$ and the estimates W depend on the definition of Diff (Diff$_A$ is an asymmetric version of Diff). Table 9.1 shows the definitions of Diff and Diff$_A$.

The first two columns represent the coverage of literal L over the instances $\mathbf{x}_k$ and $\mathbf{x}_l$, respectively. The coverage denotes the truth value of some partially built clause Cl' with literal L included when the head of the clause is instantiated with instance $\mathbf{x}_k$ or $\mathbf{x}_l$. Note that since $\mathbf{x}_k$ is always from T^+ (see Algorithm 9.4.1), the Diff$_A$ function gives the preference to literals covering the positive instances.

The good performance of the learning system that uses this version of ReliefF was empirically confirmed in many learning problems in ILP [20].

Table 9.1: Definitions of the Diff and Diff$_A$ functions.

$L(\mathbf{x}_k)$	$L(\mathbf{x}_l)$	Diff$(L, \mathbf{x}_k, \mathbf{x}_l)$	Diff$_A(L, \mathbf{x}_k, \mathbf{x}_l)$
0	0	0	0
0	1	1	0
1	0	1	1
1	1	0	0

9.4.2 Cost-Sensitive ReliefF

While historically the majority of machine learning research in classification has been focused on reducing the classification error, there also exists a corpus of work on cost-sensitive classification, where all errors are not equally important (see an overview in [6]). In general, differences in the importance of errors are handled through the cost of misclassification.

We assume that costs can be presented with the cost matrix $\mathbf{C}$, where $C_{c,u}$ is the cost (could also be benefit) associated with the prediction that an instance belongs to the class u where in fact it belongs to the class c. The optimal prediction for an instance $\mathbf{x}$ is the class u that minimizes the expected loss:

$$L(\mathbf{x}, y = u) = \sum_{c=1}^{C} P(y = c|\mathbf{x})C_{c,u}$$

where $P(y = c|\mathbf{x})$ is the probability of the class c given instance $\mathbf{x}$. The task of a learner is therefore to estimate these conditional probabilities. Feature evaluation measures need not be cost-sensitive for decision tree building, as shown by [1, 6]. However, cost-sensitivity is a desired property of an algorithm that tries to rank or weight features according to their importance. We present the best solutions for a cost-sensitive ReliefF from [25].

There are different techniques for incorporating cost information into learning. The key idea is to use the expected cost of misclassifying an instance with class c and then change the probability estimates:

$$\varepsilon_c = \frac{1}{1 - p(y = c)} \sum_{\substack{u=1 \\ u \neq c}}^{C} p(y = u)C_{c,u} \quad p'(y = c) = \frac{p(y = c)\varepsilon_c}{\sum_{u=1}^{C} p(y = u)\varepsilon_u} \quad (9.12)$$

Using probabilities (9.12) in the impurity based functions Gain (9.1) and Gini (9.2), we get their cost-sensitive variations. Similarly, we can use (9.12) in ReliefF; we only have to replace the 9th line in Algorithm 9.3.1 with

   else W[i] = W[i] + $p'_y/(1 - p'_{y_k})$∗ diff(i,$\mathbf{x}_k$,x[j, y])/($m * n$);

If we use just the information from a cost matrix and do not take prior probabilities into account, similarly to (9.12), we can compute the average cost of misclassifying an instance that belongs to the class c and the prior probability of class value:

$$\alpha_c = \frac{1}{C-1} \sum_{\substack{u=1 \\ u \neq c}}^{C} C_{c,u} \qquad \bar{p}(y=c) = \frac{\alpha_c}{\sum_{u=1}^{C} \alpha_u} \qquad (9.13)$$

The use of $\bar{p}(y=c)$ instead of $p(y=c)$ in the 9th line in Algorithm 9.3.1 also enables ReliefF to successfully use cost information. For two-class problems, ReliefF, ReliefF with p', and ReliefF with $\bar{p}$ are identical.

9.4.3   Evaluation of Ordered Features at Value Level

A context sensitive algorithm for evaluation of ordinal features was proposed in [27]. The *ordEval* algorithm exploits the information hidden in ordering of feature and class values and provides a separate score for each value of the feature. Similarly to ReliefF, the contextual information is exploited via the selection of nearest instances. The ordEval outputs probabilistic factors corresponding to the effect an increase/decrease of a feature value has on the class value. The difference to ReliefF is in handling each feature value separately and in differentiating between the positive and negative changes of the feature and their impact on the class value.

To present the algorithm we need some definitions. Let $\mathbf{x}_R$ be a randomly selected instance and x_S its most similar instance. Let j be the value of feature F_i at instance $\mathbf{x}_R$. We observe the necessary changes of the class value and features (F_i in particular) that would change $\mathbf{x}_S$ to $\mathbf{x}_R$. If these changes are positive (increase of class and/or feature values), let

- $P(y_{i,j}^p)$ be a probability that the class value of $\mathbf{x}_R$ is larger than the class value of its most similar neighbor $\mathbf{x}_S$. $P(y_{i,j}^p)$ is therefore the probability that the positive change in a similar instance's class value is needed to get from $\mathbf{x}_S$ to $\mathbf{x}_R$.

- $P(F_{i,j}^p)$ be a probability that j (the value of F_i at $\mathbf{x}_R$) is larger than the value of F_i at its most similar neighbor $\mathbf{x}_S$. By estimating $P(F_{i,j}^p)$, we gather evidence of the probability that the similar instance $\mathbf{x}_S$ has a lower value of F_i and the change of $\mathbf{x}_S$ to $\mathbf{x}_R$ is positive.

- $P(y^p F_{i,j}^p)$ be a probability that both the class and j (the value of F_i at $\mathbf{x}_R$) are larger than the class and feature value of its most similar neighbor $\mathbf{x}_S$. With $P(y^p F_{i,j}^p)$ we estimate the probability that positive change in both the class and F_i value of a similar instance $\mathbf{x}_S$ is needed to get the values of $\mathbf{x}_R$.

Similarly we define $P(y_{i,j}^n)$, $P(F_{i,j}^n)$, and $P(y^n F_{i,j}^n)$ for negative changes that would turn $\mathbf{x}_S$ into $\mathbf{x}_R$ (decrease of class and/or feature values).

The output of the algorithm are conditional probabilities called upward and downward reinforcement factors, which measure the upward/downward trends exhibited in the data. The upward reinforcement of the i-th feature's value j is

$$U_{i,j} = P(y_{i,j}^p | F_{i,j}^p) = \frac{P(y^p F_{i,j}^p)}{P(F_{i,j}^p)} \tag{9.14}$$

This factor reports the probability that a positive class change is caused by a positive feature change. This intuitively corresponds to the effect the positive change in the feature value has on the class. Similarly for downward reinforcement:

$$D_{i,j} = P(y_{i,j}^n | F_{i,j}^n) = \frac{P(y^n F_{i,j}^n)}{P(F_{i,j}^n)} \tag{9.15}$$

$D_{i,j}$ reports the effect the decrease of a feature value has on the decrease of the class value. Analogously with numerical features, we could say that U and D are similar to the partial derivatives of the prediction function.

Algorithm ordEval reliably estimates (9.14) and (9.15), borrowing from ReliefF and RReliefF many implementation details (sampling, context, treatment of distance, updates).

The ordEval algorithm is general and can be used for analysis of any survey with graded answers; The authors of [27] have used it as an exploratory tool on a marketing problem of customer (dis)satisfaction and also developed a visualization technique. The use of $U_{i,j}$ and $D_{i,j}$ for feature subset selection seems possible, but is still an open research question.

9.5 Interpretation

There are two complementary interpretations of the quality evaluations computed by Relief, ReliefF, and RReliefF. The first is based on the difference of probabilities from Equation (9.5), and the second explains them as the portions of the explained concept.

9.5.1 Difference of Probabilities

Equation (9.5) forms the basis for the difference of probabilities interpretation of the quality estimations of the Relief algorithms: the difference of the probability that two instances have different values of the feature F if they have different prediction values and the probability that two instances have different values of the feature if they have similar prediction values. These two probabilities contain the additional condition that the instances are close in the problem space and form an estimate of how well the values of the feature distinguish between the instances that are near to each other.

These two probabilities are mathematical transcriptions of Relief's idea: The first term rewards the feature if its values separate similar observations with different prediction values, and the second term punishes it if it does not separate similar observations with similar prediction values. As it turned out, this interpretation is nontrivial for human comprehension. Negated similarity (different values) and subtraction of the probabilities are difficult to comprehend for human experts.

9.5.2 Portion of the Explained Concept

The behavior of Relief, ReliefF, and RReliefF when the number of the instances approached infinity, i.e., when the problem space is densely covered with the instances, was analyzed in [26] and proved that Relief's quality estimates can be interpreted as the ratio between the number of the explained changes in the concept and the number of examined instances.

We say that feature F is responsible for the change of y_k (the predicted value of the instance $\mathbf{x}_k$) to the predicted value $b(y_k)$ if the change of its values is one of the minimal number of changes required for changing the predicted value from y_k to $b(y_k)$. We denote this responsibility by $r_F(y_k, b(y_k))$. As the number of instances M goes to infinity, the quality evaluation $W(F)$ computed from m instances $\mathbf{x}_k$ from the sample $\mathbf{S}$ ($|\mathbf{S}| = m$) for each feature converges to the ratio between the number of changes in the predicted values the feature is responsible for and the cardinality m of the sample:

$$\lim_{M \to \infty} W(F) = \frac{1}{m} \sum_{k=1}^{m} r_F(y_k, b(y_k)) \qquad (9.16)$$

Note that as $M \to \infty$, the problem space is densely covered with instances; therefore, the nearest hit comes from the same characteristic region as the randomly selected instance and its contribution in Algorithm 9.2.1 is 0.

We interpret Relief's weights $W(F)$ as the contribution (responsibility) of each feature to the explanation of the predictions. The actual quality evaluations for the features in the given problem are approximations of these ideal weights, which occur only with an abundance of data.

For ReliefF this property is somehow different. Recall that in this algorithm we search nearest misses from each of the classes and weight their contributions with prior probabilities. This weighting is also reflected in the feature evaluation when $M \to \infty$. Let $p(y = c)$ represent the prior probability of the class value c, and under the same conditions as for Relief, $r_F(y_k, b_u(y_k))$ be the responsibility of feature F for the change of y_k to the class u. Then ReliefF behaves as

$$\lim_{M \to \infty} W(F) = \frac{1}{m} \sum_{c=1}^{C} \sum_{\substack{u=1 \\ u \neq c}}^{C} \frac{p(y = c)p(y = u)}{1 - p(y = c)} \sum_{k=1}^{m} r_F(y_k, b_u(y_k)). \qquad (9.17)$$

We can therefore explain the quality estimate as the ratio between the number of class value changes the feature is responsible for and the number of examined instances, weighted with the prior probabilities of class values. In two-class problems, formulas (9.17) and (9.16) are equivalent (because $\text{diff}(F, \mathbf{x}_k, \mathbf{x}_l) = \text{diff}(F, \mathbf{x}_l, \mathbf{x}_k)$).

The interpretation of the quality estimates with the ratio of the explained changes in the concept is true for RReliefF as well, as it also computes Equation (9.5); however, the updates are proportional to the size of the difference in the prediction value.

We have noticed that in various applications (medicine, ecology) trees produced with (R)ReliefF algorithms are more comprehensible for human experts. Splits selected by them seem to mimic humans' partition of the problem, which we explain with the interpretation of Relief's weights as the portion of the explained concept.

9.6 Implementation Issues

9.6.1 Time Complexity

The time complexity of Relief and its extension ReliefF is $O(mMN)$, where m is the number of iterations in the main loop of the algorithm. For the calculation of each nearest hit and miss we need $O(MN)$ steps. Greater m implies more reliable evaluation of the feature's qualities but also greater time complexity. If we set $m = M$, we get the most reliable quality estimates and the highest time complexity. This is often unacceptably slow; therefore, for large M, we set $m \ll M$, typically $m \in [30...200]$.

The time complexity of RReliefF is equal to that of basic Relief, i.e., $O(mMN)$. The most time-consuming operation is searching for n nearest instances. We need to calculate M distances, which can be done in $O(MN)$ steps. Building the heap (full binary tree where each subtree contains the minimal element in the root) requires $O(M)$ steps, and n nearest instances can be extracted from the heap in $O(n \log M)$ steps. In practice this is always less than $O(MN)$.

If we use a k-d tree to implement the search for nearest instances, we can reduce the complexity of all three algorithms to $O(NM \log M)$ [22]. In practice, using k-d trees to select nearest instances only makes sense with reasonably small feature dimensionality ($N < 20$).

9.6.2 Active Sampling

When dealing with datasets with a huge numbers of instances, feature selection methods typically perform a random sampling. Reference [18] introduces the concept of active feature selection, and applies selective sampling based on data variance to ReliefF. The authors reduce the required number of training instances and achieve considerable time savings without performance deterioration. The idea of the approach is first to split the instances according to their density with the help of a k-d tree and then, instead of randomly choosing an instance from the whole training set, select a random representative from each leaf (bucket) of the k-d tree.

9.6.3 Parallelization

Relief algorithms are computationally more complex than some other (myopic) feature estimation measures. However, they also have a possible advantage that they can be naturally split into several independent tasks, which is a prerequisite for the successful parallelization of an algorithm. Each iteration of the algorithm is a natural candidate for a separate process, which would turn Relief into a fine-grained parallel algorithm. With the arrival of microprocessors with multiple cores, this will be an easy speedup.

9.7 Applications

(R)ReliefF has been applied in a variety of different ways in machine learning and data mining. It is implemented in many data mining tools, including Core, Weka, Orange, and R. Core (http://lkm.fri.uni-lj.si/rmarko/software) is the most complete and efficient implementation in C++, containing most of the extensions described in this chapter. Weka [30] contains Java code of ReliefF and RReliefF that can be used for feature subset selection. Orange [4] contains ReliefF and RReliefF, which can be used for many tasks within this versatile learning environment. In R [9], the ReliefF is available in a dprep package (http://math.uprm.edu/~edgar/dprep.html).

Besides the usual application for filter subset selection, (R)ReliefF was used for wrapper feature subset selection, feature ranking, feature weighing, building tree-based models and associative rules, feature discretization, controlling the search in genetic algorithms, literal ranking in ILP, and constructive induction.

9.7.1 Feature Subset Selection

Original Relief was designed for filter feature subset selection [12]. However, any algorithm from the (R)ReliefF family can also be efficiently used within the wrapper method: After ranking the features, the wrapper is used to select the appropriate size of the feature subset [7]. The usual filter way of using (R)ReliefF in a data mining process is to evaluate the features, select the appropriate subset, and then run one or more machine learning algorithms on the resulting dataset. To select the set of the most important features, [12] introduced the significance threshold θ. If the weight of a given feature is below θ, it is considered unimportant and is excluded from the resulting set. Bounds for θ were proposed, i.e., $0 < \theta \leq \frac{1}{\sqrt{\alpha m}}$, where α is the probability of accepting an irrelevant feature as relevant and m is the number of iterations used. The upper bound for θ is very loose and in practice much smaller values can be used.

Another interesting application is the feature subset selection in bagging: When K-NN classifiers are used it is important to reduce the number of features in order to provide efficient classification [10].

Many researchers have reported good performance of the (R)ReliefF family of algorithms in comparison with other feature selection methods, for example [5, 7, 28], however, with respect to the time complexity, the myopic measures are of course faster.

9.7.2 Feature Ranking

Feature ranking is needed when one has to decide the order of features in a certain data mining process. For example, if one needs to manually examine plots of features or pairs of features, in many applications it is practically impossible to examine all the plots. Therefore, only the most promising are examined. ReliefF seems to be a good choice in domains with strong interdependencies between features [3], although one must bear in mind its sensitivity to the context of redundant and irrelevant features. Feature ranking is important for guiding the search in various machine learning tasks where an exhaustive search is too complex and a heuristic search is required. Feature ranking dictates the order of features by which the algorithms search the space. Examples of such algorithms are building of decision trees, genetic algorithms, and constructive induction. Comparisons of ReliefF's feature ranking with that of other methods have confirmed ReliefF's good performance [28].

9.7.3 Feature Weighing

Feature weighing is an important component of any lazy learning scheme. Feature weights adjust the metric and therefore strongly influence the performance of lazy learning. Feature weighting is an assignment of a weight to each feature and can be viewed as a generalization of feature subset selection in the sense that it does not assign just binary weights (include-exclude) to each feature but rather an arbitrary real number. If (R)ReliefF algorithms are used in this fashion, then we do not need a significance threshold but rather use their weights directly. ReliefF was tested as the feature weighting method in lazy learning [29] and was found to be very useful. ReliefF was also applied to feature weighing in clustering [17].

9.7.4   Building Tree-Based Models

Commonly used feature estimators in decision trees are Gini-index and Gain ratio in classification and the mean squared error in regression [21, 1]. These estimators are myopic and cannot detect conditional dependencies between features and also have inappropriate bias concerning multi-valued features [14]. ReliefF was successfully employed in classification [16] and RReliefF in regression problems [23]. (R)ReliefF algorithms perform as well as myopic measures if there are no conditional dependencies among the features and surpass them if there are strong dependencies. When faced with an unknown dataset, it is unreasonable to assume that it contains no strong conditional dependencies and rely only on myopic feature estimators. Furthermore, using an impurity-based estimator near the fringe of the decision tree leads to non-optimal splits concerning accuracy, and a switch to accuracy has been suggested as a remedy. It was shown [24] that ReliefF in decision trees as well as RReliefF in regression trees do not need such switches as they contain them implicitly.

9.7.5   Feature Discretization

Discretization divides the values of the numerical feature into a number of intervals. Each interval can then be treated as one value of the new discrete feature. Discretization of features can reduce the learning complexity and help to understand the dependencies between the features and the target concept. There are several methods that can be used to discretize numerical features.

A usual top-down algorithm for the discretization of features starts with one interval and iteratively divides one subinterval into two subintervals. At each step the algorithm searches for a boundary that, when added to the current set of boundaries, maximizes the heuristic estimate of the discretized feature. The algorithm assumes that the heuristic feature quality measure increases until a (local) optima is reached. Therefore, information gain (9.1) and Gini-index gain (9.2) in classification and the expected change of variance (see Section 9.3) in regression are useless as they monotonously increase with the number of intervals. Appropriate measures are non-monotonic, such as MDL [14] and (R)ReliefF.

The main advantage of (R)ReliefF is its non-myopic behavior. Therefore, using ReliefF leads to a non-myopic discretization of numerical features. It was shown that conditionally dependent features may have important boundaries, which cannot be detected by myopic measures. The regressional version RReliefF can be used to discretize features in regression problems.

9.7.6   Association Rules and Genetic Algorithms

The use of ReliefF together with an association rules-based classifier [11] is also connected with feature subset selection. The adaptation of the algorithm to association rules changes the diff function in a similar way as in ILP (see Section 9.4.1).

In genetic algorithms, in order to speed up the learning process, one can define operators that are less random and take into account the importance of features. It was shown that ReliefF improves the efficiency of genetic searches by providing estimates of features that are then used to control the genetic operators [19].

9.7.7   Constructive Induction

In constructive induction one needs to develop new features from the existing ones describing the data. Due to the combinatorial explosion of possible combinations of features it is necessary to limit the search to the most promising subset of features. As (R)ReliefF implicitly detects dependencies between features, which are most important when constructing new features, it can be used to effectively guide the search in constructive induction. We employed (R)ReliefF algorithms to guide the constructive induction process during the growing of the tree models. Only the most promising features were selected for construction, and various operators were applied on them (conjunction, disjunction, summation, product). The results were good and in some domains the obtained constructs provided additional insight into the domain [2].

One of the most promising approaches to constructive induction is based on the function decomposition. Algorithm HINT [31] uses ReliefF to effectively guide the search through the space of all possible feature hierarchies.

9.8   Conclusion

ReliefF in classification and RReliefF in regression exploit the context of other features through distance measures and can detect highly conditionally dependent features. We have described the basic idea and showed the relation between myopic impurity measures and Relief. Then we extended Relief to a more realistic variant ReliefF, which is able to deal with incomplete data, with multi-class problems, and is robust with respect to noise. Afterwards, the basic idea was also extended to regressional problems and implemented in the Regressional ReliefF (RReliefF). The relation between Gini-index and variance is analogous to the relation between ReliefF and RReliefF. Various extensions of the (R)ReliefF family of algorithms, like the evaluation of literals in inductive logic programming, cost-sensitive feature evaluation with ReliefF, and the ordEval algorithm for the evaluation of features with ordered values, show the general applicability of the basic idea. The (R)ReliefF family of algorithms has been used in many different machine learning subproblems and applications. Besides, the comprehensive interpretability of (R)ReliefF's estimates makes it even more attractive. Although some authors claim that (R)ReliefF is computationally demanding, our discussion shows that this is not an inherent property and that efficient implementations exist.

References

[1] L. Breiman, J. H. Friedman, R. A. Olshen, and C. J. Stone. *Classification and Regression Trees*. Wadsforth International Group, 1984.

[2] A. Dalaka, B. Kompare, M. Robnik-Šikonja, and S. Sgardelis. Modeling the effects of environmental conditions on apparent photosynthesis of Stipa bromoides by machine learning tools. *Ecological Modelling*, 129:245–257, 2000.

[3] J. Demsar, G. Leban, and B. Zupan. Freeviz - an intelligent visualization approach for class-labeled multidimensional data sets. In J. Holmes and N. Peek, editors, *Proceedings of IDAMAP2005*, pages 61–66, 2005.

[4] J. Demsar, B. Zupan, M. Kattan, N. Aoki, and J. Beck. Orange and decisions-at-hand: Bridging medical data mining and decision support. In *Proceedings IDAMAP 2001*, pages 87–92, 2001.

[5] T. G. Dietterich. Machine learning research: Four current directions. *AI Magazine*, 18(4):97–136, 1997.

[6] C. Elkan. The foundations of cost-sensitive learning. In *Proc. of the Seventeenth Int. Joint Conf. on Artificaial Intelligence (IJCAI'01)*, 2001.

[7] M. A. Hall and G. Holmes. Benchmarking attribute selection techniques for discrete class data mining. *IEEE Trans. on Data and Knowledge Engineering*, 15(6):1437–1447, 2003.

[8] S. J. Hong. Use of contextual information for feature ranking and discretization. *IEEE Trans. on Knowledge and Data Engineering*, 9(5):718–730, 1997.

[9] R. Ihaka and R. Gentleman. R: A language for data analysis and graphics. *Journal of Computational and Graphical Statistics*, 5(3):299–314, 1996.

[10] Y. Jiang, J. Ling, G. Li, H. Dai, and Z. Zhou. Dependency bagging. In *Proc. Rough Sets, Fuzzy Sets, Data Mining, and Granular Computing*, pages 491–500. Springer, Berlin, 2005.

[11] V. Jovanoski and N. Lavrač. Feature subset selection in association rules

learning systems. In M. Grobelnik and D. Mladenič, editors, *Prooc. Analysis, Warehousing and Mining the Data*, pages 74–77, 1999.

[12] K. Kira and L. Rendell. A practical approach to feature selection. In D. Sleeman and P. Edwards, editors, *Int. Conf. on Machine Learning*, pages 249–256, Aberdeen, Scotland, Morgan Kaufmann, 1992.

[13] I. Kononenko. Estimating attributes: Analysis and extensions of RELIEF. In L. D. Raedt and F. Bergadano, editors, *European Conf. on Machine Learning*, pages 171–182, Catania, Italy, Springer Verlag, New York, 1994.

[14] I. Kononenko. On biases in estimating multivalued attributes. In *Proc. IJCAI-95*, pages 1034–1040, Montreal, August 20–25, 1995.

[15] I. Kononenko and M. Kukar. *Machine Learning and Data Mining: Introduction to Principles and Algorithms*. Horwood Publ., 2007.

[16] I. Kononenko, E. Šimec, and M. Robnik-Šikonja. Overcoming the myopia of inductive learning algorithms with RELIEFF. *Appl.Int.*, 7:39–55, 1997.

[17] J. Li, X. Gao, and L. Jiao. A new feature weighted fuzzy clustering algorithm. In *Rough Sets, Fuzzy Sets, Data Mining, and Granular Computing*, pages 412–420. Springer, Berlin, 2005.

[18] H. Liu, H. Motoda, and L. Yu. A selective sampling approach to active feature selection. *Artificial Intelligence*, 159(1-2):49–74, 2004.

[19] J. J. Liu and J. T.-Y. Kwok. An extended genetic rule induction algorithm. In *Evolutionary Computation, 2000. Proceedings of the 2000 Congress on*, pages 458–463, 2000. La Jolla, CA, July 16-19 2000.

[20] U. Pompe and I. Kononenko. Linear space induction in first order logic with ReliefF. In G. Riccia, R. Kruse, and R. Viertl, editors, *Mathematical and Statistical Methods in Artificial Intelligence*. Springer Verlag, New York, 1995.

[21] J. R. Quinlan. *C4.5: Programs for Machine Learning*. Morgan Kaufmann, 1993.

[22] M. Robnik Šikonja. Speeding up Relief algorithm with k-d trees. In F. Solina and B. Zajc, editors, *Proceedings of Electrotehnical and Computer Science Conference (ERK'98)*, pages B:137–140, 1998.

[23] M. Robnik Šikonja and I. Kononenko. An adaptation of Relief for attribute estimation in regression. In D. H. Fisher, editor, *Machine Learning: Proceedings of ICML'97*, pages 296–304, San Francisco, 1997.

[24] M. Robnik Šikonja and I. Kononenko. Attribute dependencies, understandability and split selection in tree based models. In I. Bratko and S. Džeroski, editors, *Machine Learning: ICML'99*, pages 344–353, 1999.

[25] M. Robnik-Šikonja. Experiments with cost-sensitive feature evaluation. In N. Lavrač et al., editor, *ECML2003*, pages 325–336, 2003.

[26] M. Robnik-Šikonja and I. Kononenko. Theoretical and empirical analysis of ReliefF and RReliefF. *Machine Learning Journal*, 53:23–69, 2003.

[27] M. Robnik-Šikonja and K. Vanhoof. Evaluation of ordinal attributes at value level. *Data Mining and Knowledge Discovery*, 14:225–243, 2007.

[28] R. Ruiz, J. C. Riquelme, and J. S. Aguilar-Ruiz. Fast feature ranking algorithm. In *Knowledge-Based Intelligent Information and Engineering Systems: KES 2003*, pages 325–331. Springer, Berlin, 2003.

[29] D. Wettschereck, D. W. Aha, and T. Mohri. A review and empirical evaluation of feature weighting methods for a class of lazy learning algorithms. *Artificial Intelligence Review*, 11:273–314, 1997.

[30] I. H. Witten and E. Frank. *Data Mining: Practical Machine Learning Tools and Techniques*. Morgan Kaufmann, 2nd edition, 2005.

[31] B. Zupan, M. Bohanec, J. Demsar, and I. Bratko. Learning by discovering concept hierarchies. *Artificial Intelligence*, 109(1-2):211–242, 1999.

Chapter 10

Weighting Method for Feature Selection in K-Means

Joshua Zhexue Huang

The University of Hong Kong

Jun Xu

The University of Hong Kong

Michael Ng

Hong Kong Baptist University

Yunming Ye

Harbin Institute of Technology, China

10.1   Introduction

The k-means type of clustering algorithms [13, 16] are widely used in real-world applications such as marketing research [12] and data mining due to their efficiency in processing large datasets. One unavoidable task of using k-means in real applications is to determine a set of features (or attributes). A common practice is to select features based on business domain knowledge and data exploration. This manual approach is difficult to use, time consuming, and frequently cannot make a right selection. An automated method is needed to solve the feature selection problem in k-means.

In this chapter, we introduce a recent development of the k-means algorithm that can automatically determine the important features in the k-means clustering process [14]. This new algorithm is called W-k-means. In this algorithm a new step is added to the standard k-means clustering process to calculate the feature weights from the current partition of data in each iteration. The weight of a feature is determined by the sum of the within-cluster dispersions of the feature. The larger the sum, the smaller the feature weight. The weights produced by the W-k-means algorithm measure the importance of the corresponding features in clustering. The small weights reduce or eliminate the effect of insignificant (or noisy) features. Therefore, the feature weights can be used in feature selection. Since the k-means clustering process is not fundamentally changed in W-k-means, the efficiency and convergency of the clustering process remain.

A further extension of this approach is to calculate a weight for each feature in each cluster [4]. This is called subspace k-means clustering because the important features in each cluster identify the subspace in which the cluster is discovered. Since the subsets of important features are different in different clusters, subspace clustering is achieved. Subspace clustering has wide applications in text clustering, bio-informatics, and customer behavior analysis, where high-dimensional data are involved. In this chapter, subspace k-means clustering is also discussed.

10.2   Feature Weighting in k-Means

Given a dataset X with M records and N features, the k-means clustering algorithm [16] searches for a partition of X into k clusters that minimizes the sum of the within-cluster dispersions of all features. The clustering process is conducted as follows:

1. Randomly select k distinct records as the initial cluster centers.

2. For each record in X, calculate the distances between the record and each cluster center, and assign the record to the cluster with the shortest distance.

3. Repeat the above step until all records have been assigned to clusters. For each cluster, compute a new cluster center as the mean (average) of the feature values.

4. Compare the new cluster centers with the previous centers. If the new centers are the same as the previous centers, stop the clustering process; otherwise, go back to Step 2.

In the above standard k-means clustering process, all features are treated the same in the calculation of the distances between the data records and the cluster centers. The importance of different features is not distinguishable. The formal presentation of the k-means clustering algorithm can be found in [13].

To identify the importance of different features, a weight can be assigned to each feature in the distance calculation. As such, the feature with a large weight will have more impact on determining the cluster a record is assigned to. Since the importance of a feature is determined by its distribution in the dataset, the feature weights are data dependent.

To automatically determine the feature weights, we add one step to the standard k-means clustering process to calculate the feature weights from the current partition of the data in each iteration. During the clustering process, weights are updated automatically until the clustering process converges. Then, the final weights of the features can indicate which features are important in clustering the data and which are not.

Formally, the process is to minimize the following objective function:

$$P(U, Z, W) = \sum_{l=1}^{k} \sum_{i=1}^{M} \sum_{j=1}^{N} u_{i,l} w_j^{\beta} d(x_{i,j}, z_{l,j}) \qquad (10.1)$$

subject to

$$\begin{cases} \sum_{l=1}^{k} u_{i,l} = 1, \ \ 1 \leq i \leq M \\ u_{i,l} \in \{0,1\}, 1 \leq i \leq M, 1 \leq l \leq k \\ \sum_{j=1}^{N} w_j = 1, \ \ 0 \leq w_j \leq 1 \end{cases} \qquad (10.2)$$

where

- U is an $M \times k$ partition matrix, $u_{i,l}$ is a binary variable, and $u_{i,l} = 1$ indicates that record i is allocated to cluster l.

- $Z = \{Z_1, Z_2, ..., Z_k\}$ is a set of k vectors representing the k-cluster centers.

- $W = [w_1, w_2, ..., w_N]$ is a set of weights.

- $d(x_{i,j}, z_{l,j})$ is a distance or dissimilarity measure between object i and the center of cluster l on the jth feature. If the feature is numeric, then

$$d(x_{i,j}, z_{l,j}) = (x_{i,j} - z_{l,j})^2 \qquad (10.3)$$

If the feature is categorical, then

$$d(x_{i,j}, z_{l,j}) = \begin{cases} 0 \ (x_{i,j} = z_{l,j}) \\ 1 \ (x_{i,j} \neq z_{l,j}) \end{cases} \qquad (10.4)$$

- β is a parameter.

The above optimization problem can be solved by iteratively solving the following three minimization problems:

1. P_1: Fix $Z = \hat{Z}$ and $W = \hat{W}$; solve the reduced problem $P(U, \hat{Z}, \hat{W})$.

2. P_2: Fix $U = \hat{U}$ and $W = \hat{W}$; solve the reduced problem $P(\hat{U}, Z, \hat{W})$.

3. P_3: Fix $U = \hat{U}$ and $Z = \hat{Z}$; solve the reduced problem $P(\hat{U}, \hat{Z}, W)$.

P_1 is solved by

$$\begin{cases} u_{i,l} = 1 \text{ if } \sum_{j=1}^{N} w_j^{\beta} d(x_{i,j}, z_{l,j}) \leq \sum_{j=1}^{N} w_j^{\beta} d(x_{i,j}, z_{t,j}) \text{ for } 1 \leq t \leq k \\ u_{i,t} = 0 \text{ for } t \neq l \end{cases} \tag{10.5}$$

and P_2 is solved for the numeric features by

$$z_{l,j} = \frac{\sum_{i=1}^{M} u_{i,l}\, x_{i,j}}{\sum_{i=1}^{M} u_{i,l}} \qquad \text{for } 1 \leq l \leq k \text{ and } 1 \leq j \leq N \tag{10.6}$$

If the feature is categorical, then

$$z_{l,j} = a_j^r \tag{10.7}$$

where a_j^r is the mode of the feature values in cluster l [13].

The solution to P_3 is given in the following theorem.

Theorem 1. Let $U = \hat{U}$ and $Z = \hat{Z}$ be fixed,
(i) When $\beta > 1$ or $\beta \leq 0$, $P(\hat{U}, \hat{Z}, W)$ is minimized iff

$$\hat{w}_j = \begin{cases} 0 & \text{if } D_j = 0 \\ \dfrac{1}{\sum\limits_{t=1}^{h} \left[\dfrac{D_j}{D_t}\right]^{\frac{1}{\beta-1}}} & \text{if } D_j \neq 0 \end{cases} \tag{10.8}$$

where

$$D_j = \sum_{l=1}^{k} \sum_{i=1}^{M} \hat{u}_{i,l} d(x_{i,j}, z_{l,j}) \tag{10.9}$$

and h is the number of features with $D_j \neq 0$.
(ii) When $\beta = 1$, $P(\hat{U}, \hat{Z}, W)$ is minimized iff

$$\hat{w}_{j'} = 1 \quad \text{and} \quad \hat{w}_j = 0, \quad j \neq j'$$

where $D_{j'} \leq D_j$ for all j.

The proof is given in [14].

Theorem 1 shows that, given a data partition, a larger weight is assigned to a feature with a smaller sum of the within-cluster dispersions and a smaller weight to a feature with a larger sum of the within-cluster dispersions. Therefore, the feature weight is reversely proportional to the sum of the within-cluster dispersions of the feature.

The real weight w_j^β of feature x_j in the distance calculation (see (1.5)) is also dependent on the value of β. In using W-k-means, we can choose either $\beta < 0$ or $\beta > 1$ for the following reasons:

- When $\beta = 0$, W-k-means is equivalent to k-means.

- When $\beta = 1$, w_j is equal to 1 for the smallest value of D_j. The other weights are equal to 0. Although the objective function is minimized, the clustering is made by the selection of one variable. It may not be desirable for high-dimensional clustering problems.

- When $0 < \beta < 1$, the larger D_j, the larger w_j, and similarly for w_j^β. This is against the variable weighting principal, so we cannot choose $0 < \beta < 1$.

- When $\beta > 1$, the larger D_j, the smaller w_j and the smaller w_j^β. The effect of variable x_j with large D_j is reduced.

- When $\beta < 0$, the larger D_j, the larger w_j. However, w_j^β becomes smaller and has less weighting to the variable in the distance calculation because of negative β.

10.3 W-k-Means Clustering Algorithm

The algorithm to solve (10.1) is an extension to the standard k-means algorithm [13, 21].

Algorithm - (The W-k-means algorithm)

Step 1. Randomly choose an initial $Z^0 = \{Z_1, Z_2, ..., Z_k\}$ and randomly generate a set of initial weights $W^0 = [w_1^0, w_2^0, ..., w_N^0]$ ($\sum_{j=1}^{N} w_j = 1$). Determine U^0 such that $P(U^0, Z^0, W^0)$ is minimized. Set $t = 0$;

Step 2. Let $\hat{Z} = Z^t$ and $\hat{W} = W^t$, solve problem $P(U, \hat{Z}, \hat{W})$ to obtain U^{t+1}. If $P(U^{t+1}, \hat{Z}, \hat{W}) = P(U^t, \hat{Z}, \hat{W})$, output $(U^t, \hat{Z}, \hat{W})$ and stop; otherwise, go to Step 3;

Step 3. Let $\hat{U} = U^{t+1}$ and $\hat{W} = W^t$, solve problem $P(\hat{U}, Z, \hat{W})$ to obtain Z^{t+1}. If $P(\hat{U}, Z^{t+1}, \hat{W}) = P(\hat{U}, Z^t, \hat{W})$, output $(\hat{U}, Z^t, \hat{W})$ and stop; otherwise, go to Step 4;

Step 4. Let $\hat{U} = U^{t+1}$ and $\hat{Z} = Z^{t+1}$, solve problem $P(\hat{U}, \hat{Z}, W)$ to obtain W^{t+1}. If $P(\hat{U}, \hat{Z}, W^{t+1}) = P(\hat{U}, \hat{Z}, W^t)$, output $(\hat{U}, \hat{Z}, W^t)$ and stop; otherwise, set $t = t + 1$ and go to Step 2.

Theorem 2. The above algorithm converges to a local minimal solution in a finite number of iterations.

The proof is given in [14].

Since the W-k-means algorithm is an extension to the k-means algorithm by adding a new step to calculate the variable weights in the iterative process, it does not seriously affect the scalability in clustering large data; therefore, it is suitable for data mining applications. The computational complexity of the algorithm is $O(tNMk)$, where t is the total number of iterations required for performing Step 2, Step 3 and Step 4; k is the number of clusters; N is the number of features; and M is the number of records.

10.4   Feature Selection

One of the drawbacks of the standard k-means algorithm is that it treats all features equally when deciding the cluster memberships. This is not desirable if the data contain a large number of diverse features. A cluster structure in a high-dimensional dataset is often confined to a subset of features rather than the entire feature set. Inclusion of all features can only obscure the discovery of the cluster structure.

The W-k-means clustering algorithm can be used to select the subset of features for clustering in real-world applications. In doing so, the clustering work can be divided in the following steps. The first step is to use W-k-means to cluster the dataset or a sample of the dataset to produce a set of weights. The second step is to select a subset of features according to the weight values and remove the unselected features from the dataset. The third step is to use W-k-means or another clustering algorithm to cluster the dataset to produce the final clustering result.

Figure 10.1 shows a dataset with three features (x_1, x_2, x_3) and two clusters in the subset of features (x_1, x_2). Feature x_3 is noise in a uniform distribution. We can see the two clusters in the plot of Figure 10.1(a) but cannot see any cluster structure in the plots of Figure 10.1(b) and Figure 10.1(c). If we did not know that the two clusters were existing in the subset of features (x_1, x_2), we would find it difficult to discover them from the dataset using the standard k-means algorithm. However, we can use W-k-means to cluster this dataset and obtain the weights of the three features as 0.47, 0.40, and 0.13, respectively. From these weights, we can easily identify the first two features (x_1, x_2) as important features. After removing the data of feature x_3, we can run the standard k-means algorithm to discover the two clusters from the subset of the features (x_1, x_2) as shown in Figure 10.1(d).

In fact, we can get the final result of Figure 10.1(d) directly from the first run of W-k-means in this simple example. Real datasets often have features in the hundreds and records in the hundreds of thousands, such as the customer datasets in large banks. In such situations, several runs of W-k-means are needed to identify the subset of important features.

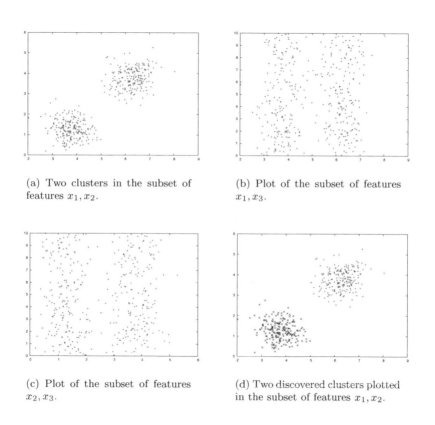

(a) Two clusters in the subset of features x_1, x_2.

(b) Plot of the subset of features x_1, x_3.

(c) Plot of the subset of features x_2, x_3.

(d) Two discovered clusters plotted in the subset of features x_1, x_2.

FIGURE 10.1: Feature selection from noise data.

10.5   Subspace Clustering with k-Means

Subspace clustering refers to the process of identifying clusters from subspaces of data, with each subspace being defined by a subset of features. Different clusters are identified from different subspaces of data. Subspace clustering is required when clustering high-dimensional data such as those in text mining, bio-informatics, and e-commerce.

Subspace clustering can be achieved by feature weighting in k-means. Instead of assigning a weight to each feature for the entire dataset, we assign a weight to each feature in each cluster. As such, if there are N features and k clusters, we will obtain $N \times k$ weights. This is achieved by rewriting the objective function (10.1) as follows:

$$P(U, Z, W) = \sum_{l=1}^{k}\sum_{i=1}^{M}\sum_{j=1}^{N} u_{i,l} w_{l,j}^{\beta} d(x_{i,j}, z_{l,j}) \qquad (10.10)$$

subject to

$$\begin{cases} \sum_{l=1}^{k} u_{i,l} = 1, \ \ 1 \le i \le M \\ u_{i,l} \in \{0,1\}, 1 \le i \le M, \ 1 \le l \le k \\ \sum_{j=1}^{N} w_{lj} = 1, \ \ 0 \le w_{lj} \le 1 \end{cases} \qquad (10.11)$$

where W is a $k \times N$ weight matrix and the other notations are the same as in (10.1).

In a similar fashion, (10.10) can be reduced to three subproblems that are solved iteratively.

The subproblem P_1 is solved by

$$\begin{cases} u_{i,l} = 1 \text{ if } \sum_{j=1}^{N} w_{lj}^{\beta} d(x_{i,j}, z_{l,j}) \le \sum_{j=1}^{N} w_{lj}^{\beta} d(x_{i,j}, z_{t,j}) \text{ for } 1 \le t \le k \\ u_{i,t} = 0 \text{ for } t \ne l \end{cases} \qquad (10.12)$$

The subproblem P_2 is solved with (10.6) or (10.7), depending on the data types.

The solution to the subproblem P_3 is given in the following theorem.

Theorem 3. Let $U = \hat{U}$ and $Z = \hat{Z}$ be fixed. When $\beta > 1$ or $\beta \le 0$, $P(\hat{U}, \hat{Z}, W)$ is minimized iff

$$\hat{w}_{lj} = \frac{1}{\sum\limits_{t=1}^{N} \left[\frac{D_{lj}}{D_{lt}}\right]^{\frac{1}{\beta-1}}} \tag{10.13}$$

where

$$D_{lj} = \sum_{i=1}^{M} \hat{u}_{i,l} d(x_{i,j}, z_{l,j}) \tag{10.14}$$

and N is the number of features with $D_{lj} > 0$.

In subspace clustering, if $D_{lj} = 0$, we cannot simply assign a weight 0 to feature j in cluster l. $D_{lj} = 0$ means all values of feature j are the same in cluster l. In fact, $D_{lj} = 0$ indicates that feature j may be an important feature in identifying cluster l. $D_{lj} = 0$ often occurs in real-world data such as text data and supplier transaction data. To solve this problem, we can simply add a small constant σ to the distance function to make $\hat{w}_{lj}$ always computable, i.e.,

$$D_{lj} = \sum_{i=1}^{M} \hat{u}_{i,l} (d(x_{i,j}, z_{l,j}) + \sigma) \tag{10.15}$$

In practice, σ can be chosen as the average dispersion of all features in the dataset. It can be proved that the subspace k-means clustering process converges [4].

10.6 Text Clustering

A typical application of subspace clustering is text mining. In text clustering, text data are usually represented in the vector space model (VSM). A set of documents is converted to a matrix where each row indicates a document and each column represents a term or word in the vocabulary of the document set. Table 10.1 is a simplified example of text data representation in VSM. Each column corresponds to a term and each line represents a document. Each entry value is the frequency of the corresponding term in the related document.

If a set of text documents contains several classes, the documents related to a particular class, for instance *sport*, are categorized by a particular subset of terms, corresponding to a subspace of the vocabulary space. Different document classes are categorized by different subsets of terms, i.e., different subspaces. For example, the subset of terms describing the *sport* class is different from the subset of terms describing the *music* class. As such, k-means subspace clustering becomes useful for text data because different clusters can be identified from different subspaces through the weights of the terms.

TABLE 10.1: A simple example
of text representation.

	t_0	t_1	t_2	t_3	t_4
x_0	1	2	3	0	6
x_1	2	3	1	0	6
x_2	3	1	2	0	6
x_3	0	0	1	3	2
x_4	0	0	2	1	3
x_5	0	0	3	2	1

TABLE 10.2: Summary of the six text datasets.

Dataset	Source	n_d	Dataset	Source	n_d
A2	alt.atheism	100	B2	talk.politics.mideast	100
	comp.graphics	100		talk.politics.misc	100
A4	comp.graphics	100	B4	comp.graphics	100
	rec.sport.baseball	100		comp.os.ms-windows	100
	sci.space	100		rec.autos	100
	talk.politics.mideast	100		sci.electronics	100
A4-U	comp.graphics	120	B4-U	comp.graphics	120
	rec.sport.baseball	100		comp.os.ms-windows	100
	sci.space	59		rec.autos	59
	talk.politics.mideast	20		sci.electronics	20

Besides, the weights can also be used to select the key words for semantic
representations of clusters.

10.6.1   Text Data and Subspace Clustering

Table 10.2 lists the six datasets built from the popular *20-Newsgroups*
collection.[1] The six datasets have different characteristics in sparsity, dimensionality, and class distribution. The classes and the number of documents
in each class are given in the columns "Source" and "n_d." The classes in the
datasets A2 and A4 are semantically apart, while the classes in the datasets
B2 and B4 are semantically close. Semantically close classes have more overlapping words. The number of documents in the datasets A4-U and B4-U are
different, indicating unbalanced class distributions.

These datasets were preprocessed using the *Bow* toolkit.[2] The preprocessing
steps included removing the headers, the stop words, and the words that
occurred in less than three documents or greater than the average number of
documents in each class, as well as stemming the remaining words with the
Porter stemming function. The standard *tf · idf* term weighting was used to
represent the document vector.

Table 10.3 shows the comparisons of accuracy in clustering these datasets
with the subspace *k*-means, the standard *k*-means, and four subspace clustering algorithms: PROCLUS [1], HARP [23], COSA [10], and LAC [8]. The

TABLE 10.3:   Comparisons of accuracies of the subspace k-means with the standard k-mean and other four subspace clustering algorithms.

	A2	B2	A4	B4	A4-U	B4-U
Subspace k-means	0.9599	0.9043	0.9003	0.8631	0.9591	0.9205
Standard k-means	0.895	0.735	0.6	0.5689	0.95	0.8729
PROCLUS	0.7190	0.6604	0.6450	0.4911	0.5239	0.5739
HARP	0.8894	0.6020	0.5073	0.3840	0.4819	0.3364
COSA	0.5781	0.5413	0.3152	0.3621	0.4159	0.3599
LAC	0.9037	0.7981	0.6721	0.5816	0.9473	0.7363

	weight intervals	word number
0~1:	(0,1e-08]	8
1~2:	(1e-08,1e-07]	280
2~3:	(1e-07,1e-06]	433
3~4:	(1e-06,1e-05]	188
4~5:	(1e-05,1)	32

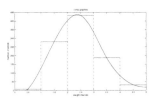

FIGURE 10.2: Distribution of words in different ranges of weights.

accuracy is calculated as the number of correctly classified documents divided by the total number of documents in a dataset. We can see that the subspace k-means performed better than the standard k-means and the other four subspace clustering algorithms on all datasets. This is due to the subspace nature of the text clusters, so the subspace k-means is more suitable in text clustering.

10.6.2   Selection of Key Words

Another advantage of using the subspace k-means in text clustering is that the weights produced by the algorithm can be used to identify the important terms or words in each cluster. The higher the weight value in a cluster, the more important the term feature in discovering the cluster. We can divide the range of the weight values into intervals and plot the distribution of features against the weight intervals as shown in Figure 10.2.

After the distribution is obtained, we remove the terms with the extremely large weights because they correspond to the terms with zero frequency in the cluster, i.e., the term did not occur in the cluster. Few terms with the extremely large weights correspond to the terms with equal frequency in each document of the cluster. Such terms can be easily identified in postprocessing.

Taking dataset B4 as an example, after preprocessing we got 1,322 feature words. We used the subspace k-means to cluster it into four clusters. Each cluster has more than 300 words with zero frequency. These words were removed from the clusters.

Figure 10.2 shows distribution of the remaining words in cluster *Computer*

Graphics of the dataset B4 against the weight intervals. Since we limit the sum of the weights for all features in a cluster to 1, the weights for most words are relatively small. Using a weight threshold, we identified 220 words with relatively larger weights. This is less than 17% of the total words. These are the words categorizing the cluster. From these words, we need to identify a few that will enable use to interpret the cluster.

Figure 10.3 show the plots of the term weights in four clusters. The horizontal axis is the index of the 220 words and the vertical lines indicate the values of the weights. We can observe that each cluster has its own subset of key words because the lines do not have big overlaps in different clusters. The classes *Computer Graphics* and *Microsoft Windows* overlap a little, which indicates that the semantics of the two classes are close to each other. Similarly, the classes *Autos* and *Electronics* are close.

We extracted 10 words from each cluster, which had the largest weights and were nouns. They are listed on the right side in Figure 10.3. We can see that these noun words indeed represent the semantic meaning of the clusters. For example, the words *graphic, color, image*, and *point* are good descriptions of the cluster *Computer Graphics*. Comparing the word distribution on the left, these words are identifiable from their large weight values. This shows that the weights, together with the word function, are useful in selecting the key words for representing the meanings of clusters.

We can also observe that some words have large weights in more than one cluster. For example, the word *request* has large weight values in two classes, *Computer Graphics* and *Microsoft Windows*. Such words indicate that the two classes are semantically close.

10.7   Related Work

Feature selection has been an important research topic in cluster analysis [5, 6, 7, 9, 10, 11, 12, 17, 18, 19, 20].

Desarbo et al. [7] introduced the first method for variable weighting in k-means clustering in the SYNCLUS algorithm. The SYNCLUS process is divided into two stages. Starting from an initial set of weights, SYNCLUS first uses the k-means clustering to partition the data into k clusters. It then estimates a new set of optimal weights by optimizing a weighted mean-square, stress-like cost function. The two stages iterate until they converge to an optimal set of weights. The algorithm is time consuming computationally [12], so it cannot process large datasets.

De Soete [5, 6] proposed a method to find optimal variable weights for ultrametric and additive tree fitting. This method was used in hierarchical clustering methods to solve variable weighting problems. Since the hierar-

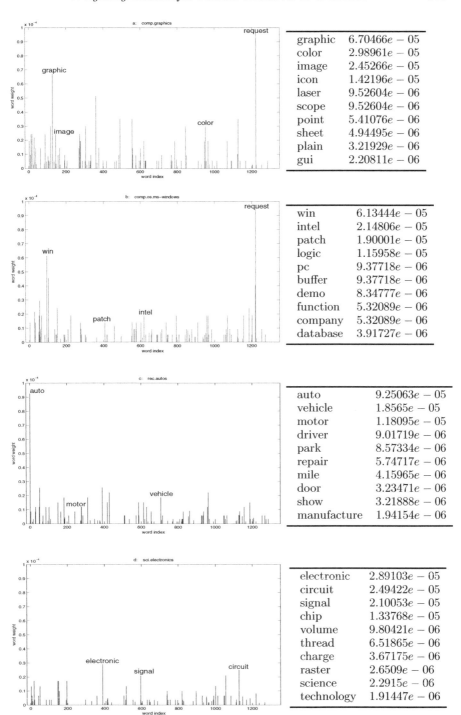

FIGURE 10.3: The noun words with large weights extracted from each cluster of the dataset B4. We can see that these words indeed represent the semantic meaning of the corresponding clusters.

chical clustering methods are computationally complex, De Soete's method cannot handle large datasets. Makarenkov and Legendre [18] extended De Soete's method to optimal variable weighting for k-means clustering. The basic idea is to assign each variable a weight w_i in calculating the distance between two objects and find the optimal weights by optimizing the cost function $L_p(w_1, w_2, ..., w_p) = \sum_{k=1}^{K} (\sum_{i,j=1}^{n_k} d_{ij}^2 / n_k)$. Here, K is the number of clusters, n_k is the number of objects in the kth cluster, and d_{ij} is the distance between the ith and the jth objects. The Polak-Ribiere optimization procedure is used in minimization, which makes the algorithm very slow. The simulation results in [18] show that the method is effective in identifying important variables but not scalable to large datasets.

Modha and Spangler [20] very recently published a new method for variable weighting in k-means clustering. This method aims to optimize variable weights in order to obtain the best clustering by minimizing the ratio of the average within-cluster distortion over the average between-cluster distortion, referred to as the generalized Fisher ratio Q. To find the minimal Q, a set of feasible weight groups was defined. For each weight group, the k-means algorithm was used to generate a data partition and Q was calculated from the partition. The final clustering was determined as the partition having the minimal Q. This method of finding optimal weights from a predefined set of variable weights may not guarantee that the predefined set of weights would contain the optimal weights. Besides, it is also a practical problem to determine the predefined set of weights for high-dimensional data.

Friedman and Meulman [10] recently published a method to cluster objects on subsets of attributes. Instead of assigning a weight to each variable for the entire dataset, their approach is to compute a weight for each variable in each cluster. As such, $p * L$ weights are computed in the optimization process, where p is the total number of variables and L is the number of clusters. Since the objective function is a complicated, highly non-convex function, a direct method to minimize it has not been found. An approximation method is used to find clusters on different subsets of variables by combining conventional distance-based clustering methods with a particular distance measure. Friedman and Meulman's work is related to the problem of subspace clustering [3]. Scalability is a concern because their approximation method is based on the hierarchical clustering methods.

Projected clustering is another method for feature selection of high-dimensional data. *PROCLUS* is the first algorithm [1]. It starts with a set of initial cluster centers discovered from a small data sample. The initial centers are made as far apart from each other as possible. For each center, a set of data points within a distance δ to the center is identified as the center *locality* L_i. Here, δ is the minimal distance between the center and other centers. For each L_i, the average distance between the points in L_i and the center is computed in each dimension. The subset of dimensions whose average distances are smaller than the average distance of all dimensions is considered as the candidate subspace for cluster i. After all candidate subspaces are identified, the clusters are discovered from the subspaces using the distance measures on subsets of dimensions. A few extensions have been made recently [2, 15, 22].

10.8   Discussions

k-means clustering is an important technique in data mining and many other real-world applications. In current practice, when using k-means, feature selection is either done manually using business domain knowledge or carried out in separate steps using statistical methods or data exploration. This is time consuming and difficult to make a right selection. Automated feature selection by feature weighting within the clustering process provides an easy solution. When handling very large data, a sample can be first clustered and features with large weights selected as the dimensions for clustering the whole dataset. Since the k-means clustering process is not changed much, this k-means feature weighting algorithm is efficient in clustering large data. Comparatively, other feature weighting methods for clustering as mentioned in the previous section are not scalable to large data.

Subspace clusters in high-dimensional data is a common phenomenon in many real-world applications, such as text mining, bio-informatics, e-business, supply chain management, and production scheduling/planning in manufacturing. In this chapter, we have demonstrated that the featuring weighting method in k-means can be extended to subspace clustering and the experimental results on text data are satisfactory. However, some further research problems remain. One is how to specify parameters β and σ when using this algorithm. To understand this, a sensitivity study needs to be conducted. The other one is a well-known problem: how to specify k, the number of clusters. To investigate this problem, a subspace cluster validation method needs to be developed. In the next step, we will work on solutions to these problems.

Acknowledgment

Michael Ng and Yunming Ye's work was supported by the National Natural Science Foundation of China (NSFC) under grant No.60603066.

Notes

1 http://kdd.ics.uci.edu/databases/20newsgroups/20newsgroups.html.

2 http://www.cs.cmu.edu/mccallum/bow.

References

[1] C. Aggarwal, C. Procopiuc, J. Wolf, P. Yu, and J. Park. Fast algorithms for projected clustering. In *Proc. of ACM SIGMOD*, pages 61–72, 1999.

[2] C. Aggarwal and P. Yu. Finding generalized projected clusters in high dimensional spaces. In *Proc. of ACM SIGMOD*, pages 70–81, 2000.

[3] R. Agrawal, J. Gehrke, D. Gunopulos, and P. Raghavan. Automatic subspace clustering of high dimensional data for data mining applications. In *Proc. of ACM SIGMOD*, pages 94–105, 1998.

[4] Y. Chan, W. K. Ching, M. K. Ng, and J. Z. Huang. An optimization algorithm for clustering using weighted dissimilarity measures. *Pattern Recognition*, 37(5):943–952, 2004.

[5] G. De Soete. Optimal variable weighting for ultrametric and additive tree clustering. *Quality and Quantity*, 20(3):169–180, 1986.

[6] G. De Soete. OVWTRE: A program for optimal variable weighting for ultrametric and addtive tree fitting. *Journal of Classification*, 5(1):101–104, 1988.

[7] W. S. Desarbo, J. D. Carroll, L. A. Clark, and P. E. Green. Synthesized clustering: A method for amalgamating clustering bases with differential weighting variables. *Psychometrika*, 49(1):57–78, 1984.

[8] C. Domeniconi, D. Papadopoulos, D. Gunopulos, and S. Ma. Subspace clustering of high dimensional data. In *Proc. of SIAM International Conference on Data Mining*, 2004.

[9] E. Fowlkes, R. Gnanadesikan, and J. Kettenring. Variable selection in clustering. *Journal of Classification*, 5(2):205–228, 1988.

[10] J. H. Friedman and J. J. Meulman. Clustering objects on subsets of attributes with discussion. *Journal of the Royal Statistical Society: Series B*, 66(4):815–849, 2004.

[11] R. Gnanadesikan, J. Kettenring, and S. Tsao. Weighting and selection of variables for cluster analysis. *Journal of Classification*, 12(1):113–136, 1995.

[12] P. E. Green, J. Carmone, and J. Kim. A preliminary study of optimal variable weighting in k-means clustering. *Journal of Classification*, 7(2):271–285, 1990.

[13] Z. Huang. Extensions to the k-means algorithms for clustering large data sets with categorical values. *Data Ming and Knowledge Discovery*, 2(3):283–304, 1998.

[14] Z. Huang, M. K. Ng, H. Rong, and Z. Li. Automated variable weighting in k-means type clustering. *IEEE Transactions on Pattern Analayis and Machine Intelligence*, 27(5):657–668, 2005.

[15] M. L. Liu. Iterative projected clustering by subspace mining. *IEEE Transactions on Knowledge and Data Engineering*, 17(2):176–189, 2005.

[16] J. MacQueen. Some methods for classification and analysis of multivariate observation. In *Proc. of the 5th Berkeley Symposium on Mathematical Statistica and Probability*, pages 281–297, 1967.

[17] V. Makarenkov and P. Leclerc. An algorithm for the fitting of a tree metric according to a weighted least-squares criterion. *Journal of Classification*, 16(1):3–26, 1999.

[18] V. Makarenkov and P. Leclerc. Optimal variable weighting for ultrametric and additive trees and k-means partitioning: methods and software. *Journal of Classification*, 18(2):245–271, 2001.

[19] G. Milligan. A validation study of a variable weighting algorithm for cluster analysis. *Journal of Classification*, 6(1):53–71, 1989.

[20] D. S. Modha and W. S. Spangler. Feature weighting in k-means clustering. *Machine Learning*, 52(3):217–237, 2003.

[21] S. Selim and M. Ismail. K-means-type algorithms: a generalized convergence theorem and characterization of local optimality. *IEEE Transactions on Pattern Analysis and Machine Intelligence*, 6(1):81–87, 1984.

[22] J. Yang, W. Wang, H. Wang, and P. Yu. δ-clusters: capturing subspace correlation in a large data set. In *Proc. of ICDE*, pages 517–528, 2002.

[23] K. Y. Yip, D. W. Cheung, and M. K. Ng. A practical projected clustering algorithm. *IEEE Transactions on knowledge and data engineering*, 16(11):1387–1397, 2004.

Chapter 11

Local Feature Selection for Classification

Carlotta Domeniconi

George Mason University

Dimitrios Gunopulos

University of California Riverside

11.1 Introduction

In a classification problem, we are given C classes and M training observations. The training observations consist of N feature measurements $\mathbf{x} = (x_1, \cdots, x_N)^T \in \Re^N$ and the known class labels $y = 1, \ldots, C$. The goal is to predict the class label of a given query $\mathbf{x}_0$.

The K nearest neighbor classification method [10, 14] is a simple and appealing approach to this problem: It finds the K nearest neighbors of $\mathbf{x}_0$ in the training set, and then predicts the class label of $\mathbf{x}_0$ as the most frequent one occurring in the K neighbors. Such a method produces continuous and overlapping, rather than fixed, neighborhoods and uses a different neighborhood for each individual query so that all points in the neighborhood are close to the query, to the extent possible. It is based on the assumption of smoothness of the target functions, which translates to locally constant class posterior probabilities for a classification problem. That is, $f_j(\mathbf{x}+\delta\mathbf{x}) \simeq f_j(\mathbf{x})$ for $||\delta\mathbf{x}||$ small enough, where $\{f_j(\mathbf{x})\}_{j=1}^C = \{P(j|\mathbf{x})\}_{j=1}^C$. Then,

$$f_j(\mathbf{x}_0) \simeq \frac{1}{|N(\mathbf{x}_0)|} \sum_{\mathbf{x}\in N(\mathbf{x}_0)} f_j(\mathbf{x}) \tag{11.1}$$

where $N(\mathbf{x}_0)$ is a neighborhood of $\mathbf{x}_0$ that contains points $\mathbf{x}$ in the N-

dimensional space that are "close" to $\mathbf{x}_0$. $|N(\mathbf{x}_0)|$ denotes the number of points in $N(\mathbf{x}_0)$. Given the training data $\{(\mathbf{x}_n, y_n)\}_{n=1}^M$, this motivates the estimates

$$\hat{f}(j|\mathbf{x}_0) = \frac{\sum_{n=1}^{M} 1(\mathbf{x}_n \in N(\mathbf{x}_0))1(y_n = j)}{\sum_{n=1}^{M} 1(\mathbf{x}_n \in N(\mathbf{x}_0))} \qquad (11.2)$$

where $1(\cdot)$ is an indicator function such that it returns 1 when its argument is true, and 0 otherwise.

A particular nearest neighbor method is defined by how the neighborhood $N(\mathbf{x}_0)$ is specified. K nearest neighbor methods (K-NN) define the region at $\mathbf{x}_0$ to be the one that contains exactly the K closest training points to $\mathbf{x}_0$ according to a p-norm distance metric on the Euclidean space of the input measurement variables:

$$D_p(\mathbf{x}_0, \mathbf{x}) = \{\sum_{i=1}^{N} |[W(\mathbf{x}_0)(\mathbf{x}_0 - \mathbf{x})]_i|^p\}^{1/p} \qquad (11.3)$$

The resulting neighborhood is determined by the value of K and by the choice of the distance measure, which in turn depends on a norm $p > 0$ and a metric defined by the matrix $W(\mathbf{x}_0) \in \Re^{N \times N}$.

The K nearest neighbor method has nice asymptotic properties. In particular, it has been shown [5] that the one nearest neighbor (1-NN) rule has an asymptotic error rate that is at most twice the Bayes error rate, independent of the distance metric used. The nearest neighbor rule becomes less appealing with finite training samples, however. This is due to the curse of dimensionality [4]. Severe bias can be introduced in the nearest neighbor rule in a high-dimensional input feature space with finite samples. As such, the choice of a distance measure becomes crucial in determining the outcome of a nearest neighbor classification. The commonly used Euclidean distance measure, while simple computationally, implies that the input space is isotropic or homogeneous. However, the assumption for isotropy is often invalid and generally undesirable in many practical applications. Figure 11.1 illustrates a case in point, where class boundaries are parallel to the coordinate axes. For query $\mathbf{a}$, dimension X is more relevant, because a slight move along the X axis may change the class label, while for query $\mathbf{b}$, dimension Y is more relevant. For query $\mathbf{c}$, however, both dimensions are equally relevant. This implies that distance computation does not vary with equal strength or in the same proportion in all directions in the feature space emanating from the input query. Capturing such information, therefore, is of great importance to any classification procedure in high-dimensional settings.

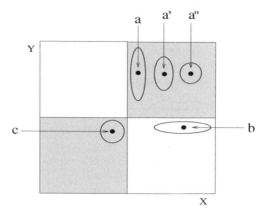

FIGURE 11.1: Feature relevance varies with query locations.

11.2 The Curse of Dimensionality

Related to the question of rate of convergence of the 1-NN rule is the one on how well the rule works in finite-sample settings. The asymptotic results rely on the fact that the bias of the estimate of each $f_j(\mathbf{x})$,

$$bias \hat{f}_j(\mathbf{x}) = f_j(\mathbf{x}) - E[\hat{f}_j(\mathbf{x})] \qquad (11.4)$$

becomes arbitrarily small. This is because the region $N(\mathbf{x}_0)$ will only contain training points $\mathbf{x}$ arbitrarily close to $\mathbf{x}_0$ (provided that $f_j(\mathbf{x})$ is continuous at $\mathbf{x}_0$ and $K/M \to 0$). In a finite setting, if the number of training data M is large and the number of input features N is small, then the asymptotic results may still be valid. However, for a moderate to large number of input variables, the sample size required for their validity is usually beyond feasibility.

This phenomenon is known as the *curse of dimensionality* [4]. It refers to the fact that in high-dimensional spaces data become extremely sparse and are far apart from each other. To get a quantitive idea of this phenomenon, consider a random sample of size M drawn from a uniform distribution in the N-dimensional unit hypercube. The expected diameter of a $K = 1$ neighborhood using Euclidean distance is proportional to $M^{-1/N}$, which means that for a given N, the diameter of the neighborhood containing the closest training point shrinks as $M^{-1/N}$ for increasing M. Table 11.1 shows the length d of the diameter for various values of N and M. For example, for $N = 20$, if $M = 10^4$, the length d of the diameter is 1.51; if $M = 10^6$, $d = 1.20$; if $M = 10^{10}$, $d = 0.76$. Considering that the entire range of each variable is 1, we note that even for a moderate number of input variables, very large training sample sizes are required to make a $K = 1$ nearest neighborhood relatively

TABLE 11.1:
Expected length of the
diameter d of a $K = 1$
neighborhood for
various values of N
and M.

N	M	$d(N, M)$
4	100	0.42
4	1000	0.23
6	100	0.71
6	1000	0.48
10	1000	0.91
10	10^4	0.72
20	10^4	1.51
20	10^6	1.20
20	10^{10}	0.76

small. The proportion (diameter $\sim M^{-1/N}$) discussed here for $p = 2$ inflicts all $p > 0$ norms.

The fact that the data become so sparse in high-dimensional spaces has the consequence that the bias of the estimate can be quite large even for $K = 1$ and very large datasets. This high bias effect due to the curse of dimensionality can be reduced by taking into consideration the fact that the class probability functions may not vary with equal strength in all directions in the feature space emanating from the query point $\mathbf{x}_0$. This can be accomplished by choosing a metric $W(\mathbf{x}_0)$ (11.3) that credits the highest influence to those directions along which the class probability functions are not locally constant, and have correspondingly less influence on other directions. As a result, the class conditional probabilities tend to be approximately constant in the resulting modified neighborhood, whereby better classification can be obtained, as we will see later.

From the above discussion it should be clear that, in finite settings, the choice of the metric $W(\mathbf{x}_0)$ can strongly affect performance, and therefore the choice of a distance measure becomes crucial in determining the outcome of a nearest neighbor classification.

11.3   Adaptive Metric Techniques

Pattern classification faces a difficult challenge in finite settings and high-dimensional spaces due to the curse of dimensionality. It becomes crucial in estimating different degrees of relevance that input features may have in various locations in feature space. In this section we discuss relevant work in the literature on flexible metric computations.

11.3.1   Flexible Metric Nearest Neighbor Classification

Friedman [8] describes an adaptive approach for pattern classification that combines some of the best features of K-NN learning and recursive partitioning. The resulting hybrid method inherits the flexibility of recursive partitioning to adapt the shape of a region $N(\mathbf{x}_0)$ as well as the ability of nearest neighbor techniques to keep the points within the region close to the point being predicted. The method is capable of producing nearly continuous probability estimates with the region $N(\mathbf{x}_0)$ centered at $\mathbf{x}_0$, and the shape of the region separately customized for each individual prediction point. In the following we describe the method proposed in [8] in more detail.

Consider an arbitrary function $f(\mathbf{x})$ of N arguments $(x_1, \cdots, x_N)$. In the absence of values for any of the argument variables, the least-squares estimate for $f(\mathbf{x})$ is just the expected value $Ef = \int f(\mathbf{x})p(\mathbf{x})d\mathbf{x}$, over the joint probability density of its arguments. Suppose now that the value of just one of the argument variables x_i were known, say $x_i = z$. The least-squares prediction for $f(\mathbf{x})$ in this case would be the expected value of $f(\mathbf{x})$, under the restriction that x_i assumes the known value z: $E[f|x_i = z] = \int f(\mathbf{x})p(\mathbf{x}|x_i = z)d\mathbf{x}$. The improvement in the squared prediction error $I_i^2(z)$ associated with knowing the value z of the ith input variable $x_i = z$ is therefore

$$I_i^2(z) = (Ef - E[f|x_i = z])^2 \qquad (11.5)$$

$I_i^2(z)$ measures how much we gain by knowing that $x_i = z$. It reflects the influence of the ith input variable on the variation of $f(\mathbf{x})$ at the particular point $x_i = z$. Note that if $Ef = E[f|x_i = z]$, then $f(\mathbf{x})$ is independent of x_i at the particular point $x_i = z$, and accordingly $I^2(z) = 0$.

Consider an arbitrary point $\mathbf{z} = (z_1, \cdots, z_N)$ in the N-dimensional input space. A measure of the relative influence, *relevance*, of the ith input variable x_i to the variation of $f(\mathbf{x})$ at $\mathbf{x} = \mathbf{z}$ is given by

$$r_i^2(\mathbf{z}) = \frac{I_i^2(z_i)}{\sum_{k=1}^{N} I_k^2(z_k)} \qquad (11.6)$$

In [8], Friedman proposes an algorithm, called *machete*, that uses the local relevance measure (11.6) to define a splitting procedure centered at the prediction point, overcoming some of the limitations of the static splitting of recursive partitioning. As with recursive partitioning, the machete begins with the entire input measurement space R_0 and divides it into two regions by a split on one of the input variables. However, the manner in which the splitting variable is selected, and the nature of the split itself, are quite different. The input variable used for splitting is the one that maximizes the estimated relevance as evaluated at the point $\mathbf{z}$ to be predicted:

$$i^*(\mathbf{z}) = \arg \max_{1 \le i \le N} \hat{r}_i^2(\mathbf{z}) \qquad (11.7)$$

Thus, for the same training data, different input variables can be selected for this first split at different prediction points $\mathbf{z}$, depending on how the relevance of each input variable changes with location in feature space. The space is then split on the i^*th input variable so that the i^*th component of $\mathbf{z}$, z_{i^*}, is centered within the resulting subinterval that contains it. In particular, the training data are sorted in increasing order on $|z_{i^*} - x_{ni^*}|$ and the new region $R_1(\mathbf{z})$ is

$$R_1(\mathbf{z}) = \{\mathbf{x}_n \mid |z_{i^*} - x_{ni^*}| \le d(M_1)\} \qquad (11.8)$$

where $d(M_1)$ is a distance value such that $R_1(\mathbf{z})$ contains $M_1 < M$ training observations.

As with all recursive methods, the entire machete procedure is defined by successively applying its splitting procedure to the result of the previous split. The algorithm stops when there are K training observations left in the region under consideration, with K being one of the input parameters of the machete.

In [8], Friedman also proposes a generalization of the machete algorithm, called *scythe*, in which the input variables influence each split in proportion to their estimated local relevance, rather than according to the winner-take-all strategy of the machete.

The major limitation concerning the machete/scythe method is that, like recursive partitioning methods, it applies a "greedy" strategy. Since each split is conditioned on its "ancestor" split, minor changes in an early split, due to any variability in parameter estimates, can have a significant impact on later splits, thereby producing different terminal regions. This makes the predictions highly sensitive to the sampling fluctuations associated with the random nature of the process that produces the training data, and therefore may lead to high variance predictions.

We performed a comparative study (see Section 11.5) that shows that while machete/scythe demonstrates performance improvement over recursive partitioning, simple K-NN still remains highly competitive.

11.3.2 Discriminant Adaptive Nearest Neighbor Classification

In [9], Hastie and Tibshirani propose a discriminant adaptive nearest neighbor classification method (DANN) based on linear discriminant analysis. The method computes a local distance metric as a product of properly weighted within and between sum of squares matrices. The authors also describe a method to perform global dimensionality reduction, by pooling the local dimension information over all points in the training set [9].

The goal of linear discriminant analysis (LDA) is to find an orientation in feature space on which the projected training data are well separated. This is obtained by maximizing the difference between the class means relative to some measure of the standard deviations for each class. The difference between the class means is estimated by the *between-class scatter matrix B*, and the measure of the standard deviations for each class is given by the *within-class scatter matrix W*. Both matrices are computed by using the given training data. Once the data are rotated and scaled for the best separation of classes, a query point is classified to the class of the closest centroid, with a correction for the class prior probabilities.

In [9], the authors estimate B and W locally at the query point, and use them to form a metric that behaves locally like the LDA metric. The metric proposed is $\Sigma = W^{-1}BW^{-1}$, which has the effect of crediting larger weights to directions in which the centroids are more spread out than to those in which they are close. First the metric Σ is initialized to the identity matrix. A nearest neighborhood of K_m points around the query point $\mathbf{x}_0$ is identified using the metric Σ. Then, the weighted within and between sum of squares matrices W and B are calculated using the points in the neighborhood of $\mathbf{x}_0$. The result is a new metric $\Sigma = W^{-1}BW^{-1}$ for use in a nearest neighbor classification rule at $\mathbf{x}_0$. The algorithm can be either a single-step procedure, or a larger number of iterations can be carried on.

The authors also show that the resulting metric used in DANN approximates the weighted *Chi-squared* distance:

$$D(\mathbf{x}, \mathbf{x}_0) = \sum_{j=1}^{C} \frac{[P(j|\mathbf{x}) - P(j|\mathbf{x}_0)]^2}{P(j|\mathbf{x}_0)} \qquad (11.9)$$

which measures the distance between the query point $\mathbf{x}_0$ and its nearest neighbor $\mathbf{x}$, in terms of their class posterior probabilities. The approximation, derived by a Taylor series expansion, holds only under the assumption of Gaussian class densities with equal covariance matrices.

While sound in theory, DANN may be limited in practice. The main concern is that in high dimensions, we may never have sufficient data to fill in $N \times N$ matrices. Also, the fact that the distance metric computed by DANN approximates the weighted *Chi-squared* distance (11.9) only when class densities are Gaussian and have the same covariance matrix may cause a performance degradation in situations where data do not follow Gaussian distributions or are corrupted by noise, which is often the case in practice. This hypothesis is validated in our experimental results (Section 11.5).

11.3.3 Adaptive Metric Nearest Neighbor Algorithm

In [7], a technique (ADAMENN) based on the *Chi-squared* distance was introduced to compute local feature relevance. ADAMENN uses the *Chi-squared* distance to estimate to which extent each dimension can be relied on to predict class posterior probabilities. A detailed description of the method follows.

11.3.3.1 Chi-Squared Distance

Consider a query point with feature vector $\mathbf{x}_0$. Let $\mathbf{x}$ be the nearest neighbor of $\mathbf{x}_0$ computed according to a distance metric $D(\mathbf{x}, \mathbf{x}_0)$. The goal is to find a metric $D(\mathbf{x}, \mathbf{x}_0)$ that minimizes $E[r(\mathbf{x}_0, \mathbf{x})]$, where $r(\mathbf{x}_0, \mathbf{x}) = \sum_{j=1}^{C} \Pr(j|\mathbf{x}_0)(1 - \Pr(j|\mathbf{x}))$. Here C is the number of classes, and $\Pr(j|\mathbf{x})$ is the class conditional probability at $\mathbf{x}$. That is, $r(\mathbf{x}_0, \mathbf{x})$ is the finite sample error risk given that the nearest neighbor to $\mathbf{x}_0$ by the chosen metric is $\mathbf{x}$. Equivalently, the following function can be minimized:

$$E(r^*(\mathbf{x}_0) - r(\mathbf{x}_0, \mathbf{x}))^2 \qquad (11.10)$$

where $r^*(\mathbf{x}_0) = \sum_{j=1}^{C} \Pr(j|\mathbf{x}_0)(1 - \Pr(j|\mathbf{x}_0))$ is the theoretical infinite sample risk at $\mathbf{x}_0$. By substituting this expression and that for $r(\mathbf{x}_0, \mathbf{x})$ into (11.10), we obtain the following metric that minimizes (11.10) [11]: $D(\mathbf{x}_0, \mathbf{x}) = (\sum_{j=1}^{C} \Pr(j|\mathbf{x}_0)(\Pr(j|\mathbf{x}) - \Pr(j|\mathbf{x}_0)))^2$. The idea behind this metric is that if the value of $\mathbf{x}$ for which $D(\mathbf{x}_0, \mathbf{x})$ is small is selected, then the expectation (11.10) will be minimized.

This metric is linked to the theory of the two-class case developed in [13]. However, a major concern with the above metric is that it has a cancellation effect when all classes are equally likely [11]. This limitation can be avoided by considering the *Chi-squared* distance [9] $D(\mathbf{x}, \mathbf{x}_0) = \sum_{j=1}^{C} [\Pr(j|\mathbf{x}) - \Pr(j|\mathbf{x}_0)]^2$, which measures the distance between the query $\mathbf{x}_0$ and the point $\mathbf{x}$, in terms of the difference between the class posterior probabilities at the two points. Furhermore, by multiplying it by $1/\Pr(j|\mathbf{x}_0)$ we obtain the following weighted *Chi-squared* distance:

$$D(\mathbf{x}, \mathbf{x}_0) = \sum_{j=1}^{C} \frac{[\Pr(j|\mathbf{x}) - \Pr(j|\mathbf{x}_0)]^2}{\Pr(j|\mathbf{x}_0)} \qquad (11.11)$$

Note that in comparison to the *Chi-squared* distance, the weights $1/\Pr(j|\mathbf{x}_0)$ in (11.11) have the effect of increasing the distance of $\mathbf{x}_0$ to any point $\mathbf{x}$ whose most probable class is unlikely to include $\mathbf{x}_0$. That is, if $j^* = \arg\max_j \Pr(j|\mathbf{x})$, we have $\Pr(j^*|\mathbf{x}_0) \approx 0$. As a consequence, it becomes highly improbable for any such point to be a nearest neighbor candidate.

Equation (11.11) computes the distance between the true and estimated posteriors. The goal is to estimate the relevance of feature i by computing its ability to predict the class posterior probabilities locally at the query point. This is accomplished by considering the expectation of $\Pr(j|\mathbf{x})$ conditioned at a location along feature dimension i. Then, the *Chi-squared* distance (11.11) tells us the extent to which dimension i can be relied on to predict $\Pr(j|\mathbf{x})$. Thus, Equation (11.11) provides a foundation upon which to develop a theory of feature relevance in the context of pattern classification.

11.3.3.2 Local Feature Relevance

Based on the above discussion, the computation of local feature relevance proceeds as follows. We first notice that $\Pr(j|\mathbf{x})$ is a function of $\mathbf{x}$. Therefore, we can compute the conditional expectation of $\Pr(j|\mathbf{x})$, denoted by $\overline{\Pr}(j|x_i = z)$, given that x_i assumes value z, where x_i represents the ith component of $\mathbf{x}$. That is, $\overline{\Pr}(j|x_i = z) = E[\Pr(j|\mathbf{x})|x_i = z] = \int \Pr(j|\mathbf{x})p(\mathbf{x}|x_i = z)d\mathbf{x}$. Here $p(\mathbf{x}|x_i = z)$ is the conditional density of the other input variables defined as $p(\mathbf{x}|x_i = z) = p(\mathbf{x})\delta(x_i - z)/\int p(\mathbf{x})\delta(x_i - z)d\mathbf{x}$, where $\delta(x - z)$ is the Dirac delta function having the properties $\delta(x - z) = 0 \quad if \quad x \neq z$ and $\int_{-\infty}^{\infty} \delta(x - z)dx = 1$. Let

$$r_i(\mathbf{z}) = \sum_{j=1}^{C} \frac{[\Pr(j|\mathbf{z}) - \overline{\Pr}(j|x_i = z_i)]^2}{\overline{\Pr}(j|x_i = z_i)} \tag{11.12}$$

$r_i(\mathbf{z})$ represents the ability of feature i to predict the $\Pr(j|\mathbf{z})$s at $x_i = z_i$. The closer $\overline{\Pr}(j|x_i = z_i)$ is to $\Pr(j|\mathbf{z})$, the more information feature i carries for predicting the class posterior probabilities locally at $\mathbf{z}$.

We can now define a measure of feature relevance for $\mathbf{x}_0$ as

$$\bar{r}_i(\mathbf{x}_0) = \frac{1}{K} \sum_{\mathbf{z} \in N(\mathbf{x}_0)} r_i(\mathbf{z}) \tag{11.13}$$

where $N(\mathbf{x}_0)$ denotes the neighborhood of $\mathbf{x}_0$ containing the K nearest training points, according to a given metric. $\bar{r}_i$ measures how well on average the class posterior probabilities can be approximated along input feature i within a local neighborhood of $\mathbf{x}_0$. Small $\bar{r}_i$ implies that the class posterior probabilities will be well approximated along dimension i in the vicinity of $\mathbf{x}_0$. Note that $\bar{r}_i(\mathbf{x}_0)$ is a function of both the test point $\mathbf{x}_0$ and the dimension i, thereby making $\bar{r}_i(\mathbf{x}_0)$ a local relevance measure in dimension i.

The relative relevance, as a weighting scheme, can then be given by $w_i(\mathbf{x}_0) = \frac{R_i(\mathbf{x}_0)^t}{\sum_{l=1}^{N} R_l(\mathbf{x}_0)^t}$, where $t = 1, 2$, giving rise to linear and quadratic weightings respectively, and $R_i(\mathbf{x}_0) = \max_j\{\bar{r}_j(\mathbf{x}_0)\} - \bar{r}_i(\mathbf{x}_0)$. In [7], the following exponential weighting scheme was proposed:

$$w_i(\mathbf{x}_0) = \exp(cR_i(\mathbf{x}_0))/\sum_{l=1}^{N} \exp(cR_l(\mathbf{x}_0)) \tag{11.14}$$

where c is a parameter that can be chosen to maximize (minimize) the influence of $\bar{r}_i$ on w_i. When $c = 0$ we have $w_i = 1/N$, which has the effect of ignoring any difference among the $\bar{r}_i$'s. On the other hand, when c is large, a change in $\bar{r}_i$ will be exponentially reflected in w_i. The exponential weighting is more sensitive to changes in local feature relevance and in general gives rise to better performance improvement. In fact, it is more stable because it prevents neighborhoods from extending infinitely in any direction, i.e., zero weight. This, however, can occur when either linear or quadratic weighting is used. Thus, Equation (11.14) can be used to compute the weight associated with each feature, resulting in the weighted distance computation:

$$D(\mathbf{x}, \mathbf{y}) = \sqrt{\sum_{i=1}^{N} w_i (x_i - y_i)^2} \qquad (11.15)$$

The weights w_i enable the neighborhood to elongate less important feature dimensions and, at the same time, to constrict the most influential ones. Note that the technique is *query-based* because the weights depend on the query [1].

Since both $\Pr(j|\mathbf{z})$ and $\overline{\Pr}(j|x_i = z_i)$ in (11.12) are unknown, we must estimate them using the training data $\{\mathbf{x}_n, y_n\}_{n=1}^{M}$ in order for the relevance measure (11.13) to be useful in practice. Here $y_n \in \{1, \cdots, C\}$. The quantity $\Pr(j|\mathbf{z})$ is estimated by considering a neighborhood $N_1(\mathbf{z})$ centered at $\mathbf{z}$:

$$\hat{\Pr}(j|\mathbf{z}) = \frac{\sum_{n=1}^{M} 1(\mathbf{x}_n \in N_1(\mathbf{z}))1(y_n = j)}{\sum_{n=1}^{M} 1(\mathbf{x}_n \in N_1(\mathbf{z}))} \qquad (11.16)$$

where $1(\cdot)$ is an indicator function such that it returns 1 when its argument is true, and 0 otherwise.

To compute $\overline{\Pr}(j|x_i = z) = E[\Pr(j|\mathbf{x})|x_i = z]$, we introduce an additional variable g_j such that $g_j|\mathbf{x} = 1$ if $y = j$, and 0 otherwise, where $j \in \{1, \cdots, C\}$. We then have $\Pr(j|\mathbf{x}) = E[g_j|\mathbf{x}]$, from which it is not hard to show that $\overline{\Pr}(j|x_i = z) = E[g_j|x_i = z]$. However, since there may not be any data at $x_i = z$, the data from the neighborhood of z along dimension i are used to estimate $E[g_j|x_i = z]$, a strategy suggested in [8]. In detail, by noticing $g_j = 1(y = j)$, the estimate can be computed from

$$\hat{\overline{\Pr}}(j|x_i = z_i) = \frac{\sum_{\mathbf{x}_n \in N_2(\mathbf{z})} 1(|x_{ni} - z_i| \le \Delta_i)1(y_n = j)}{\sum_{\mathbf{x}_n \in N_2(\mathbf{z})} 1(|x_{ni} - z_i| \le \Delta_i)} \qquad (11.17)$$

where $N_2(\mathbf{z})$ is a neighborhood centered at $\mathbf{z}$ (larger than $N_1(\mathbf{z})$), and the value of Δ_i is chosen so that the interval contains a fixed number L of points: $\sum_{n=1}^{M} 1(|x_{ni} - z_i| \le \Delta_i)1(\mathbf{x}_n \in N_2(\mathbf{z})) = L$. Using the estimates in (11.16) and in (11.17), we obtain an empirical measure of the relevance (11.13) for each input variable i.

11.3.3.3 The ADAMENN Algorithm

The adaptive metric nearest neighbor algorithm (ADAMENN) has six adjustable tuning parameters: K_0: the number of neighbors of the test point (query); K_1: the number of neighbors in $N_1(\mathbf{z})$ (11.16); K_2: the size of the neighborhood $N_2(\mathbf{z})$ for each of the K_0 neighbors (11.17); L: the number of points within the Δ intervals; K: the number of neighbors in the final nearest

Given a test point $\mathbf{x}_0$, and input parameters K_0, K_1, K_2, L, K, and c:

1. Initialize w_i in (11.15) to $1/N$, for $i = 1, \ldots, N$.

2. Compute the K_0 nearest neighbors of $\mathbf{x}_0$ using the weighted distance metric (11.15).

3. For each dimension i, $i = 1, \ldots, N$, compute relevance estimate $\bar{r}_i(\mathbf{x}_0)$ (11.13) using Equations (11.16) and (11.17).

4. Update $\mathbf{w}$ according to (11.14).

5. Iterate steps 2, 3, and 4 (zero and five times in our implementation).

6. At completion of iterations, use $\mathbf{w}$, hence (11.15), for K nearest neighbor classification at the test point $\mathbf{x}_0$.

FIGURE 11.2: The ADAMENN algorithm

neighbor rule; and c: the positive factor for the exponential weighting scheme (11.14).

Cross-validation can be used to determine the optimal values of the parameters. Note that K is common to all nearest neighbor rules. K_0 is used to reduce the variance of the estimates; its value should be a small fraction of M, e.g., $K_0 = \max(0.1M, 20)$. Often a smaller value is preferable for K_1 to avoid biased estimates. K_2 and L are common to the machete and scythe algorithms described in [8]. The values of K_2 and L determine the bias and variance trade-offs for the estimation of $E[g_j|x_i = z]$. The way these estimates are used does not require a high accuracy. As a consequence, ADAMENN performance is basically insensitive to the values chosen for K_2 and L, provided they are not too small (close to one) or too large (close to M). The value of c should increase as the input query moves close to the decision boundary, so that highly stretched neighborhoods will result; c can be chosen empirically in practice. Arguably we have introduced a few more parameters that might potentially cause overfitting. However, it is important to realize that one of the parameters (K_0) plays the role of averaging or smoothing. Because it helps reduce variance, we can afford to have a few parameters that adapt to avoid bias, without incurring the risk of overfitting.

At the beginning, the estimation of the $\bar{r}_i$ values in (11.13) is accomplished by using a weighted distance metric (11.15) with w_i, $\forall i = 1, \ldots, N$, being initialized to $1/N$. Then, the elements w_i of $\mathbf{w}$ are updated according to $\bar{r}_i$ values via (11.14). The update of $\mathbf{w}$ can be iterated. At the completion of the iterations, the resulting $\mathbf{w}$ is plugged into (11.15) to compute nearest neighbors at the test point $\mathbf{x}_0$. An outline of the ADAMENN algorithm is shown in Figure 11.2.

11.4   Large Margin Nearest Neighbor Classifiers

The previously discussed techniques have been proposed to try to minimize bias in high dimensions by using locally adaptive mechanisms. The "lazy learning" approach used by these methods, while appealing in many ways, requires a considerable amount of on-line computation, which makes it difficult for such techniques to scale up to large datasets. In this section we discuss a locally adaptive metric classification method that, although still founded on a query-based weighting mechanism, computes off-line the information relevant to define local weights [6].

The technique uses support vector machines (SVMs) as a guidance for the process of defining a local flexible metric. SVMs have been successfully used as a classification tool in a variety of areas [12], and the maximum margin boundary they provide has been proved to be optimal in a structural risk minimization sense. While the solution provided by SVMs is theoretically sound, SVMs maximize the margin in feature space. However, the feature space does not always capture the structure of the input space. As noted in [2], the large margin in the feature space does not necessarily translate into a large margin in the input space. In fact, it is argued that sometimes SVMs give a very small margin in the input space, because the metric of the feature space is usually quite different from that of the input space [2]. Such a situation is undesirable. The approach discussed here overcomes this limitation. In fact, it can be shown that the proposed weighting scheme increases the margin, and therefore the separability of classes, in the transformed space where classification is performed (see Section 11.4.4).

The solution provided by SVMs guides the extraction of local information in a neighborhood around the query. This process produces highly stretched neighborhoods along boundary directions when the query is close to the boundary. As a result, the class conditional probabilities tend to be constant in the modified neighborhood, whereby better classification performance can be achieved. The amount of elongation-constriction decays as the query moves farther from the vicinity of the decision boundary. This phenomenon is exemplified in Figure 11.1 by queries a, $a^{'}$, and $a^{''}$.

Cross validation is avoided by using a principled technique for setting the procedural parameters of the method. The approach to efficient and automatic settings of parameters leverages the sparse solution provided by SVMs. As a result, the algorithm has only one adjustable tuning parameter, namely, the number K of neighbors in the final nearest neighbor rule. This parameter is common to all nearest neighbor techniques.

11.4.1   Support Vector Machines

In this section we introduce the main concepts of learning with support vector machines (SVMs). Again, we are given M observations. Each observation consists of a pair: a vector $\mathbf{x}_i \in \Re^N$, $i = 1, \ldots, M$, and the associated class label $y_i \in \{-1, 1\}$.

In the simple case of two linearly separable classes, a support vector machine selects, among the infinite number of linear classifiers that separate the data, the classifier that minimizes an upper bound on the generalization error. The SVM achieves this goal by computing the classifier that satifies the maximum margin property, i.e., the classifier whose decision boundary has the maximum minimum distance from the closest training point.

If the two classes are non-separable, the SVM looks for the hyperplane that maximizes the margin and that, at the same time, minimizes an upper bound of the error. The trade-off between margin and upper bound of the misclassification error is driven by a positive constant C that has to be chosen beforehand. The corresponding decision function is then obtained by considering the $sign(f(\mathbf{x}))$, where $f(\mathbf{x}) = \sum_i \alpha_i y_i \mathbf{x}_i^T \mathbf{x} - b$, and the coefficients α_i are the solutions of a convex quadratic problem, defined over the hypercube $[0, C]^l$. The parameter b is also computed from the data. In general, the solution will have a number of coefficients α_i equal to zero, and since there is a coefficient α_i associated to each data point, only the data points corresponding to non-zero α_i will influence the solution. These points are the support vectors. Intuitively, the support vectors are the data points that lie at the border between the two classes, and a small number of support vectors indicates that the two classes can be well separated.

This technique can be extended to allow for non-linear decision surfaces. This is done by mapping the input vectors into a higher dimensional feature space, $\phi : \Re^N \to \Re^{N'}$, and by formulating the linear classification problem in the feature space. Therefore, $f(\mathbf{x})$ can be expressed as $f(\mathbf{x}) = \sum_i \alpha_i y_i \phi^T(\mathbf{x}_i) \phi(\mathbf{x}) - b$.

If one were given a function $K(\mathbf{x}, \mathbf{y}) = \phi^T(\mathbf{x}) \phi(\mathbf{y})$, one could learn and use the maximum margin hyperplane in feature space without having to compute explicitly the image of points in $\Re^{N'}$. It has been proved (Mercer's Theorem) that for each continuous positive definite function $K(\mathbf{x}, \mathbf{y})$ there exists a mapping ϕ such that $K(\mathbf{x}, \mathbf{y}) = \phi^T(\mathbf{x}) \phi(\mathbf{y})$, $\forall \mathbf{x}, \mathbf{y} \in \Re^N$. By making use of such function K (*kernel function*), the equation for $f(\mathbf{x})$ can be rewritten as

$$f(\mathbf{x}) = \sum_i \alpha_i y_i K(\mathbf{x}_i, \mathbf{x}) - b \tag{11.18}$$

11.4.2   Feature Weighting

The maximum margin boundary found by the SVM is used here to determine local discriminant directions in the neighborhood around the query. The normal direction to local decision boundaries identifies the orientation along which data points between classes are well separated. The gradient vector computed at points on the boundary allows one to capture such information, and to use it for measuring local feature relevance and weighting features accordingly. The resulting weighting scheme improves upon the solution computed by the SVM by increasing the margin in the space transformed by the weights. Here are the major thrusts of the proposed method.

SVMs classify patterns according to the $sign(f(\mathbf{x}))$. Clearly, in the case of a non-linear feature mapping ϕ, the SVM classifier gives a non-linear boundary $f(\mathbf{x}) = 0$ in the input space. The gradient vector $\mathbf{n_d} = \nabla_{\mathbf{d}} f$, computed at any point $\mathbf{d}$ on the level curve $f(\mathbf{x}) = 0$, points to the direction perpendicular to the decision boundary in the input space at $\mathbf{d}$. As such, the vector $\mathbf{n_d}$ identifies the orientation in the input space onto which the projected training data are well separated in the neighborhood around $\mathbf{d}$. Therefore, the orientation given by $\mathbf{n_d}$, and any orientation close to it, carries highly discriminant information for classification. As a result, this information can be used to define a local measure of feature relevance.

Let $\mathbf{x_0}$ be a query point whose class label we want to predict. Suppose $\mathbf{x_0}$ is close to the boundary, which is where class conditional probabilities become locally non-uniform, and therefore estimating local feature relevance becomes crucial. Let $\mathbf{d}$ be the closest point to $\mathbf{x_0}$ on the boundary $f(\mathbf{x}) = 0$:

$$\mathbf{d} = \arg\min_{\mathbf{p}} \|\mathbf{x_0} - \mathbf{p}\|, \quad subject \ to \ the \ constraint \ f(\mathbf{p}) = 0 \quad (11.19)$$

Then we know that the gradient $\mathbf{n_d}$ identifies a discriminant direction.

As a consequence, the subspace spanned by the orientation $\mathbf{n_d}$ intersects the decision boundary and contains changes in class labels. Therefore, when applying a nearest neighbor rule at $\mathbf{x_0}$, we desire to stay close to $\mathbf{x_0}$ along the $\mathbf{n_d}$ direction, because that is where it is likely to find points similar to $\mathbf{x_0}$ in terms of the class conditional probabilities. Distances should be increased (due to large weight) along $\mathbf{n_d}$ and directions close to it, thus excluding points along $\mathbf{n_d}$ that are away from $\mathbf{x_0}$. The farther we move from the $\mathbf{n_d}$ direction, the less discriminant the correspondending orientation. This means that class labels are unlikely to change along those orientations, and distances should be reduced (due to small weight), thus including points that are likely to be similar to $\mathbf{x_0}$ in terms of the class conditional probabilities.

Formally, we can measure how close a direction $\mathbf{t}$ is to $\mathbf{n_d}$ by considering the dot product $\mathbf{n_d^T t}$. In particular, denoting $\mathbf{e}_j$ the canonical unit vector along input feature j, for $j = 1, \ldots, N$, we can define a measure of relevance for feature j, locally at $\mathbf{x}_0$ (and therefore at $\mathbf{d}$), as

$$R_j(\mathbf{x}_0) \equiv |\mathbf{e}_j^T \mathbf{n_d}| = |n_{\mathbf{d},j}| \tag{11.20}$$

where $\mathbf{n_d} = (n_{\mathbf{d},1}, \ldots, n_{\mathbf{d},N})^T$.

The measure of relative feature relevance, as a weighting scheme, can then be given by

$$w_j(\mathbf{x}_0) = (R_j(\mathbf{x}_0))^t / \sum_{i=1}^{N} (R_i(\mathbf{x}_0))^t \tag{11.21}$$

where t is a positive integer, giving rise to polynomial weightings. The following exponential weighting scheme is used in [6]:

$$w_j(\mathbf{x}_0) = exp(AR_j(\mathbf{x}_0)) / \sum_{i=1}^{n} exp(AR_i(\mathbf{x}_0)) \tag{11.22}$$

where A is a parameter that can be chosen to maximize (minimize) the influence of R_j on w_j. When $A = 0$ we have $w_j = 1/N$, thereby ignoring any difference between the R_j's. On the other hand, when A is large a change in R_j will be exponentially reflected in w_j. Thus, (11.22) can be used as weights associated with features for weighted distance computation:

$$D(\mathbf{x}, \mathbf{y}) = \sqrt{\sum_{i=1}^{N} w_i (x_i - y_i)^2}. \tag{11.23}$$

11.4.3 Large Margin Nearest Neighbor Classification

We desire that the parameter A in the exponential weighting scheme (11.22) increases as the distance of $\mathbf{x}_0$ from the boundary decreases. By using the knowledge that support vectors are mostly located around the boundary surface, we can estimate how close a query point $\mathbf{x}_0$ is to the boundary by computing its distance from the closest non-bounded support vector:

$$B_{\mathbf{x}_0} = \min_{\mathbf{s}_i} \|\mathbf{x}_0 - \mathbf{s}_i\| \tag{11.24}$$

where the minimum is taken over the non bounded $(0 < \alpha_i < C)$ support vectors $\mathbf{s}_i$. Following the same principle described in [3], the spatial resolution around the boundary is increased by enlarging the volume elements locally in neighborhoods of support vectors.

Then, we can achieve the goal by setting

$$A = \max\{D - B_{\mathbf{x}_0}, 0\} \tag{11.25}$$

where D is a constant ("meta") parameter input to the algorithm. In practice, D can be set equal to the approximated average distance between the training points $\mathbf{x}_k$ and the boundary:

$$D = \frac{1}{M} \sum_{\mathbf{x}_k} \{\min_{\mathbf{s}_i} \|\mathbf{x}_k - \mathbf{s}_i\|\} \qquad (11.26)$$

By doing so, the value of A nicely adapts to each query point according to its location with respect to the boundary. The closer $\mathbf{x}_0$ is to the decision boundary, the greater impact R_j will have on distance computation (when $B_{\mathbf{x}_0} > D$, $A = 0$, and therefore $w_j = 1/N$).

Input: Decision boundary $f(\mathbf{x}) = 0$ produced by an SVM; query point $\mathbf{x}_0$ and parameter K.

1. Compute the closest point $\mathbf{d}$ to $\mathbf{x}_0$ on the boundary (11.19).

2. Compute the gradient vector $\mathbf{n_d} = \nabla_{\mathbf{d}} f$.

3. Set feature relevance values $R_j(\mathbf{x}_0) = |n_{\mathbf{d},j}|$ for $j = 1, \ldots, N$.

4. Estimate the distance of $\mathbf{x}_0$ from the boundary as: $B_{\mathbf{x}_0} = \min_{\mathbf{s}_i} \|\mathbf{x}_0 - \mathbf{s}_i\|$.

5. Set $A = \max\{D - B_{\mathbf{x}_0}, 0\}$, where D is defined as in equation (11.26).

6. Set $\mathbf{w}$ according to (11.21) or (11.22).

7. Use the resulting $\mathbf{w}$ for K nearest neighbor classification at the query point $\mathbf{x}_0$.

FIGURE 11.3: The LaMaNNa algorithm

We observe that this principled technique for setting the parameters of our method takes advantage of the sparse representation of the solution provided by the SVM. In fact, for each query point $\mathbf{x}_0$, in order to compute $B_{\mathbf{x}_0}$ we only need to consider the support vectors, whose number is typically small compared to the total number of training examples. Furthermore, D can be computed off-line and used in subsequent on-line classification.

The resulting locally flexible metric nearest classification algorithm based on SVMs is summarized in Figure 11.3. We call our algorithm LaMaNNa (**La**rge **Ma**rgin **N**earest **N**eighbor algorithm) to highlight the fact that the algorithm operates in a space with enlarged margins, as formally shown in the next section. The algorithm has only one adjustable tuning parameter, namely, the number K of neighbors in the final nearest neighbor rule. This parameter is common to all nearest neighbor classification techniques.

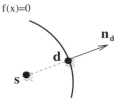

f(x)=0

FIGURE 11.4: Perpendicular distance and gradient vector.

11.4.4 Weighting Features Increases the Margin

We define the input space margin as the minimal distance from the training points to the classification boundary in the input space [2]. More specifically, let $s \in \Re^N$ be a sample point, and d (defined in (11.19)) the (nearest) foot of the perpendicular on the separating surface $f(x) = 0$ from s (see Figure 11.4). We define the input space margin as

$$IM = \min_{s} D(s, d) = \min_{s} \sqrt{\frac{1}{N} \sum_{i=1}^{N} (s_i - d_i)^2} \qquad (11.27)$$

where s is in the training set, and equal weights are assigned to the feature dimensions. In the following we show that the weighting schemes implemented by LaMaNNa increase the margin in the space transformed by the weights. For lack of space we omit the proofs. The interested reader should see [6].

Consider the gradient vector $n_d = \nabla_d f = (\frac{\partial}{\partial x_1} f_d, \ldots, \frac{\partial}{\partial x_N} f_d)$ computed with respect to x at point d. Our local measure of relevance for feature j is then given by

$$R_j(s) = |e_j^T n_d| = |n_{d,j}|$$

and $w_j(s)$ is defined as in (11.21) or (11.22), with $\sum_{j=1}^{N} w_j(s) = 1$.

Let

$$D_w^2(s, d) = \sum_{i=1}^{N} w_i(s)(s_i - d_i)^2 \qquad (11.28)$$

be the squared weighted Euclidean distance between s and d. The main result is summarized in the following theorem.

Theorem 1
Let $s \in \Re^N$ be a sample point and $d \in \Re^N$ the nearest foot of the perpendicular on the separating surface $f(x) = 0$. Define $D^2(s, d) = \frac{1}{N} \sum_{i=1}^{N} (s_i - d_i)^2$ and $D_w^2(s, d) = \sum_{i=1}^{N} w_i(s)(s_i - d_i)^2$, where $w_i(x_0)$ are the weights computed

according to (11.21) or (11.22). Then

$$D^2(\mathbf{s}, \mathbf{d}) \leq D_{\mathbf{w}}^2(\mathbf{s}, \mathbf{d})$$

Using this result, it can be shown that the weighting schemes increase the margin in the transformed space. Let

$$\mathbf{s}^* = \arg \min_{\mathbf{s}} D(\mathbf{s}, \mathbf{d})$$

Then

$$IM = \sqrt{\frac{1}{N} \sum_{i=1}^{N} (s_i^* - d_i^*)^2}$$

We have the following result.

Corollary 1
$IM \leq D_{\mathbf{w}}(\mathbf{s}^*, \mathbf{d}^*)$.

Theorem 1 shows that $D^2(\mathbf{s}, \mathbf{d}) \leq D_{\mathbf{w}}^2(\mathbf{s}, \mathbf{d})$. Now we show that the equality holds only when $w_i = \frac{1}{N} \forall i$. This result guarantees an effective increase of the margin in the transformed space whenever differential weights are credited to features (according to the given weighting schemes), as stated in Corollary 3.

Corollary 2
$D^2(\mathbf{s}, \mathbf{d}) = D_{\mathbf{w}}^2(\mathbf{s}, \mathbf{d})$ *if and only if* $w_i = \frac{1}{n} \forall i$.

And finally, from Corollaries 1 and 2, we obtain:

Corollary 3
$IM = D_{\mathbf{w}}(\mathbf{s}^*, \mathbf{d}^*)$ *if and only if* $w_i = \frac{1}{N} \forall i$.

11.5 Experimental Comparisons

In the following we compare the previously discussed classification techniques using real data. In the experiments, we also included the RBF-SVM classifier with radial basis kernels, the simple K-NN method using the Euclidean distance measure, and the C4.5 decision tree method.

In our experiments we used seven different real datasets. They are all taken from the UCI Machine Learning Repository at http://www.cs.uci.edu/~mlearn/MLRepository. For the Iris, Sonar, Liver, and Vote data we performed leave-one-out cross-validation to measure performance, since the number of available data is limited for these datasets. For the Breast, OQ-letter, and Pima data we randomly generated five independent training sets of size 200. For each of these, an additional independent test sample consisting of 200 observations was generated. Table 11.2 shows the cross-validated error rates for the eight methods under consideration on the seven real data. Procedural parameters

TABLE 11.2: Average classification error rates for real data.

	Iris	Sonar	Liver	Vote	Breast	OQ	Pima
LaMaNNa	4.0	11.0	28.1	**2.6**	3.0	3.5	**19.3**
RBF-SVM	4.0	12.0	**26.1**	3.0	3.1	3.4	21.3
ADAMENN	**3.0**	9.1	30.7	3.0	3.2	**3.1**	20.4
Machete	5.0	21.2	27.5	3.4	3.5	7.4	20.4
Scythe	4.0	16.3	27.5	3.4	2.7	5.0	20.0
DANN	6.0	**7.7**	30.1	3.0	**2.2**	4.0	22.2
K-NN	6.0	12.5	32.5	7.8	2.7	5.4	24.2
C4.5	8.0	23.1	38.3	3.4	4.1	9.2	23.8

(including K) for each method were determined empirically through cross validation over training data.

LaMaNNa achieves the best performance in 2/7 of the real datasets; in one case it shows the second best performance, and in the remaining four its error rate is still quite close to the best one.

It seems natural to quantify this notion of robustness, that is, how well a particular method m performs on average across the problems taken into consideration. Following Friedman [8], we capture robustness by computing the ratio b_m of the error rate e_m of method m and the smallest error rate over all methods being compared in a particular example:

$$b_m = e_m / \min_{1 \leq k \leq 8} e_k$$

Thus, the best method m^* for that example has $b_{m^*} = 1$, and all other methods have larger values $b_m \geq 1$, for $m \neq m^*$. The larger the value of b_m, the worse the performance of the m-th method is in relation to the best one for that example, among the methods being compared. The distribution of the b_m values for each method m over all the examples, therefore, seems to be a good indicator concerning its robustness. For example, if a particular method has an error rate close to the best in every problem, its b_m values should be densely distributed around the value 1. Any method whose b value distribution deviates from this ideal distribution reflects its lack of robustness.

Figure 11.5 plots the distribution of b_m for each method over the seven real datasets. The dark area represents the lower and upper quartiles of the distribution that are separated by the median. The outer vertical lines show the entire range of values for the distribution. The outer vertical lines for the LaMaNNa method are not visible because they coincide with the limits of the lower and upper quartiles. The spread of the error distribution for LaMaNNa is narrow and close to one. The spread for ADAMENN has a similar behavior, with the outer bar reaching a slightly higher value. The results clearly demonstrate that LaMaNNa (and ADAMENN) obtained the most robust performance over the datasets.

The poor performance of the Machete and C4.5 methods might be due to

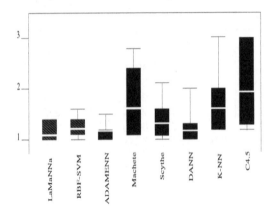

FIGURE 11.5: Performance distributions for real data.

the greedy strategy they employ. Such a recursive peeling strategy removes at each step a subset of data points permanently from further consideration. As a result, changes in an early split, due to any variability in parameter estimates, can have a significant impact on later splits, thereby producing different terminal regions. This makes predictions highly sensitive to the sampling fluctuations associated with the random nature of the process that produces the traning data, thus leading to high variance predictions. The Scythe algorithm, by relaxing the winner-take-all splitting strategy of the Machete algorithm, mitigates the greedy nature of the approach, and thereby achieves better performance.

In [9], the authors show that the metric employed by the DANN algorithm approximates the weighted Chi-squared distance, given that class densities are Gaussian and have the same covariance matrix. As a consequence, we may expect a degradation in performance when the data do not follow Gaussian distributions and are corrupted by noise, which is likely the case in real scenarios like the ones tested here.

We observe that LaMaNNa avoids expensive cross-validation by using a principled technique for setting the procedural parameters. The approach to efficient and automatic settings of parameters leverages the sparse solution provided by SVMs. As a result, LaMaNNa has only one adjudtable tuning parameter, the number K of neighbors in the final nearest neighbor rule. This parameter is common to all nearest neighbor techniques. On the other hand, the competing techniques have multiple parameters whose values must be determined through cross-validation: ADAMENN has *six* parameters; Machete/Scythe each has *four* parameters; and DANN has *two* parameters.

The LaMaNNa technique offers accuracy improvements over the RBF-SVM algorithm alone. The reason for such performance gain may rely on the effect of the local weighting scheme on the margin in the transformed space, as shown in Section 11.4.4. Assigning large weights to input features close to the gradient direction, locally in neighborhoods of support vectors, corresponds to an increase in the spatial resolution along those orientations, and therefore to improve the separability of classes.

11.6 Conclusions

Pattern classification faces a difficult challenge in finite settings and high-dimensional spaces due to the curse of dimensionality. In this chapter we have presented and compared techniques to address data exploration tasks such as classification. All methods design adaptive metrics or parameter estimates that are local in input space in order to dodge the curse of dimensionality phenomenon. Such techniques have been demonstrated to be effective for the achievement of accurate predictions.

References

[1] D. Aha. Lazy learning. *Artificial Intelligence Review*, 11:1–5, 1997.

[2] S. Akaho. Svm maximizing margin in the input space. In *6th Kernel Machines Workshop on Learning Kernels, Proceedings of Neural Information Processing Systems, (NIPS)*, 2002.

[3] S. Amari and S. Wu. Improving support vector machine classifiers by modifying kernel functions. *Neural Networks*, 12:783–789, 1999.

[4] R. Bellman. *Adaptive Control Processes*. Princeton Univ. Press, Princeton, NJ, 1961.

[5] T. Cover and P. Hart. Nearest neighbor pattern classification. *IEEE Trans. on Information Theory*, 13:21–27, 1967.

[6] C. Domeniconi, D. Gunopulos, and J. Peng. Large margin nearest neighbor classifiers. *IEEE Trans. on Neural Networks*, 16:899–909, 2005.

[7] C. Domeniconi, J. Peng, and D. Gunopulos. Locally adaptive metric nearest neighbor classification. *IEEE Trans. on Pattern Analysis and Machine Intelligence*, 24:1281–1285, 2002.

[8] J. Friedman. Flexible metric nearest neighbor classification. Technical Report, Dept. of Statistics, Stanford University, 1994.

[9] T. Hastie and R. Tibshirani. Discriminant adaptive nearest neighbor clas-

sification. *IEEE Trans. on Pattern Analysis and Machine Intelligence*, 18:607–615, 1996.

[10] G. McLachlan. *Discriminant Analysis and Statistical Pattern Recognition*. Wiley, New York, 1992.

[11] J. Myles and D. Hand. The multi-class metric problem in nearest neighbor discrimination rules. *Pattern Recognition*, 23:1291–1297, 1990.

[12] J. Shawe-Taylor and N. Cristianini. *Kernel Methods for Pattern Analysis*. Cambridge University Press, New York, 2004.

[13] R. Short and K. Fukunaga. Optimal distance measure for nearest neighbor classification. *IEEE Transactions on Information Theory*, 27:622–627, 1981.

[14] C. Stone. Nonparametric regression and its applications (with discussion). *Ann. Statist.*, 5, 1977.

Chapter 12

Feature Weighting through Local Learning

Yijun Sun

University of Florida

12.1 Introduction

Feature selection is one of the fundamental problems in machine learning. The role of feature selection is critical, especially in applications involving many irrelevant features. Yet, compared to classifier design (e.g., SVM and AdaBoost), much rigorous theoretical treatment to feature selection is needed. Most feature selection algorithms rely on heuristic searching and thus cannot provide any guarantee of optimality. This is largely due to the difficulty in defining an objective function that can be easily optimized by well-established optimization techniques. It is particularly true for wrapper methods when a nonlinear classifier is used to evaluate the goodness of selected feature subsets. This problem can to some extent be alleviated by using a feature-weighting strategy, which assigns to each feature a real-valued number, instead of a binary one, to indicate its relevance to a learning problem. Among the existing feature weighting algorithms, the Relief algorithm [10] is considered one of the most successful ones due to its simplicity and effectiveness [5]. However, it is unclear to date what objective function Relief optimizes. In this chapter, we first prove that Relief implements an online algorithm that solves a convex optimization problem with a margin-based objective function. The margin is defined based on a 1-NN classifier. Therefore, compared with filter methods, Relief usually performs better due to the performance feedback of a nonlinear classifier when searching for useful features; and compared with conventional wrapper methods, by optimizing a convex problem, Relief avoids *any* exhaustive or heuristic combinatorial search and thus can be implemented very efficiently. The new interpretation clearly explains the simplicity and effectiveness of Relief.

The new interpretation of Relief enables us to identify and address some weaknesses of the algorithm. One major drawback of Relief is that the nearest neighbors are defined in the original feature space, which are highly unlikely to be the ones in the weighted space. Moreover, Relief lacks a mechanism to deal with outlier data. In the presence of a large number of irrelevant features and mislabeling, the solution quality of Relief can be severely degraded. To mitigate these problems, in Section 12.3, we propose a new feature weighting algorithm, referred to as I-Relief, by following the principle of the Expectation-Maximization (EM) algorithm [4]. I-Relief treats the nearest neighbors and identity of a pattern as hidden random variables, and iteratively estimates feature weights until convergence. We provide a convergence theorem for I-Relief, which shows that under certain conditions I-Relief converges to a unique solution irrespective of the initial starting points. We also extend I-Relief to multiclass problems. In Section 12.4, by using the fact that Relief optimizes a margin-based objective function, we propose a new multiclass Relief algorithm using a new multiclass margin definition. We also consider online learning for I-Relief. The new proposed I-Relief algorithms are based on batch learning. In the case where there exists a large number of training samples, online learning is computationally much more attractive. We develop an online I-Relief algorithm in Section 12.5, wherein a convergence theorem is also provided. To verify the effectiveness of the newly proposed algorithms and confirm the established theoretical results, we conduct some experiments in Section 12.7 on six UCI datasets and six microarray datasets. We finally conclude this chapter in Section 12.8.

Relief Algorithm

(1) **Initialization:** $\mathcal{D} = \{(\mathbf{x}_n, y_n)\}_{n=1}^{N}$, $w_i = 0$, $1 \leq i \leq I$, T;

(2) **for** $t = 1 : T$

    (3) Randomly select a pattern $\mathbf{x}$ from $\mathcal{D}$;

    (4) Find the nearest hit $\text{NH}(\mathbf{x})$ and miss $\text{NM}(\mathbf{x})$ of $\mathbf{x}$;

    (5) **for** $i = 1 : I$

        (6) Compute: $w_i = w_i + |\mathbf{x}^{(i)} - \text{NM}^{(i)}(\mathbf{x})| - |\mathbf{x}^{(i)} - \text{NH}^{(i)}(\mathbf{x})|$;

    (7) **end**

(8) **end**

FIGURE 12.1: Pseudo-code of Relief.

12.2 Mathematical Interpretation of Relief

We first present a brief review of Relief. The pseudo-code of Relief is presented in Fig. 12.1. Let $\mathcal{D} = \{(\mathbf{x}_n, y_n)\}_{n=1}^{N} \in \mathbb{R}^I \times \{\pm 1\}$ denote a training dataset, where N is the sample size and I is the data dimensionality. The key idea of Relief is to iteratively estimate the feature weights according to their ability to discriminate between neighboring patterns. In each iteration, a pattern $\mathbf{x}$ is randomly selected and then two nearest neighbors of $\mathbf{x}$ are found, one from the same class (termed the *nearest hit* or NH) and the other from the different class (termed the *nearest miss* or NM). The weight of the i-th feature is then updated as $w_i = w_i + |\mathbf{x}^{(i)} - \mathrm{NM}^{(i)}(\mathbf{x})| - |\mathbf{x}^{(i)} - \mathrm{NH}^{(i)}(\mathbf{x})|$, for $\forall i \in \mathbb{N}_I$.

We provide below a mathematical interpretation for the seemingly heuristic Relief algorithm. Following the margin definition in [7], we define the margin for pattern $\mathbf{x}_n$ as $\rho_n = d(\mathbf{x}_n - \mathrm{NM}(\mathbf{x}_n)) - d(\mathbf{x}_n - \mathrm{NH}(\mathbf{x}_n))$, where $d(\cdot)$ is a distance function. For the moment, we define $d(\mathbf{x}) = \sum_i |x_i|$, which is consistent with the distant function used in the original Relief algorithm. Other distance functions can also be used. Note that $\rho_n > 0$ if only if $\mathbf{x}_n$ is correctly classified by 1-NN. One natural idea is to scale each feature such that the averaged margin in a weighted feature space is maximized:

$$
\begin{aligned}
& \max_{\mathbf{w}} \sum_{n=1}^{N} \rho_n(\mathbf{w}) \\
= {} & \max_{\mathbf{w}} \sum_{n=1}^{N} \left(\sum_{i=1}^{I} w_i |\mathbf{x}_n^{(i)} - \mathrm{NM}^{(i)}(\mathbf{x}_n)| - \sum_{i=1}^{I} w_i |\mathbf{x}_n^{(i)} - \mathrm{NH}^{(i)}(\mathbf{x}_n)| \right) \\
& \text{s.t. } \|\mathbf{w}\|_2^2 = 1, \mathbf{w} \geqslant 0 ,
\end{aligned}
\tag{12.1}
$$

where $\rho_n(\mathbf{w})$ is the margin of $\mathbf{x}_n$ computed with respect to $\mathbf{w}$. The constraint $\|\mathbf{w}\|_2^2 = 1$ prevents the maximization from increasing without bound, and $\mathbf{w} \geqslant 0$ ensures that the learned weight vector induces a distance measure. By defining $\mathbf{z} = \sum_{n=1}^{N} |\mathbf{x}_n - \mathrm{NM}(\mathbf{x}_n)| - |\mathbf{x}_n - \mathrm{NH}(\mathbf{x}_n)|$, where $|\cdot|$ is the point-wise absolute operator, Equation (12.1) can be simplified as

$$
\begin{aligned}
\max_{\mathbf{w}} \ & \mathbf{w}^T \mathbf{z}, \\
\text{s.t. } & \|\mathbf{w}\|_2^2 = 1, \mathbf{w} \geqslant 0
\end{aligned}
\tag{12.2}
$$

By using the Lagrangian technique, the solution can be expressed as $\mathbf{w} = \frac{1}{2\lambda}(\mathbf{z} + \boldsymbol{\zeta})$, where λ and $\boldsymbol{\zeta} \geqslant 0$ are the Lagrangian multipliers. With the Karush-Kuhn-Tucker condition [3], namely, $\sum_i \zeta_i w_i = 0$, it is easy to verify the following three cases: (1) $z_i = 0 \Rightarrow \zeta_i = 0 \Rightarrow w_i = 0$; (2) $z_i > 0 \Rightarrow z_i + \zeta_i > 0 \Rightarrow w_i > 0 \Rightarrow \zeta_i = 0$; and (3) $z_i < 0 \Rightarrow \zeta_i > 0 \Rightarrow w_i = 0 \Rightarrow z_i = -\zeta_i$. It immediately follows that the optimum solution can be calculated in a closed form as $\mathbf{w} = (\mathbf{z})^+ / \|(\mathbf{z})^+\|_2$, where $(\mathbf{z})^+ = [\max(z_1, 0), \cdots, \max(z_I, 0)]^T$.

By comparing the expression of $\mathbf{w}$ with the update rule of Relief, we conclude that Relief is an online solution to the optimization scheme in Eq. (12.1). This is true except when $w_i = 0$ for $z_i \leq 0$, which usually corresponds to irrelevant features. From the above analysis, we note that Relief learns discriminant information locally through a highly nonlinear 1-NN classifier and solves a simple convex problem globally with a closed-form solution. In this sense, Relief combines the merits of both filter and wrapper methods, which clearly explains its simplicity and effectiveness.

Other distance functions can also be used. If Euclidean distance is used, the resulting algorithm is Simba [7]. However, Simba returns many local maxima, for which the mitigation offered in Simba is to restart the algorithm from several starting points. Hence, the acquisition of the global minimum is not guaranteed through its invocation.

12.3 Iterative Relief Algorithm

Two major drawbacks of Relief become clear from above analysis: First, the nearest neighbors are defined in the original feature space, which are highly unlikely to be the ones in the weighted space; second, the objective function optimized by Relief is actually the average margin. In the presence of outliers, some margins can take large negative values. In a highly noisy data case with a large amount of irrelevant features or mislabelling, the aforementioned two issues can become so severe that the performance of Relief may be greatly deteriorated. A heuristic algorithm, called ReliefF [11], has been proposed to address the first problem. ReliefF averages K, instead of just one, nearest neighbors in computing the sample margins. Empirical studies have shown that ReliefF can achieve significant performance improvement over the original Relief. As for the second problem, to our knowledge, no such algorithm exists. In this section, we propose an analytic solution capable of handling these two issues simultaneously.

12.3.1 Algorithm

We first define two sets, $\mathcal{M}_n = \{i : 1 \leq i \leq N, y_i \neq y_n\}$ and $\mathcal{H}_n = \{i : 1 \leq i \leq N, y_i = y_n, i \neq n\}$, associated with each pattern $\mathbf{x}_n$. Suppose now that we have known, for each pattern $\mathbf{x}_n$, its nearest hit and miss, the indices of which are saved in the set $\mathcal{S}_n = \{(s_{n1}, s_{n2})\}$, where $s_{n1} \in \mathcal{M}_n$ and $s_{n2} \in \mathcal{H}_n$. For example, $s_{n1} = 1$ and $s_{n2} = 2$ mean that the nearest miss and hit of $\mathbf{x}_n$ are $\mathbf{x}_1$ and $\mathbf{x}_2$, respectively. We also denote $\mathbf{o} = [o_1, \cdots, o_N]^{\mathrm{T}}$ as a set of binary parameters, such that $o_n = 0$ if $\mathbf{x}_n$ is an outlier, or $o_n = 1$ otherwise. Then the objective function we want to optimize may be formulated as $C(\mathbf{w}) = \sum_{\{n=1,o_n=1\}}^{N} \left(\|\mathbf{x}_n - \mathbf{x}_{s_{n1}}\|_{\mathbf{w}} - \|\mathbf{x}_n - \mathbf{x}_{s_{n2}}\|_{\mathbf{w}} \right)$, which can be easily optimized by using the conclusion drawn in Section 12.2. Of course, we do not know the set $\mathcal{S} = \{\mathcal{S}_n\}_{n=1}^{N}$ and the vector $\mathbf{o}$. However, if we assume the elements of $\{\mathcal{S}_n\}_{n=1}^{N}$ and $\mathbf{o}$ are random variables, we can proceed by deriving the probability distributions of the unobserved data. We first make a guess on the weight vector $\mathbf{w}$. By using the pairwise distances that have been computed when searching for the nearest hits and misses, the probability of the i-th data point being the nearest miss of $\mathbf{x}_n$ can be defined as

$$P_m(i|\mathbf{x}_n, \mathbf{w}) = \frac{f(\|\mathbf{x}_n - \mathbf{x}_i\|_{\mathbf{w}})}{\sum_{j \in \mathcal{M}_n} f(\|\mathbf{x}_n - \mathbf{x}_j\|_{\mathbf{w}})}$$

Similarly, the probability of the i-th data point being the nearest hit of $\mathbf{x}_n$ is

$$P_h(i|\mathbf{x}_n, \mathbf{w}) = \frac{f(\|\mathbf{x}_n - \mathbf{x}_i\|_{\mathbf{w}})}{\sum_{j \in \mathcal{H}_n} f(\|\mathbf{x}_n - \mathbf{x}_j\|_{\mathbf{w}})}$$

and the probability of $\mathbf{x}_n$ being an outlier can be defined as:

$$P_o(o_n = 0|\mathcal{D}, \mathbf{w}) = \frac{\sum_{i \in \mathcal{M}_n} f(\|\mathbf{x}_n - \mathbf{x}_i\|_{\mathbf{w}})}{\sum_{\mathbf{x}_i \in \mathcal{D} \backslash \mathbf{x}_n} f(\|\mathbf{x}_n - \mathbf{x}_i\|_{\mathbf{w}})} \qquad (12.3)$$

where $f(\cdot)$ is a kernel function. One commonly used example is $f(d) = \exp(-d/\sigma)$, where the kernel width σ is a user-defined parameter. Throughout the chapter, the exponential kernel is used. Other kernel functions can also be used, and the descriptions of their properties can be found in [1].

Now we are ready to derive the following iterative algorithm. Although we adopt the idea of the EM algorithm that treats unobserved data as random variables, it should be noted that the following method is not an EM algorithm since the objective function is not a likelihood. For brevity of notation, we define $\alpha_{i,n} = P_m(i|\mathbf{x}_n, \mathbf{w}^{(t)}), \beta_{i,n} = P_h(i|\mathbf{x}_n, \mathbf{w}^{(t)}), \gamma_n = 1 - P_o(o_n = 0|\mathcal{D}, \mathbf{w}^{(t)}), \mathcal{W} = \{\mathbf{w} : \|\mathbf{w}\|_2 = 1, \mathbf{w} \geq 0\}, \mathbf{m}_{n,i} = |\mathbf{x}_n - \mathbf{x}_i|$ if $i \in \mathcal{M}_n$, and $\mathbf{h}_{n,i} = |\mathbf{x}_n - \mathbf{x}_i|$ if $i \in \mathcal{H}_n$.

Step 1: After the t-th iteration, the Q function is calculated as

$$Q(\mathbf{w}|\mathbf{w}^{(t)}) = \mathrm{E}_{\{S,\mathbf{o}\}}[C(\mathbf{w})]$$
$$= \sum_{n=1}^{N} \gamma_n \Big(\sum_{i \in \mathcal{M}_n} \alpha_{i,n} \|\mathbf{x}_n - \mathbf{x}_i\|_{\mathbf{w}} - \sum_{i \in \mathcal{H}_n} \beta_{i,n} \|\mathbf{x}_n - \mathbf{x}_i\|_{\mathbf{w}} \Big)$$
$$= \sum_{n=1}^{N} \gamma_n \Big(\underbrace{\sum_j w_j \sum_{i \in \mathcal{M}_n} \alpha_{i,n} m_{n,i}^j}_{\bar{m}_n^j} - \underbrace{\sum_j w_j \sum_{i \in \mathcal{H}_n} \beta_{i,n} h_{n,i}^j}_{\bar{h}_n^j} \Big) \qquad (12.4)$$
$$= \mathbf{w}^T \sum_{n=1}^{N} \gamma_n (\bar{\mathbf{m}}_n - \bar{\mathbf{h}}_n) = \mathbf{w}^T \boldsymbol{\nu}$$

Step 2: The re-estimation of $\mathbf{w}$ in the $(t+1)$-th iteration is

$$\mathbf{w}^{(t+1)} = \arg \max_{\mathbf{w} \in \mathcal{W}} Q(\mathbf{w}|\mathbf{w}^{(t)}) = (\boldsymbol{\nu})^+ / \|(\boldsymbol{\nu})^+\|_2$$

The above two steps iterate alternatively until convergence, i.e., $\|\mathbf{w}^{(t+1)} - \mathbf{w}^{(t)}\| < \theta$.

We name the above algorithm as iterative Relief, or I-Relief for short. Since P_m, P_h, and P_o return us with reasonable probability estimates and the re-estimation of $\mathbf{w}$ is a convex optimization problem, we expect a good convergence behavior and reasonable performance from I-Relief. We provide a convergence analysis below.

12.3.2   Convergence Analysis

We begin by studying the asymptotic behavior of I-Relief. If $\sigma \to +\infty$, we have $\lim_{\sigma \to +\infty} P_m(i|\mathbf{x}_n, \mathbf{w}) = 1/|\mathcal{M}_n|$ for $\forall \mathbf{w} \in \mathcal{W}$ since $\lim_{\sigma \to +\infty} f(d) = 1$. On the other hand, if $\sigma \to 0$, by assuming that for $\forall n, d_{i,n} \triangleq \|\mathbf{x}_i - \mathbf{x}_n\|_\mathbf{w} \neq d_{j,n}$ if $i \neq j$, it can be shown that $\lim_{\sigma \to 0} P_m(i|\mathbf{x}_n, \mathbf{w}) = 1$ if $d_{in} = \min_{j \in \mathcal{M}_n} d_{jn}$ and 0 otherwise. $P_h(i|\mathbf{x}_n, \mathbf{w})$ and $P_o(n|\mathbf{w})$ can be computed similarly. We observe that if $\sigma \to 0$, I-Relief is equivalent to iterating the original Relief (NM = NH = 1) provided that outlier removal is not considered. In our experiments, we rarely observe that the resulting algorithm converges. On the other hand, if $\sigma \to +\infty$, I-Relief converges in one step because the term $\boldsymbol{\nu}$ in Eq. (12.4) is a constant vector for any initial feature weights. This suggests that the convergence behavior of I-Relief and the convergent rates are fully controlled by the choice of the kernel width. In the following, we present a proof by using the Banach fixed point theorem. We first state the theorem without proof. For detailed proofs, we refer the interested reader to [12].

DEFINITION 12.1   *Let $\mathcal{U}$ be a subset of a norm space $\mathcal{Z}$, and $\|\cdot\|$ is a norm defined in $\mathcal{Z}$. An operator $T : \mathcal{U} \to \mathcal{Z}$ is called a contraction operator if there exists a constant $q \in [0,1)$ such that $\|T(x) - T(y)\| \leq q\|x - y\|$ for $\forall x, y \in \mathcal{U}$. q is called the contraction number of T.*

DEFINITION 12.2   *An element of a norm space $\mathcal{Z}$ is called a fixed point of $T : \mathcal{U} \to \mathcal{Z}$ if $T(x) = x$.*

THEOREM 12.1 (Banach Fixed Point Theorem)
Let T be a contraction operator mapping a complete subset $\mathcal{U}$ of a norm space $\mathcal{Z}$ into itself. Then the sequence generated as $x^{(t+1)} = T(x^{(t)})$, $t = 0, 1, 2, \cdots$ with arbitrary $x^{(0)} \in \mathcal{U}$ converges to the unique fixed point x^* of T. Moreover,

the following error bounds hold:

$$\|x^{(t)} - x^*\| \leq \frac{q^t}{1-q}\|x^{(1)} - x^{(0)}\|$$
$$\text{and } \|x^{(t)} - x^*\| \leq \frac{q}{1-q}\|x^{(t)} - x^{(t-1)}\|$$ (12.5)

In order to apply the fixed point theorem to prove the convergence of I-Relief, the gist is to identify the contraction operator in I-Relief and check if all conditions in Theorem 12.1 are met. To this end, let $\mathcal{P} = \{\mathbf{p} : \mathbf{p} = [P_m, P_h, P_o]\}$ and we specify the two steps of I-Relief in a functional form as $A1 : \mathcal{W} \rightarrow \mathcal{P}, A1(\mathbf{w}) = \mathbf{p}$ and $A2 : \mathcal{P} \rightarrow \mathcal{W}, A2(\mathbf{p}) = \mathbf{w}$. By indicating the functional composition by a circle ($\circ$), I-Relief can be written as $\mathbf{w}^{(t)} = (A2 \circ A1)(\mathbf{w}^{(t-1)}) \triangleq T(\mathbf{w}^{(t-1)})$, where $T : \mathcal{W} \rightarrow \mathcal{W}$. Since $\mathcal{W}$ is a closed subset of a norm space $\mathcal{R}^I$ and complete, T is an operator mapping a complete subset $\mathcal{W}$ into itself. However, it is difficult to directly verify that T is a contraction operator satisfying Definition 12.1. Noting that for $\sigma \rightarrow +\infty$, I-Relief converges with one step, we have $\lim_{\sigma \rightarrow +\infty} \|T(\mathbf{w}_1, \sigma) - T(\mathbf{w}_2, \sigma)\| = 0$ for $\forall \mathbf{w}_1, \mathbf{w}_2 \in \mathcal{W}$. Therefore, in the limit, T is a contraction operator with contraction constant $q = 0$, that is, $\lim_{\sigma \rightarrow +\infty} q(\sigma) = 0$. Therefore, for $\forall \varepsilon > 0$, there exists a $\bar{\sigma}$ such that $q(\sigma) \leq \varepsilon$ whenever $\sigma > \bar{\sigma}$. By setting $\varepsilon < 1$, the resulting operator T is a contraction operator. Combining the above arguments, we establish the following convergence theorem for I-Relief.

THEOREM 12.2

Let I-Relief be defined as above. There exists a $\bar{\sigma}$ such that $\lim_{t \rightarrow +\infty} \|\mathbf{w}^{(t)} - \mathbf{w}^{(t-1)}\| = 0$ for $\forall \sigma > \bar{\sigma}$. Moreover, for a fixed $\sigma > \bar{\sigma}$, I-Relief converges to the unique solution for any initial weight $\mathbf{w}^{(0)} \in \mathcal{W}$.

Theorem 12.2 ensures the convergence of I-Relief but does not tell us how large a kernel width should be. In our experiment, we find that with a relatively large σ value, say $\sigma > 0.5$, the convergence is guaranteed. Also, the error bound in Ineq. (12.5) tells us that the smaller the contraction number q, the tighter the error bound and hence the larger the convergence rate. Since it is difficult to explicitly express q as a function of σ, it is difficult to prove that q monotonically decreases with σ. However, in general, a larger kernel width yields a larger convergence rate, which is experimentally confirmed in Section 12.7.3. It is also worthwhile to emphasize that, unlike other machine learning algorithms, such as neural networks, the convergence and the solution of I-Relief are not affected by the initial value if the kernel width is fixed. We experimentally find that setting the initial feature weights all to be $1/I$ can only lead to a slight but negligible improvement of the convergence rate compared to a randomly generated initial value.

12.4 Extension to Multiclass Problems

The original Relief algorithm can only handle binary problems. ReliefF overcomes this limitation by modifying the weight update rule as

$$w_i = w_i + \sum_{\{c \in \mathcal{Y}, c \neq y(\mathbf{x})\}} \frac{P(c)}{1 - P(y(\mathbf{x}))} |\mathbf{x}^{(i)} - \text{NM}_c^{(i)}(\mathbf{x})| - |\mathbf{x}^{(i)} - \text{NH}^{(i)}(\mathbf{x})| \quad (12.6)$$

where $\mathcal{Y} = \{1, \cdots, C\}$ is the label space, $\text{NM}_c(\mathbf{x})$ is the nearest miss of $\mathbf{x}$ from class c, and $P(c)$ is the *a priori* probability of class c. By using the conclusions drawn in Section 12.2, it can be shown that ReliefF is equivalent to defining a sample margin as

$$\rho = \sum_{\{c \in \mathcal{Y}, c \neq y(\mathbf{x})\}} \frac{P(c)}{1 - P(y(\mathbf{x}))} d(\mathbf{x} - \text{NM}_c(\mathbf{x})) - d(\mathbf{x} - \text{NH}(\mathbf{x})) \quad (12.7)$$

Note that a positive sample margin does not necessarily imply a correct classification. The extension of ReliefF to the iterative version is quite straightforward, and therefore we skip the detailed derivations here. We name the resulting algorithm as I-Relief-1.

From the commonly used margin definition for multiclass problems, however, it is more natural to define a margin as

$$\begin{aligned} \rho &= \min_{\{c \in \mathcal{Y}, c \neq y(\mathbf{x})\}} d(\mathbf{x} - \text{NM}_c(\mathbf{x})) - d(\mathbf{x} - \text{NH}(\mathbf{x})) \\ &= \min_{\{\mathbf{x}_i \in \mathcal{D} \backslash \mathcal{D}_{y(\mathbf{x})}\}} d(\mathbf{x} - \mathbf{x}_i) - d(\mathbf{x} - \text{NH}(\mathbf{x})) \end{aligned} \quad (12.8)$$

where $\mathcal{D}_c$ is a subset of $\mathcal{D}$ containing only the patterns from class c. Compared to the first definition, this definition regains the property that a positive sample margin corresponds to a correct classification. The derivation of the iterative version of multiclass Relief using the new margin definition, which we call I-Relief-2, is straightforward.

12.5 Online Learning

I-Relief is based on batch learning, i.e., feature weights are updated after seeing all of the training data. In case the amount of training data is enormous, or we do not have the luxury of seeing all of the data when starting training, online learning is computationally much more attractive than batch learning. In this section, we derive an online algorithm for I-Relief. Convergence analysis is also presented.

Recall that in I-Relief one needs to compute $\boldsymbol{\nu} = \sum_{n=1}^{N} \gamma_n(\bar{\mathbf{m}}_n - \bar{\mathbf{h}}_n)$. Analogously, in online learning, after the T-th iteration, we may consider computing $\boldsymbol{\nu}^{(T)} = \frac{1}{T}\sum_{t=1}^{T}\gamma^{(t)}(\bar{\mathbf{m}}^{(t)} - \bar{\mathbf{h}}^{(t)})$. Denote $\boldsymbol{\pi}^{(t)} = \gamma^{(t)}(\bar{\mathbf{m}}^{(t)} - \bar{\mathbf{h}}^{(t)})$. It is easy to show that $\boldsymbol{\nu}^{(T)} = \boldsymbol{\nu}^{(T-1)} + \frac{1}{T}(\boldsymbol{\pi}^{(T)} - \boldsymbol{\nu}^{(T-1)})$. By defining $\eta^{(T)} = 1/T$ as a learning rate, the above formulation states that the current estimate can be simply computed as a linear combination of the previous estimate and the current observation. Moreover, it suggests that other learning rates are possible. One simple example is to set $\eta^{(T)} = 1/aT$ with $a \in (0, 1]$. Below we establish the convergence property of online I-Relief. We first present a useful lemma without proof.

LEMMA 12.1

Let $\{a_n\}$ be a bounded sequence, i.e., for $\forall n$, $M_1 \leq a_n \leq M_2$. If $\lim_{n \to +\infty} a_n = a^*$, then $\lim_{n \to +\infty} \frac{1}{n}\sum_{i=1}^{n} a_i = a^*$.

THEOREM 12.3

Online I-Relief converges when the learning rate is appropriately selected. If both algorithms converge, I-Relief and online I-Relief converge to the same solution.

PROOF The proof of the first part of the theorem can be easily done by recognizing that the above formulation has the same form as the Robbins-Moron stochastic approximation algorithm [13]. The conditions on the learning rate $\eta^{(t)}$: $\lim_{t \to +\infty} \eta^{(t)} = 0$, $\sum_{t=1}^{+\infty} \eta^{(t)} = +\infty$, and $\sum_{t=1}^{+\infty}(\eta^{(t)})^2 < +\infty$ ensure the convergence of online I-Relief. $\eta^{(t)} = 1/t$ meets the above conditions.

Now we prove the second part of the theorem. To eliminate the randomness, instead of randomly selecting a pattern from $\mathcal{D}$, we divide the data into blocks, denoted as $\mathcal{B}^{(m)} = \mathcal{D}$. Online I-Relief successively performs online learning over $\mathcal{B}^{(m)}, m = 1, 2, \cdots$. For the m-th block, denote $\tilde{\boldsymbol{\pi}}^{(m)} = \frac{1}{N}\sum_{t=(m-1)\times N+1}^{m \times N}\boldsymbol{\pi}^{(t)}$. After running over M blocks of data, we have $\boldsymbol{\nu}^{(M \times N)} = \frac{1}{M \times N}\sum_{t=1}^{M \times N}\boldsymbol{\pi}^{(t)} = \frac{1}{M}\sum_{m=1}^{M}\tilde{\boldsymbol{\pi}}^{(m)}$. From the proof of the first part, we know that $\lim_{t \to +\infty}\boldsymbol{\nu}^{(t)} = \boldsymbol{\nu}^*$. It follows that $\lim_{m \to +\infty}\tilde{\boldsymbol{\pi}}^{(m)} = \tilde{\boldsymbol{\pi}}^*$. Using Lemma 12.1, we have $\lim_{M \to +\infty}\boldsymbol{\nu}^{(M \times N)} = \tilde{\boldsymbol{\pi}}^* = \boldsymbol{\nu}^*$. The last equality is due to the fact that a convergent sequence cannot have two limits.

We prove the convergence of online I-Relief to I-Relief by using the uniqueness of the fixed point for a contraction operator. Recall that if the kernel width is appropriately selected, $T : \mathcal{W} \to \mathcal{W}$ is a contraction operator for I-Relief, i.e., $T(\mathbf{w}^*) = \mathbf{w}^*$. We then construct an operator $\tilde{T}$:

$\mathcal{W} \rightarrow \mathcal{W}$ for online I-Relief, which, in the m-th iteration, uses $\tilde{\mathbf{w}}^{(m-1)} = (\boldsymbol{\nu}^{((m-1) \times N)})^+ / \|(\boldsymbol{\nu}^{((m-1) \times N)})^+\|_2$ as input, and then computes $\boldsymbol{\nu}^{(m \times N)}$ by performing online learning on $\mathcal{B}^{(m)}$ and returns $\tilde{\mathbf{w}}^{(m)} = (\boldsymbol{\nu}^{(m \times N)})^+ / \|(\boldsymbol{\nu}^{(m \times N)})^+\|_2$. Since $\lim_{t \rightarrow +\infty} \boldsymbol{\nu}^{(t)} = \boldsymbol{\nu}^* = \tilde{\boldsymbol{\pi}}^*$, it follows that as $m \rightarrow +\infty$, we have $\tilde{T}(\tilde{\mathbf{w}}^*) = \tilde{\mathbf{w}}^*$, where $\tilde{\mathbf{w}}^* = (\boldsymbol{\nu}^*)^+ / \|\boldsymbol{\nu}^*\|_2$. Therefore, $\tilde{\mathbf{w}}^*$ is the fixed point of $\tilde{T}$. The only difference between T and $\tilde{T}$ is that $\tilde{T}$ performs online learning while T does not. Since $\{\boldsymbol{\nu}^{(t)}\}$ is convergent, it is also a Cauchy sequence. In other words, as $m \rightarrow +\infty$, the difference between every pair of $\boldsymbol{\nu}$ within one block goes to zero with respect to some norms. The operator $\tilde{T}$, therefore, is identical to T in the limit. It follows that $\tilde{\mathbf{w}}^* = \mathbf{w}^*$, since otherwise there would be two fixed points for a contraction operator, which contradicts Theorem 12.1.   ⬜

12.6 Computational Complexity

One major advantage of Relief and its variations over other algorithms is their computational efficiency. The computational complexities of Relief, I-Relief and online I-Relief are $\mathcal{O}(TNI)$, $\mathcal{O}(TN^2I)$, and $\mathcal{O}(TNI)$, respectively, where T is the number of iterations, I is the feature dimensionality, and N is the number of data points. If Relief runs over the entire dataset, i.e., $T = N$, then the complexity is $\mathcal{O}(N^2I)$. In the following section, we show that online I-Relief can attain similar solutions to I-Relief after one pass of the training data. Therefore, the computational complexity of online I-Relief is of the same order as that of Relief.

12.7 Experiments

12.7.1 Experimental Setup

We conducted large-scale experiments to demonstrate the effectiveness of the proposed algorithms and to study their behavior. Since in most practical applications one typically does not know the true feature set, it is necessary to conduct experiments in a controlled manner. We performed experiments on two test-beds. The first test-bed contains six datasets: *twonorm, waveform, ringnorm, f-solar, thyroid*, and *segmentation*, all publicly available at the UCI Machine Learning Repository [2]. The data information is summarized in Table 17.2. We added 50 independently Gaussian distributed irrelevant fea-

TABLE 12.1: Data summary of six UCI and six microarray datasets.

Dataset	Train	Test	Feature	Class
twonorm	400	7000	20	2
waveform	400	4600	21	2
ringnorm	400	7000	20	2
f-solar	666	400	9	2
thyroid	140	75	5	2
segmentation	210	2100	19	7
9-tumors	60	/	5726	9
Brain-tumor2	60	/	10367	4
Leukemia-1	72	/	5327	3
Prostate-tumors	83	/	2308	4
SRBCT	102	/	10509	2
DLBCL	77	/	5469	2

tures to each pattern, representing different levels of signal-to-noise ratios[1]. In real applications, it is also possible that some patterns are mislabeled. To evaluate the robustness of each algorithm against mislabeling, we introduced noise to the training data but kept the testing data intact. The level of noise represents a percentage of randomly selected training data for which its class labels are changed.

The second test-bed contains six microarray datasets: *9-tumors* [17], *Brain-tumor2* [14], *Leukemia-1* [8], *prostate-tumors* [16], *DLBCL* [15], and *SRBCT* [9]. Except for *prostate-tumors* and *DLBCL*, the remaining four datasets are multiclass problems (from three to nine classes). One characteristic of microarray data, different from most of the classification problems we encounter, is the extremely large feature dimensionality (from 2308 to 10509) compared to the small sample numbers (from 60 to 102). The data information is presented in Table 17.2. For all of the datasets, except for a simple scaling of each feature value to be between 0 and 1 as required in Relief, no other preprocessing was performed.

We used two metrics to evaluate the performance of the feature weighting algorithms. In most applications, feature weighting is performed for selecting a small feature subset to defy the curse of dimensionality. Therefore, a natural choice of a performance metric is classification errors. The classification-error metric, however, may not be able to fully characterize algorithmic performance. We found experimentally that in some cases, including a few irrelevant features may not change classification errors significantly. Indeed, improving classification performance sometimes is not the only purpose for performing feature weighting. In applications where the acquisition of data is quite expensive, including some useless features is highly undesirable. For microarray data, including irrelevant genes may complicate subsequent research. This consideration was the main motivation for us to add 50 useless features to the original feature sets in the UCI datasets. We treat feature selection as a target recognition problem. Though the features in the original feature sets may be weakly relevant or even useless, it is reasonable to assume that the original features contain at least the same or more information than the useless ones that are added artificially. By changing a threshold, we can plot a receiver operating characteristic (ROC) curve [6] that gives us a direct view on the capabilities of each algorithm to identify useful features and at the same time rule out useless ones. However, as the classification-error metric, the ROC metric is not exclusive. Some algorithms are down-biased and tend to assign zero weights to not only useless features but also to some presumably useful features in original feature sets (c.f. Fig. 12.3), resulting in a small area under a ROC curve. Since we do not know the true status of the features in the original feature sets, in this case, we need to check classification errors to see if the studied algorithm does select all of the useful features.

12.7.2   Experiments on UCI Datasets

We first performed experiments on the UCI datasets. For binary problems, we compared I-Relief with ReliefF and Simba. For multiclass problems, we compared ReliefF with I-Relief-1 and I-Relief-2.

To make the experiment computationally feasible, we used KNN to estimate classification errors for each feature weighting algorithm. KNN is certainly not an optimal classifier for each dataset. However, the focus of the chapter is not on the optimal classification but on feature weighting. KNN provides us with a platform where we can compare different algorithms fairly with a reasonable computational cost. The number of the nearest neighbors K was estimated through a stratified 10-fold cross validation using training data. We did not spend extra effort on re-estimating K when only a subset of features were used in training and testing, rather opting to use the one estimated in the original feature space. Though the value of K is surely not optimal, we found that it is fair for each algorithm.

The kernel width σ is the only free parameter in I-Relief. We show in Section 12.7.3 that σ is not a critical parameter. Nevertheless, we estimated it through 10-fold cross validation in the experiment. One problem associated with the estimation with cross validation using classification errors as criterion is that it requires us to specify the optimal number of features used in KNN. To overcome this difficulty, the following heuristic method was used: For a given candidate of σ, feature weights were estimated, and then KNN was performed in the induced weighted feature space [18]. The optimal σ was then chosen as the one with the smallest classification error. Likewise, we found the number of NH and NM in ReliefF through cross validation, rather than presetting it to 10 as suggested in [11]. The code of Simba used in the study was downloaded from [7]. As we have discussed in Section 12.2, there are some local maxima in Simba's objective function. Simba tries to overcome this problem by performing a gradient ascent from serval different starting points. We set the number of starting points to be 5, which is the default value of Simba. Also, we set the number of passes of the training data to be 5, the default value of which is 1.

To eliminate statistical variations, each algorithm was run 20 times for each dataset. In each run, a dataset was randomly partitioned into training and testing, and 50 irrelevant features were added. The averaged testing errors of KNN as a function of the number of the top ranked features and the ROC curves of the algorithms are plotted in Fig. 12.2. (In the notation 50/10, the first number refers to the number of irrelevant features and the second one to the percentage of mislabeled samples.) As a reference, the classification errors of KNN on the clean data (without irrelevant features and mislabeling) and noisy data are reported in Table 12.2. From these experimental results, we arrive at the following observations.

(1) The performance of KNN is degraded significantly in the presence of a large amount of irrelevant features, as reported in the literature, while mislabeling has less influence on the performance of KNN than irrelevant features.

(2) From Fig. 12.2, we can see that with respect to classification errors, in nearly all of the datasets, I-Relief performs the best, ReliefF the second, and Simba the worst. For a more rigorous comparison between I-Relief and ReliefF, a Student's paired two-tailed t-test was performed. The p-value of the t-test reported in Table 12.2 represents the probability that two sets of compared samples come from distributions with equal means. The smaller the p-value, the more significant the difference of the two average values is. At the 0.03 p-value level, I-Relief wins on seven cases (*ringnorm* (50/10), *twonorm* (50/10), *thyroid* (50/0), *waveform*, and *f-solar*), and ties with ReliefF on the remaining five cases. As we argued before, the classification-error metric may not fully characterize algorithmic performance. Therefore, we checked the ROC curves plotted in Fig. 12.2. In almost all of the datasets, I-Relief has the largest area under an ROC curve, ReliefF the second, and Simba the smallest. For three cases (*ringnorm* (50/0), *heart* (50/10), and *thyroid* (50/10)) that have no significant differences in classification errors, it is clear from the ROC curves that I-Relief performs much better than ReliefF with respect to the ROC metric. This suggests that when comparing feature selection and weighting algorithms, using classification errors as the only performance metric may not be enough.

To further demonstrate the performance of each algorithm, we particularly focused on *waveform* datasets. We plotted the learned feature weights of one realization in Fig. 12.3. For ease of comparison, the maximum value of each feature weight vector is normalized to be 1. Without mislabeling, the weights learned in ReliefF are similar to those of I-Relief, but the former have larger weights on the useless features than the latter. It is interesting to note that Simba assigns zero weights to not only useless features but also to some presumably useful ones. In this case, we need to go back to the classification-error metric. Particularly, for *waveform* (50/0), we observe that the testing error of Simba becomes flat after the tenth feature since, except for these 10 features,

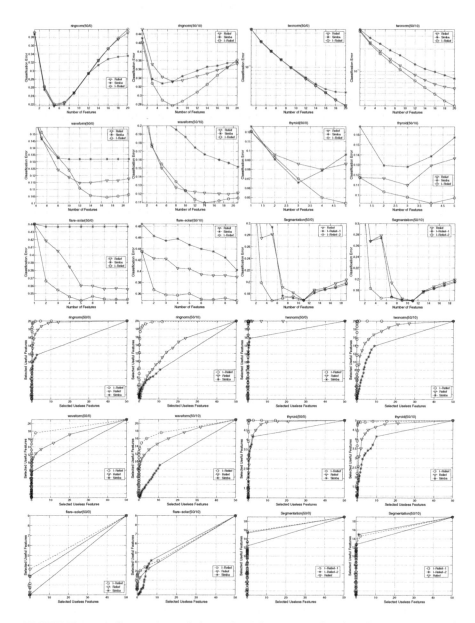

FIGURE 12.2: Comparison of three algorithms using the classification error and ROC metrics on six UCI datasets.

TABLE 12.2: The testing errors and standard deviations (%) on six UCI datasets. The last row (W/L/T) summarizes win/loss/tie in comparing Relief and I-Relief based on the 0.03 *p*-value level.

Dataset	KNN (clean data)	Mislabel	KNN (noisy data)	I-Relief	Relief	P-value
Ringnorm	39.2(1.3)	0%	45.1(1.2)	22.0(1.2)	21.7(1.1)	0.47
		10%	44.2(1.1)	**28.1(1.5)**	**34.0(4.5)**	**0.00**
Twonorm	3.1(0.2)	0%	4.8(0.6)	3.1(0.7)	3.2(0.5)	0.96
		10%	6.4(0.7)	**3.7(0.7)**	**6.2(1.3)**	**0.00**
Waveform	12.6(0.7)	0%	14.2(1.7)	**10.5(1.1)**	**11.2(1.1)**	**0.03**
		10%	14.7(1.6)	**11.2(1.2)**	**12.2(1.3)**	**0.00**
Thyroid	4.4(2.4)	0%	24.1(3.8)	**5.8(3.2)**	**8.7(4.3)**	**0.02**
		10%	26.0(4.1)	9.8(3.8)	11.3(3.6)	0.20
F-solar	34.8(2.4)	0%	34.5(2.6)	**34.5(3.3)**	**37.1(3.8)**	**0.03**
		10%	36.1(1.7)	**35.1(2.1)**	**38.7(3.7)**	**0.00**
Segment	12.5(1.4)	0%	27.9(1.7)	17.0(1.4)	17.7(1.7)	0.17
		10%	29.2(1.8)	17.3(1.4)	17.4(1.2)	0.92
					W/T/L	**=9/9/0**

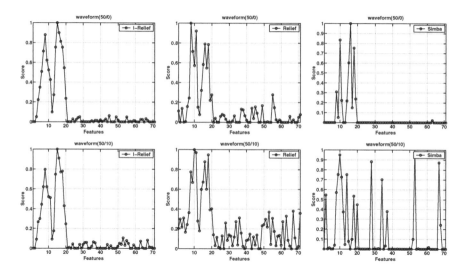

FIGURE 12.3: Feature weights learned in three algorithms on *waveform* dataset. The first 21 features are presumably useful.

the weights of the remaining features are all zero. This implies that Simba in effect does not identify all of the useful features. With 10% mislabeling, the weight quality of both ReliefF and Simba degrades significantly, whereas I-Relief performs similarly as before. For example, for *waveform* (50/10), Simba mistakenly identifies an irrelevant feature as the top feature. These observations imply that both Simba and ReliefF are not robust against label noise.

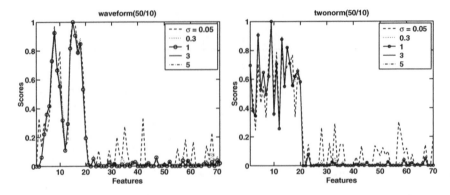

FIGURE 12.4: Feature weights learned using different σ values on the *twonorm* and *waveform* datasets.

12.7.3   Choice of Kernel Width

The kernel width σ in I-Relief can be estimated through cross validation on training data. It is well-known that the cross-validation method may result in an estimate with a large variance. Fortunately, this problem does not pose a serious concern. In this subsection, we show that σ is not a critical parameter. In Fig. 12.4, we plot the feature weights learned on *twonorm* and *waveform* using different σ values. We observe that for relatively large σ values, the resulting feature weights do not have much difference. This indicates that the performance of I-Relief is not sensitive to the choice of σ values, which makes model selection easy in practical applications.

We also conducted some experiments to confirm the convergence results established in Section 12.3.2. Plotted in Fig. 12.5(a) are the convergence rates of I-Relief with different σ values on the *waveform* dataset. We observe that the algorithm diverges when $\sigma = 0.05$ but converges in all other cases. Moreover, with the increase of σ values, the convergence becomes faster. In Fig. 12.5(b), we plotted the convergence rates of I-Relief with different initial values for a fixed kernel width. The line with stars is for the uniformly distributed initial value, and the line with circles for randomly generated initial values, both averaged from 10 runs. This experimental result confirms that I-Relief converges from *any* starting point, and using the uniform initial value does improve convergence, but the improvement is negligible.

12.7.4   Online Learning

In this subsection, we perform some experiments to verify the convergence properties of online I-Relief established in Section 12.5. The feature weights learned in I-Relief are used as a target vector. The stopping criterion θ is set to be 10^{-5} to ensure that the target vector is a good approximation of the true solution (c.f. Ineq. (12.5)). The convergence results with different

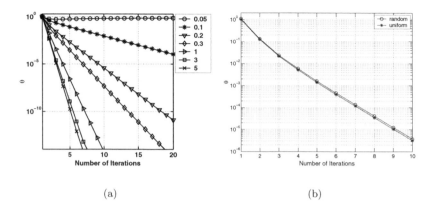

(a)   (b)

FIGURE 12.5: The convergence rates of I-Relief using (a) different σ, and (b) different initial values on the *waveform* dataset. The y-axis $\theta = \|\mathbf{w}^{(t+1)} - \mathbf{w}^{(t)}\|$.

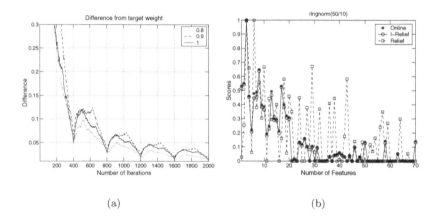

(a)   (b)

FIGURE 12.6: Convergence analysis of online I-Relief on the *ringnorm* dataset.

learning rates (different a in $\eta^{(t)} = 1/at$), averaged from 20 runs, are plotted in Fig. 12.6(a). We only present the results of *ringnorm* since the results for other datasets are almost identical. From the figure, we first observe that online I-Relief, regardless of the learning rates, converges to I-Relief, which confirms the theoretical findings in Theorem 12.3. We also find that after 400 iterations (*ringnorm* has only 400 training samples), the feature weights are already very close to the target vector. In Fig. 12.6(b), we plotted the target vector and the feature weights learned in online I-Relief (after 400 iterations). For comparison, the feature weights of ReliefF are also plotted.

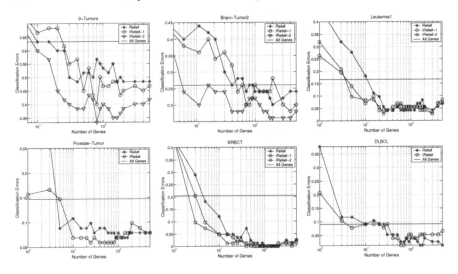

FIGURE 12.7: Classification errors on six microarray datasets.

## 12.7.5	Experiments on Microarray Data

We finally compare ReliefF to I-Relief on six microarray datasets. Due to the limited sample numbers, the leave-one-out method is used to evaluate the performance of each algorithm.

The classification errors of KNN as a function of the 500 top ranked features are plotted in Fig. 12.7. Since *Prostate-Tumor* and *DLBCL* are binary problems, I-Relief-1 is equivalent to I-Relief-2. From the figure, we observe that except for *DLBCL*, in which I-Relief performs similar to ReliefF, for the remaining five datasets, I-Relief-2 is the clear winner compared to Relief and I-Relief-1. For *Leukemia-1* and *SRBCT*, though the performances of the three algorithms all converge after 100 genes, it is clear that I-Relief is much more accurate than Relief in ranking genes. For comparison, we report the classification errors of KNN using all genes. We can see that gene selection can significantly improve the performance of KNN .

We note that the numbers of genes found by I-Relief that correspond to the minimum classification errors are all less than 200. With these small gene sets, oncologists may be able to work on them directly to infer the molecular mechanisms underlying disease causes. Also, for classification purposes, some computationally expensive methods such as wrapper methods can be used to further filter out some redundant genes. By using some sophisticated classification algorithms such as SVM, much improvement on classification performance is expected.

12.8   Conclusion

In this chapter, we have provided a mathematical interpretation of the seemingly heuristic Relief algorithm as an online method solving a convex optimization problem with a margin-based objective function. Starting from this new interpretation, we have proposed a set of new feature weighting algorithms. The key idea is to learn non-linear discriminant information of features *locally* and solve a set of convex optimization problems *globally*. Due to the existence of analytic solutions, these algorithms can be implemented extremely easily. The core module of these algorithms is just the finding of the nearest neighbors of each pattern. We have shown that the performance of our algorithms is not sensitive to the choice of the free parameter, which makes model selection easy in practical applications. Another merit of these algorithms is that the LOOCV-based objective function optimized by our algorithms provides built-in regularization to prevent from overfitting, and hence no explicit regularization is needed. We have conducted a large-scale experiment showing that our algorithms perform significantly better than Relief and Simba.

Considering the many heuristic approaches used in feature selection, we believe that the contribution of our work is not merely limited to the algorithmic aspects. The I-Relief algorithms are one of the first feature weighting methods that use the performance of a non-linear classifier as a guidance in searching for informative features and yet can be solved efficiently by using numerical analysis and optimization techniques, instead of combinatorial searching. They provide a promising direction for the future research of the difficult feature selection problem.

Notes

1 The signal-to-noise ratio refers to the ratio between the number of original features and that of the artificially added useless ones.

References

[1] C. G. Atkeson, A. W. Moore, and S. Schaal. Locally weighted learning. *Artificial Intelligence Review*, 11(15):11–73, 1997.

[2] C. Blake and C. Merz. UCI Repository of Machine Learning Databases, 1998.

[3] E. K. P. Chong and S. H. Zak. *An Introduction to Optimization.* John Wiley and Sons, New York, 2001.

[4] A. Dempster, N. Laird, and D. Rubin. Maximum likelihood from incomplete data via the EM algorithm. *Journal of the Royal Statistical Society B*, 39(1):1–38, 1977.

[5] T. G. Dietterich. Machine learning research: Four current directions. *AI Magazine*, 18(4):97–136, 1997.

[6] R. Duda, P. Hart, and D. Stork. *Pattern Classification.* John Wiley and Sons, New York, 2000.

[7] R. Gilad-Bachrach, A. Navot, and N. Tishby. Margin based feature selection - theory and algorithms. In *Proc. 21st International Conference on Machine Learning*, pages 43–50. ACM Press, 2004.

[8] T. Golub, D. Slonim, P. Tamayo, C. Huard, M. Gaasenbeek, J. Mesirov, H. Coller, M. Loh, J. Downing, M. Caligiuri, C. Bloomfield, and E. Lander. Molecular classification of cancer: class discovery and class prediction by gene expression monitoring. *Science*, 286(5439):531–537, October 1999.

[9] J. Khan, J. Wei, M. Ringner, L. Saal, M. Ladanyi, F. Westermann, F. Berthold, M. Schwab, C. Antonescu, C. Peterson, and P. Meltzer. Classification and diagnostic prediction of cancers using gene expression profiling and artificial neural networks. *Nature Medicine*, 7(6):673–679, 2001.

[10] K. Kira and L. A. Rendell. A practical approach to feature selection. In *Proc. 9th International Conference on Machine Learning*, pages 249 – 256. Morgan Kaufmann, 1992.

[11] I. Kononenko. Estimating attributes: Analysis and extensions of RELIEF. In *Proc. European Conference on Machine Learning*, pages 171–182, 1994.

[12] R. Kress. *Numerical Analysis.* Springer-Verlag, New York, 1998.

[13] H. Kushner and G. Yin. *Stochastic Approximation and Recursive Algorithms and Applications.* Springer-Verlag, New York, 2 edition, 2003.

[14] C. Nutt, D. Mani, R. Betensky, P. Tamayo, J. Cairncross, C. Ladd, U. Pohl, C. Hartmann, M. McLaughlin, T. Batchelor, P. Black, A. von Deimling, S. Pomeroy, T. Golub, and D. N. Louis. Gene expression-based classification of malignant gliomas correlates better with survival than histological classification. *Cancer Research*, 63(7):1602–1607, April 2003.

[15] M. Shipp, K. Ross, P. Tamayo, A. Weng, J. Kutok, R. Aguiar, M. Gaasen-

beek, M. Angelo, M. Reich, G. Pinkus, T. Ray, M. Koval, K. Last, A. Norton, T. Lister, J. Mesirov, D. Neuberg, E. Lander, J. Aster, and T. Golub. Diffuse large b-cell lymphoma outcome prediction by gene-expression profiling and supervised machine learning. *Nature Medicine*, 8(1):68–74, 2002.

[16] D. Singh, P. Febbo, K. Ross, D. Jackson, J. Manola, C. Ladd, P. Tamayo, A. Renshaw, A. D'Amico, J. Richie, E. Lander, M. Loda, P. Kantoff, T. Golub, and W. Sellers. Gene expression correlates of clinical prostate cancer behavior. *Cancer Cell*, 1(2):203–209, March 2002.

[17] J. Staunton, D. Slonim, H. Coller, P. Tamayo, M. Angelo, J. Park, U. Scherf, J. Lee, W. Reinhold, J. Weinstein, J. Mesirov, E. Lander, and T. Golub. Chemosensitivity prediction by transcriptional profiling. *Proc. Natl. Acad. Sci. USA*, 98(19):10787–10792, September 2001.

[18] D. Wettschereck, D. W. Aha, and T. Mohri. A review and empirical evaluation of feature weighting methods for a class of lazy learning algorithms. *Artificial Intelligence Review*, 11(1–5):273–314, 1997.

Part IV

Text Classification and Clustering

Chapter 13

Feature Selection for Text Classification

George Forman

Hewlett-Packard Labs

13.1 Introduction

Applications of text classification technology are becoming widespread. In the defense against spam email, suspect messages are flagged as potential spam and set aside to facilitate batch deletion. News articles are automatically sorted into topic channels and conditionally routed to individuals based on learned profiles of user interest. In content management, documents are categorized into multi-faceted topic hierarchies for easier searching and browsing. Shopping and auction Web sites do the same with short textual item descriptions. In customer support, the text notes of call logs are categorized with respect to known issues in order to quantify trends over time [3].These are but a few examples of how text classification is finding its way into applications. Readers are referred to the excellent survey by Sebastiani [13].

All these applications are enabled by standard machine learning algorithms, such as support vector machines (SVMs) and naïve Bayes variants, coupled with a pre-processing step that transforms the text string representation into a numeric feature vector. By far, the most common transformation is the "bag of words," in which each column of a case's feature vector corresponds to the number of times it contains a specific word of the training corpus. Strikingly, although this representation is oblivious to the order of the words in the document, it achieves satisfactory accuracy in most topic-classification applications. For intuition behind this: If the word "viagra" appears anywhere in an email message, regardless of its position, the probability that it is spam is much greater than if it had not appeared at all.

Rather than allocate every unique word in the training corpus to a distinct feature column, one can optionally perform feature selection to be more discriminating about which words to provide as input to the learning algorithm. This has two major motivations:

1. Accuracy (error rate, F-measure, ROC area, etc.): The accuracy of many learning algorithms can be improved by selecting the most predictive features. For example, naïve Bayes tends to perform poorly without feature selection in text classification settings. The purpose of feature selection is sometimes described as a need to eliminate useless noise words, but a study showed that even the lower ranked words continue to have predictive value [8]—only a small set of words are truly equally likely to occur in each class. Thus, feature selection may be viewed as selecting those words with the strongest signal-to-noise ratio. Pragmatically, the goal is to select whatever subset of features yields a highly accurate classifier.

2. Scalability: A large text corpus can easily have tens to hundreds of thousands of distinct words. By selecting only a fraction of the vocabulary as input, the induction algorithm may require a great deal less computation. This may also yield savings in storage or network bandwidth. These benefits could be an enabling factor in some applications, e.g., involving large numbers of classifiers to train or large numbers of cases.

Even so, the need for feature selection has been somewhat lessened by continuing advances in the accuracy and scalability of core machine learning algorithms. For example, Joachims recently demonstrated a new linear SVM classifier that can be trained on over 800,000 text cases with nearly 50,000 word features in less than 3 minutes on a 3.6GHz PC processor [9]. What is more, for some training sets, feature selection provides no improvement in accuracy. Hence, the additional complexity of feature selection can be omitted for many researchers who are not interested in feature selection, but simply need a fixed and easily replicable input representation.

Nevertheless, a data-mining practitioner faced with a given training set from which to produce the best possible classifier should not ignore feature selection. It can significantly boost accuracy for some datasets, and may at least produce modest improvements on average. Thus, feature selection still has a role to play for those who seek to maximize accuracy, e.g., industrial practitioners, application programmers, and contestants in data-mining competitions.

Moreover, the accuracy and scalability benefits accrue more substantially when one considers other possibilities for feature terms besides just individual words. For example, having a single feature representing the occurrence of the phrase "John Denver" can be far more predictive for some classification tasks than just having one feature for the word "John" and another for the word

"Denver." Other potentially useful features include any consecutive sequence of characters (n-grams) and, for domains that include multiple text fields (e.g., title, authors, abstract, keywords, body, and references), separate feature sets may be generated for each field or any combination of concatenated fields. It can be prohibitive simply to extend the bag of terms to include every potential feature that occurs in the training corpus. Thus, feature selection is also needed for scalability into large feature spaces. One can then search via cross validation to improve the input representation to the core induction algorithm. That is, different choices for feature generators can be tried, as well as different choices for feature selectors. In this way, the scalability improvements of feature selection can also benefit accuracy by extending the space that may be searched in a reasonable time.

In an ideal world, we might know, for any task domain, the best feature generator and feature selection method that dominates all others. However, in the research literature, no single dominant method appears. We must either choose one ourselves from among many reasonable options, or use cross validation to select one of many. If the latter, then our role becomes one of providing a sufficiently large (but not intractable) search space to cover good possibilities. This changes the game somewhat—we can propose features that might be useful, without having to assure their usefulness.

Section 13.2 describes a variety of common feature generators, which may be used to produce many potentially useful features. Section 13.3 describes the details of feature selection for binary and multi-class settings. Section 13.4 discusses the efficient evaluation of feature selection and the computational corners that may be cut for repeated evaluations. Section 13.5 illustrates the gains that the described methods can provide, both in selecting a subset of words and in selecting a good combination of feature generators. The remainder of this introduction describes the three major paradigms of feature selection, and the common characteristics of the text domain.

13.1.1   Feature Selection Phyla

There are three major paradigms of feature selection: *Filter methods* evaluate each feature independently with respect to the class labels in the training set and determine a ranking of all features, from which the top-ranked features are selected [1]. *Wrapper methods* use classic AI search methods—such as greedy hill-climbing or simulated-annealing—to search for the "best" subset of features, repeatedly evaluating different feature subsets via cross validation with a particular induction algorithm. *Embedded methods* build a usually linear prediction model that simultaneously tries to maximize the goodness-of-fit of the model and minimize the number of input features [6]. Some variants build a classifier on the full dataset, and then iteratively remove features the classifier depends on least [7]. By beginning with the full dataset, they qualify as the least scalable. Given large feature spaces, memory may be exceeded simply to realize the full feature vectors with all potential features. We will not consider such methods further. Filter methods are the simplest to implement and the most scalable. Hence, they are appropriate to treat very large feature spaces and are the focus here. They can also be used as a pre-processing step to reduce the feature dimensionality sufficiently to enable other, less scalable methods. Wrapper methods have traditionally sought specific combinations of individual features from the power set of features, but this approach scales poorly for the large number of features inherent with classifying text. Using cross validation to select among feature generators and optimize other parameters is somewhat like a wrapper method, but one that involves far fewer runs of the induction algorithm than typical wrapper feature selection.

13.1.2   Characteristic Difficulties of Text Classification Tasks

Besides the high dimensionality of the feature space, text classification problems are also characterized as frequently having a high degree of class imbalance. Consider training a text classifier to identify pages on any particular topic across the entire Web. High class skew is problematic for induction algorithms. If only 1% of the training cases are in the positive class, then the classifier can obtain 99% accuracy simply by predicting the negative class for all cases. Often the classifier must have its decision threshold adjusted in a separate post-processing phase, or else it must explicitly optimize for F-measure—which pressures it to increase recall of the positive class, without sacrificing too much precision.

One complication of high class skew is that even large training sets can end up having very few positive examples from which to characterize the positive class. Given 1% positives, a training set with 5000 randomly selected, manually labeled examples ends up with only 50 positives on average. This leads to significantly more uncertainty in the frequency estimates of words in the positive class than in the negative class. And if a predictive word is spelled "color" in half the positive cases and "colour" in the other half, then this *dispersion* of information into separate features yields more uncertainty. In technical notes or Web text, we often encounter misspellings, which may yield other variants, such as "collor." This problem is exacerbated by the fact that natural language provides many ways to express the same idea, e.g., hue, tint, shade, dye, or paint.

Another common aspect of text classification is that the large feature space typically follows a Zipf-like distribution [10]. That is, there are a few very common words, and very many words that rarely appear. By contrast, the most predictive features would be those that appear nearly always in one class, but not in the other.

Finally, text classification problems sometimes have only small amounts of training data available, perhaps more often than in other domains. This may partly be because a person's judgment is often needed to determine the topic label or interest level. By contrast, non-text classification problems may sometimes have their training sets labeled by machines, e.g., classifying which inkjet pens during manufacture ultimately fail their final quality test.

13.2   Text Feature Generators

Before we address the question of how to discard words, we must first determine what shall count as a word. For example, is "HP-UX" one word, or is it two words? What about "650-857-1501"? When it comes to programming, a simple solution is to take any contiguous sequence of alphabetic characters, or alphanumeric characters, which includes identifiers such as "ioctl32", which may sometimes be useful. By using the Posix regular expression \p{L&}+ we avoid breaking "naïve" in two, as well as many accented words in French, German, etc. But what about "win_32", "can't," or words that may be hyphenated over a line break? Like most data cleaning endeavors, the list of exceptions is endless, and one must simply draw a line somewhere and hope for an 80%-20% trade off. Fortunately, semantic errors in word parsing are usually only seen by the core learning algorithm, and it is their statistical properties that matter, not their readability or intuitiveness to people. Our purpose is to offer a range of feature generators so that the feature selector may discover the strongly predictive features. The most beneficial feature generators will vary according to the characteristics of the domain text.

13.2.1   Word Merging

One method of reducing the size of the feature space somewhat is to merge word variants together and treat them as a single feature. More importantly, this can also improve the predictive value of some features.

Forcing all letters to lowercase is a nearly ubiquitous practice. It normalizes for capitalization at the beginning of a sentence, which does not otherwise affect the word's meaning, and helps reduce the dispersion issue mentioned in the introduction. For proper nouns, it occasionally conflates with other words in the language, e.g., "Bush" or "LaTeX."

Likewise, various word stemming algorithms can be used to merge multiple related word forms. For example, "cat," "cats," "catlike," and "catty" may all be merged into a common feature. Various studies find that stemming typically benefits recall but at a cost of precision. If one is searching for "catty" and the word is treated the same as "cat," then a certain amount of precision is necessarily lost. For extremely skewed class distributions, this loss may be unsupportable.

Stemming algorithms make both over-stemming errors and under-stemming errors, but again, the semantics are less important than the feature's statistical properties. Unfortunately, stemmers must be separately designed for each natural language, and while many good stemmers are available for Romance languages, languages such as Hebrew and Arabic continue to be quite difficult to stem well. Another difficulty is that in some text classification applications, multiple natural languages are mixed together, sometimes even within a single training case. This would require a language recognizer to identify which stemming algorithm should be used on each case or each sentence. This level of complexity and slowdown is unwelcome. Simply taking the first few characters of each word may yield equivalent classification accuracy for many classification problems.

For classifying technical texts or blogs, misspellings may be common to rampant. Inserting an automatic spelling correction step into the processing pipeline is sometimes proposed, but the mistakes introduced may outweigh the purported benefit. One common problem is that out-of-vocabulary (OOV) words of the spell checker may be forced to the nearest known word, which may have quite a different meaning. This often happens with technical terms, which may be essential predictors. For misspellings that are common, the misspelled form may occur frequently enough to pose a useful feature, e.g., "volcanoe."

A common source of OOV words is abbreviations and acronyms, especially in governmental or technical texts. Where glossaries are available, the short and long forms may be merged into a single term. Although many acronym dictionaries are available online, there are many collisions for short acronyms, and they tend to be very domain-specific and even document-specific. Some research has shown success recognizing acronym definitions in text, such as "(OOV)" above, which provides a locally unambiguous definition for the term.

Online thesauruses can also be used to merge different words together, e.g., to resolve the "color" vs. "hue" problem mentioned in the introduction. Unfortunately, this approach rarely helps, as many words have multiple meanings, and so their meanings become distorted. To disambiguate word meanings correctly would require a much deeper understanding of the text than is needed for text classification. However, there are domain-specific situations where thesauruses of synonyms can be helpful. For example, if there is a large set of part numbers that correspond to a common product line, it could be very advantageous to have a single feature to represent this.

13.2.2 Word Phrases

Whereas merging related words together can produce features with more frequent occurrence (typically with greater recall and lower precision), identifying multiple word phrases as a single term can produce rarer, highly specific features (which typically aid precision and have lower recall), e.g., "John Denver" or "user interface." Rather than require a dictionary of phrases as above, a simple approach is to treat all consecutive pairs of words as a phrase term, and let feature selection determine which are useful for prediction. The recent trend to remove spaces in proper names, e.g., "SourceForge," provides the specificity of phrases without any special software consideration—perhaps motivated by the modern world of online searching.

This can be extended for phrases of three or more words with strictly decreasing frequency, but occasionally more specificity. A study by Mladenic and Grobelnik [12] found that most of the benefit is obtained by two-word phrases. This is in part because portions of the phrase may already have the same statistical properties, e.g., the four-word phrase "United States of America" is covered already by the two-word phrase "United States." In addition, the reach of a two-word phrase can be extended by eliminating common stopwords, e.g., "head of the household" becomes "head household." Stopword lists are language specific, unfortunately. Their primary benefit to classification is in extending the reach of phrases, rather than eliminating commonly useless words, which most feature selection methods can already remove in a language-independent fashion.

13.2.3 Character N-grams

The word identification methods above fail in some situations, and can miss some good opportunities for features. For example, languages such as Chinese and Japanese do not use a space character. Segmenting such text into words is complex, whereas nearly equivalent accuracy may be obtained by simply using every pair of adjacent Unicode characters as features—*n-grams*. Certainly many of the combinations will be meaningless, but feature selection can identify the most predictive ones. For languages that use the Latin character set, 3-grams or 6-grams may be appropriate. For example, n-grams would capture the essence of common technical text patterns such as "HP-UX 11.0", `while (<>) {`, `#!/bin/`, and `" :)."` Phrases of two adjacent n-grams simply correspond to (2n)-grams. Note that while the number of potential n-grams grows exponentially with n, in practice only a small fraction of the possibilities occur in actual training examples, and only a fraction of those will be found predictive.

Interestingly, the common Adobe PDF document format records the position of each character on the page, but does not explicitly represent spaces. Software libraries to extract the text from PDF use heuristics to decide where to output a space character. That is why text extracts are sometimes missing spaces between words, or have a space character inserted between every pair of letters. Clearly, these types of errors would wreak havoc with a classifier that depends on spaces to identify words. A more robust approach is for the feature generator to strip all whitespace, and generate n-grams from the resulting sequence.

13.2.4 Multi-Field Records

Although most research deals with training cases as a single string, many applications have multiple text (and non-text) fields associated with each record. In document management, these may be *title, author, abstract, keywords, body*, and *references*. In technical support, they may be *title, product, keywords, engineer, customer, symptoms, problem description*, and *solution*. The point is that multi-field records are common in applications, even though the bulk of text classification research treats only a single string. Furthermore, for classifying long strings such as arbitrary files, the first few kilobytes may be treated as a separate field and often prove sufficient for generating adequate features, avoiding the overhead of processing huge files, such as tar or zip archives.

The simplest approach is to concatenate all strings together. However, supposing the classification goal is to separate technical support cases by product, then the most informative features may be generated from the product description field alone, and concatenating all fields will tend to water down the specificity of the features.

Another simple approach is to give each field its own separate bag-of-words feature space. That is, the word "OfficeJet" in the title field would be treated as though it were unrelated to a feature for the same word in the product field. Sometimes multiple fields need to be combined while others are kept separate, and still others are ignored. These decisions are usually made manually today. Here, again, an automated search can be useful to determine an effective choice of which fields to combine and which to keep separate. This increases computation time for search, but more importantly it saves the expert's time. And it may discover better choices than would have been explored manually.

13.2.5 Other Properties

For some classification tasks, other text properties besides words or n-grams can provide the key predictors to enable high accuracy. Some types of spam use deceptions such as "4ree v!@gr@ 4 u!" to thwart word-based features, but these might easily be recognized by features revealing their abnormal word lengths and the density symbols. Likewise, to recognize Perl or awk code, the specific alphanumeric identifiers that appear are less specific than the distribution of particular keywords and special characters. Formatting information, such as the amount of whitespace, the word count, or the average number of words per line, can be key features for particular tasks.

Where task-specific features are constructed, they are often highly valuable, e.g., parsing particular XML structures that contain name-value pairs. By being task-specific, it is naturally difficult to make generally useful comments about their generation or selection. The little that is said about task-specific features in the research literature on general text classification belies their true importance in many practical applications.

13.2.6 Feature Values

Once a decision has been made about what to consider as a feature term, the occurrences of that term can be determined by scanning the texts. For some purposes, a binary value is sufficient, indicating whether the term appears at all. This representation is used by the Bernoulli formulation of the naïve Bayes classifier[11]. Many other classifiers use the term frequency $tf_{t,k}$ (the word count in document k) directly as the feature value, e.g., the multinomial naïve Bayes classifier[11].

The support vector machine (SVM) has proven highly successful in text classification. For such kernel methods, the distance between two feature vectors is typically computed as their dot product (cosine similarity), which is dominated by the dimensions with larger values. To avoid the situation where the highly frequent but non-discriminative words (such as stopwords) dominate the distance function, one can either use binary features or else weight the term frequency value $tf_{t,k}$ inversely to the feature's document frequency df_t in the corpus (the number of documents in which the word appears one or more times). In this way, very common words are downplayed. This idea, widely known as "TF.IDF," has a number of variants, one form being $tf_{t,k} \times log(\frac{M+1}{df_t+1})$. While this representation requires more computation and more storage per feature than simply using binary features, it can often lead to better accuracy for kernel methods.

If the document lengths vary widely, then a long document will exhibit larger word counts than a short document on the same topic. To make these feature vectors more similar, the $tf_{t,k}$ values may be normalized so that the length (Euclidean norm) of each feature vector equals 1.

13.3 Feature Filtering for Classification

With a panoply of choices for feature generation laid out, we now turn to feature filtering, which independently scores each feature with respect to the training class labels. The subsections below describe how this can be done for different classification settings. After the scoring is completed, the final issue is determining how many of the best features to select for the best performance. Unfortunately, the answer varies widely from task to task, so several values should be tried, including the option of using all features. This parameter can be tuned automatically via cross validation on the training data.

Cross validation may also be needed to select which feature generators to use, as well as selecting parameters for the induction algorithm, such as the well-known complexity constant C in the SVM model. The simplest to program is to optimize each parameter in its own nested loop. However, with each successive nesting a smaller fraction of the training data is being given to the induction algorithm. For example, if nested 5-fold cross validation is being used to select the feature generator, the number of features, and the complexity constant, the innermost loop trains with only half of the training set ($\frac{4}{5} \times \frac{4}{5} \times \frac{4}{5} = 51\%$). Unfortunately, for some parameters, the optimal value found in this way can be a poor choice for the full training set size. This is one reason why 10-fold cross validation, despite its computational cost, is usually preferred to 2-fold cross validation, which trains on nearly half as much of the data.

Instead, a single loop of cross validation should be married with a multi-parameter search strategy. The simplest to program is to measure the cross validation accuracy (or F-measure) at each point on a simple grid, and then select the best parameters. There is a large literature in multi-parameter optimization that has yielded methods that are typically much more efficient, if more complex to program.

13.3.1 Binary Classification

We first discuss the setting where there are two classes. Binary classification is a fundamental case, because (1) binary domain tasks are common, e.g., identifying spam email from good email, and (2) it is used as a subroutine to solve most types of multi-class tasks.

To clarify the nature of the problem, we demonstrate with an exemplary binary task: identifying papers about probabilistic machine learning methods among a collection of other computer science papers. The dataset has 1800 papers altogether, with only 50 of them in the positive class—2.8% positive. Each is represented by its title and abstract, which generate 12,500 alphanumeric words when treated as a single text string. Figure 13.1 shows the document frequency counts tp_t for each of the word features t with respect to the 50 documents in the positive class (y-axis) and separately fp_t for the 1750 documents in the negative class (x-axis), similar to an ROC graph. A feature in the topmost left corner (or bottommost right corner) would be perfectly predictive of the positive (negative) class, and would aid the induction algorithm a great deal. Unfortunately, these regions are typically devoid of features.

Common stopwords such as "of" and "the" occur frequently in both classes, and approximate the diagonal. These have no predictive value. The slightly larger points indicate which words appear on a generic list of 570 common English stopwords. Observe that the non-stopwords "paper" and "algorithm" behave like stopwords in this dataset, unsurprisingly. This illustrates that stopwords are not only language-specific, but also domain-specific.

Because of the Zipf-like distribution of words, most words occur rarely in each class. In fact, the majority of the points are plotted atop one another near the origin, belying their overwhelming density in this region. Over half of the words appear only once in the dataset and are plotted at just two

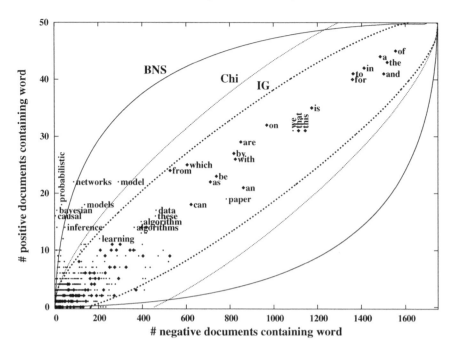

FIGURE 13.1: Word document frequencies in the positive and negative classes for a typical problem with class imbalance.

points—(1,0) and (0,1). One often removes such extremely rare words via a minimum count threshold df_{min}; in this case, $df_{min} = 2$ removes about half the features. Whether this is a good idea depends on the induction algorithm and the character of the dataset.

Filter methods typically evaluate each feature term t according to a function of its document frequency counts in the positive tp_t and negative fp_t classes. Three commonly used feature selection formulas are given in Table 13.1: Information Gain (IG), Chi-Squared (Chi), and Bi-Normal Separation (BNS).

Returning to Figure 13.1, the contour lines on the graph show the decision boundaries that these three feature selection methods would make for this dataset when the top 100 features are requested. That is, for each method, the points along its contour lines all evaluate to the same "goodness" value by its function, and there are 100 features with greater values. Naturally, they all devalue and remove the stopwords near the diagonal, without being language- or domain-specific, as stopword lists would be.

The features that are selected lie above the upper contour as well as below the matching lower contour; the contours are rotationally symmetric about the center point. Despite this symmetry, these two areas differ in character because the Zipfian word distribution focuses the selection decisions to be near the origin, and the feature selection methods each have an asymmetry. The

TABLE 13.1: Three common feature selection formulas, computed from document frequency counts in the positive and negative classes.

Name	Formula
Information Gain (IG)	$e(pos, neg) - [P_{word}e(tp, fp) + (1 - P_{word})e(fn, tn)]$ where $e(x, y) = -\text{xlx}(\frac{x}{x+y}) - \text{xlx}(\frac{y}{x+y})$, and $\text{xlx}(x) = x \log_2(x)$
Chi-Squared, χ^2 (Chi)	$g(tp, (tp + fp)P_{pos}) + g(fn, (fn + tn)P_{pos}) +$ $g(fp, (tp + fp)P_{neg}) + g(tn, (fn + tn)P_{neg})$ where $g(count, expect) = \frac{(count - expect)^2}{expect}$
Bi-Normal Separation (BNS)	$\lvert F^{-1}(tpr) - F^{-1}(fpr) \rvert$ where F^{-1} is the inverse of the Normal CDF

Notation:

pos:	number of positive cases $= tp + fn$
neg:	number of negative cases $= fp + tn$
tp:	true positives = number of positive cases containing the word
fp:	false positives = number of negative cases containing the word
fn:	false negatives
tn:	true negatives
tpr:	true positive rate $= tp/pos$
fpr:	false positive rate $= fp/neg$
P_{pos}:	percentage of positive cases $= pos/all$
P_{neg}:	percentage of negative cases $= neg/all$
P_{word}:	percentage of cases containing word $= (tp + fp)/all$

most noticeable asymmetry is that the chi-squared method results in a strong bias toward positive features; there are no features under its lower contour. Information gain selects some negative features, but still has a bias for positive features. This is more easily seen at the top, where there are no word points obscuring the place where the contour meets the top of the graph.

By contrast, the BNS decision boundary comes much closer to the origin on the x-axis. Compared to the other methods, BNS prefers many more of the negative features—in this case, only those occurring more than 50 times among the negatives and not once among the 50 positives. It is for this reason that BNS excels in improving recall, usually at a minor cost to precision. This trade off often yields an overall improvement in F-measure compared to other methods.

Why is BNS asymmetric, given that its formula is symmetric? It stems from the class skew. Since the inverse Normal cumulative distribution function (CDF) is undefined at zero, whenever there are zero occurrences of a word in a class, we must substitute a small constant, e.g., $\xi = 0.1$ occurrences. Since there are typically more negatives than positives, the minimum false positive rate $fpr = \frac{\xi}{neg}$ is smaller than the minimum true positive rate $tpr = \frac{\xi}{pos}$. In this way, a feature that occurs x times in only the majority class is correctly preferred to one that occurs x times in only the minority class.

Likewise, to avoid the undefined value $F^{-1}(1.0)$, if ever a feature occurs in every single positive (negative) case, we back off the tp (fp) count by ξ. This does not occur naturally with language texts, but text classification techniques are regularly used to treat string features of all sorts of classification problems. In industrial settings with many classes to process, it sometimes happens that there is a perfect indicator in the texts, e.g., `<meta name="Novell_ID" val="Win32">`, which may be discovered by long n-grams or phrases of alphanumeric words. Note that without feature selection, an SVM classifier will not make effective use of a few excellent features [4].

As a side note, sometimes the purpose of feature selection is just to characterize a class for user understanding rather than machine classification. In this case, ordinarily one only wants to see the positive words and phrases.

13.3.2 Multi-Class Classification

There are two major forms of multi-class classification: single-label (1-of-n) classification, where each case is known to belong in exactly one of the n classes, and multi-label (m-of-n) classification, where each case may belong to several, none, or even all classes.

In the multi-label case, the problem is naturally decomposed into n binary classification tasks: class$_i$ vs. not class$_i$. Each of these binary tasks is solved independently, and each can have its own feature selection to maximize its accuracy. In the single-label case, many induction algorithms operate by decomposing the problem into n binary tasks as above, and then making a final decision by some form of voting. Here also, feature selection can be optimized independently for each binary subtask. However, some 1-of-n induction algorithms do not perform binary decompositions, and need *multi-class feature selection* to select a single set of features that work well for the many classes. Other 1-of-n induction algorithms perform very many binary decompositions, e.g., those that search for optimal splitting hierarchies, or error-correcting code classifiers that consider $O(n^2)$ dichotomies. For these it would be impractical to perform a separate feature selection for each binary task.

Setting aside such incompatible induction algorithms, all multi-class tasks could be dealt with by binary decompositions in theory, and so there would be no need for multi-class feature selection. But practice often recants theory. The APIs for many good software products and libraries expect the transformation of text into numerical feature vectors to be performed as a pre-processing step, and there is no facility for injecting it into the inner loops where the decompositions occur. Even some m-of-n applications that can be programmed *de novo* demand multi-class feature selection for performance and scalability reasons. For example, where a centralized server must classify millions of objects on the network into multiple, orthogonal taxonomies, it can be much more efficient to determine a single, reasonably sized feature vector to send across the network than to send all the large documents themselves. As another example, a large database of unstructured, multi-field (technical support) cases is represented in memory by a cached, limited size feature vector representation. This is used for quick interactive exploration, classification, and labeling into multiple 1-of-n and m-of-n taxonomies, where the classifiers are periodically retrained in real time [3]. It would be impractical to re-extract features for each binary decomposition, or to union all the features into a very long feature vector that would be requested by all the binary feature selection subtasks.

Many multi-class feature selection schemes have been devised, and some methods such as Chi-squared naturally extend to multiple classes. However, most of them suffer from the following liability: Suppose in a typical, multi-class topic recognition problem in English, one of the classes happens to contain all German texts, which will generate many extremely predictive words. Nearly all feature selection schemes will prefer the stronger features, and myopically starve the other classes. Likewise, if one class is particularly difficult, multi-class feature selectors will tend to ignore it.

A solution to this problem is to perform feature selection for each class separately via binary decompositions, and then to determine the final ranking of features by a round-robin algorithm where each class gets to nominate its most desired features in turn [2]. This scheme was devised to improve robustness for such situations that arise in practice occasionally. Usually efforts to improve robustness come at some loss in average performance. Remarkably, this improves performance even for well-balanced research benchmarks. Why? Inevitably, some classes are easier to recognize than others, and this disparity causes most feature selection methods to slight the very classes that need more help.

13.3.3   Hierarchical Classification

Hierarchy is among the most powerful of organizing abstractions. *Hierarchical classification* includes a variety of tasks where the goal is to classify items into a set of classes that are arranged into a tree or directed acyclic graph, such as the Yahoo Web directory. In some settings, the task is a single-label problem to select 1-of-n nodes—or even restricted to the leaf classes in the case of a "virtual hierarchy." In other settings, the problem is cast as a multi-label task to select multiple interior nodes, optionally including all super-classes along the paths to the root.

Despite the offered hierarchy of the classes, these problems are sometimes treated simply as flat multi-class tasks, aggregating training examples up the tree structure for each class. Alternately, a top-down hierarchy of classifiers can be generated to match the class hierarchy. The training set for each step down the tree is composed of all the training examples under each child subtree, optionally including a set of items positioned at the interior node itself, which terminates the recursion. Although this decomposition of classes is different from a flat treatment of the problem, in either decomposition, the same single-label or multi-label feature selection methods apply to the many sub-problems. It has been suggested that each internal hierarchical classifier may be faster because each may depend on only a few features (selected by feature selection), and may be more accurate because it only considers cases within a limited context. For example, an interior node about recycling that has subtopics for glass recycling and can recycling would have a classifier under it that need only consider cases that have to do with recycling. In this way, the training sets for each of the interior classifiers may be more balanced than with a flat treatment of the problem.

13.4   Practical and Scalable Computation

We discuss briefly the matter of programming software feature selection, with some practical pointers for efficient implementation. These issues are usually not recorded in the research literature or in product documentation. Since feature selection is usually accompanied by multiple runs of cross validation to select the best number of features, it makes sense to save computation where possible, rather than run each fold from scratch.

We begin with the binary case. By dividing the training cases into F folds in advance, the true positive and false positive counts can be kept track of separately for each fold. It then becomes very fast to determine the tp, fp counts for any subset of folds using just 2F integers per feature. This makes feature-filtering methods extremely scalable, and requires only one pass through the dataset.

Furthermore, for M training cases, each fold has only M/F cases, and an 8-bit counter will often suffice. For the 1800 paper dataset—which altogether can generate over 300,000 word, phrase, 3-gram, 4-gram, and 5-gram features – we can efficiently support feature selection for 10-fold cross validation with less than 6MB of memory. This is nowadays an insignificant amount of memory and its computation takes only seconds on a PC. For 10-folds on $M \leq 640K$ cases, 100MB is sufficient for 2.6 million features. Moreover, likely half or more of the features will occur only once, and they can be discarded after one pass through the dataset, freeing up memory for inductions that follow in cross validation.

Large generated feature spaces need not be stored on the first pass through the dataset. Once the counts are made—possibly on a subset of the training cases—the best, say 100K, features are determined (if done carefully, considering the separate 10-fold views). Then a second feature-generation pass through the dataset stores only these useful features. The ensuing cross validations then work with ever decreasing subsets of the realized dataset.

In the multi-class setting, further optimization is possible. If the total number of distinct classes is C, then we can efficiently determine the tp, fp counts for "class vs. not class" binary subtasks using C+1 integers per feature (for this exposition, we ignore the orthogonal issue of the F-folds). The number of occurrences of the feature in each of the C classes is tracked separately, plus one integer tracks the total number of occurrences in the dataset. This missing fp counter is determined from the total minus the tp counter. Further, if we know the classes are mutually exclusive (1-of-n classification), then we can efficiently determine the tp, fp count for any dichotomy between sets of classes.

It is fortunate that feature selection for cross validation can be so efficient. The bottleneck is then the induction algorithm. Reducing the number of folds from 10-fold to 2-fold cuts the workload substantially, but the smaller training sets can yield different behavior that misleads the search for optimum parameters. For example, using smaller training sets on the two folds may prefer substantially fewer features than is optimal for the full training set.

Rather than reduce the data folds, early termination can be used. With only a few of the fold measurements completed, the current parameter settings may be deemed inferior to the best result found so far. For example, suppose the best parameters made 80 misclassifications on all 10 folds, and the current parameters have already committed 80 mistakes on just 3.2 of the folds. Early termination can be done more aggressively with various statistical methods. Or by being less conservative. After all, even with exhaustive evaluation of the folds, it is only a somewhat arbitrary subset of possibilities that are explored.

Concerns about computational workload for practical text applications may gradually become insignificant, considering that 80-core CPUs are within a five-year horizon and that algorithmic breakthroughs often yield super-linear improvements.

13.5   A Case Study

The overall benefit of feature selection can vary to the extremes for different datasets. For some, the accuracy can be greatly improved by selecting ∼1000 features, or for others, by selecting only a few strongly predictive features. For still others, the performance is substantially worse with anything fewer than all words. In some cases, including 5-grams among the features may make

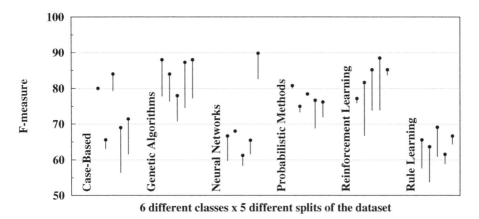

FIGURE 13.2: F-measure improvement via feature selection. The dot shows the best performance with feature selection; the other end of the whisker shows the best performance without feature selection, i.e., simply using all words.

all the difference. Because large gains are sometimes possible, text feature selection will never become obsolete—although it would be welcome if it were hidden under layers of software the way SVM solvers are today.

Nonetheless, the chapter would not be complete without an example. Figure 13.2 shows the improvement in F-measure for including feature selection vs. just giving all word features to the state-of-the-art SVM-Perf classifier. The improvement is shown for six different (mutually exclusive) classes, corresponding to different computer science topics in machine learning (each with 2.8% positives). Half the dataset was used for training, and the other half was used for testing; five such splits were evaluated and their results are shown separately. Apparently identifying papers on rule learning approaches is generally harder, but the main point is how large a gain is *sometimes* made by considering feature selection. In every case, the SVM complexity constant C was optimized from a set of five values ranging from 0.01 to 1. Where feature selection was employed, the number of features evaluated ranged from 100 upwards in steps of 1.5×, using all features. The optimal parameters were chosen according to which showed the best F-measure on the test set. Certainly this is not a large enough study or dataset to draw general conclusions, but the trend in the illustration is clear. A full scale study would also need to optimize its parameters via cross validation on the training set, rather that taking the omniscient view we have here for expediency.

Once we have paid the software complexity price to have the cross validation framework in place, we can also use it to try different feature generators. Figure 13.3 shows the further improvement in F-measure over the previous figure that is available by trying different combinations of feature generators. The three settings tried were: (1) words; (2) words plus 2-word phrases;

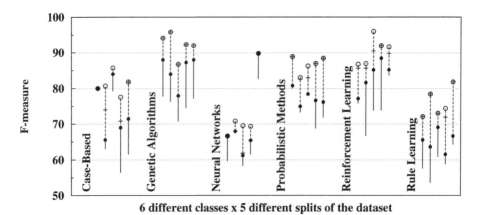

FIGURE 13.3: Further F-measure improvement for trying different feature generators. The plot from Figure 13.2 is reproduced, and the whiskers are extended up to the maximum performance for using words, + 2-word phrases, + 3-, 4-, and 5-grams.

and (3) words plus 2-word phrases, plus 3-grams, 4-grams and 5-grams. (The maximum performance without the latter is shown by the cross-hair, revealing that most of the performance gain is usually captured by 2-word phrases. Nonetheless, n-grams do sometimes improve performance.)

13.6 Conclusion and Future Work

Text classification is an elephant among blind researchers. As we approach it from different sides, we inevitably find that different strategies are called for in feature generation and feature selection. Unfortunately for the practitioner, there is much sound advice that is conflicting. A challenge for research in this decade is to develop methods and convenient software packages that consistently generate feature sets leading to good accuracy on most any training set, without requiring the user to spend their time trying different modules and parameter settings. Today, when faced with lackluster text classification performance on a new domain problem, one has to wonder whether it could be greatly improved by "tweaking the many knobs," or whether the poor performance is inherent to the task.

Cross validation for model selection and parameter tuning appears to be the straightforward solution. However, proposing a large number of potential features for a class that has few training cases can lead to overfitting the training data—generating features that are only predictive for the particular training cases studied. Small training sets are a common problem, since obtaining correct topic labels requires people's time and concentration. Even a seemingly large training set can be meager if it is divided into many classes, if the class sizes are highly imbalanced, or if the words used for a single topic are diverse. Examples of the latter include multilingual texts, many creative authors, or topics that consist of many implicit subtopics, such as sports. These are common situations in real-world datasets and pose worthy research challenges, since obtaining additional labeled training examples usually comes with a cost. One direction may be to develop priors that leverage world knowledge, e.g., gathered from many other available training sets [5].

Other open problems arise in generating and selecting useful features for classes that are not topic-based. For example, one may need to classify texts as business vs. personal, by author, or by genre (e.g., news, scientific literature, or recipes). In these situations, the specific topic words used are less predictive, and instead one may need features that represent the verb tenses used, complexity of sentences, or pertinence to company products. While there is a healthy and growing literature in authorship, genre, and sentiment classification, there are many other types of desirable and challenging classifications that have not been addressed, for example, determining the writing quality of an article containing figures, or classifying company Web sites into a multi-faceted yellow pages, such as UDDI.org. There is certainly no shortage of research opportunities.

References

[1] G. Forman. An extensive empirical study of feature selection metrics for text classification. *J. of Machine Learning Research*, 3:1289–1305, 2003.

[2] G. Forman. A pitfall and solution in multi-class feature selection for text classification. In *ICML '04: Proc. of the 21st Int'l Conf. on Machine learning*, pages 297–304. ACM Press, 2004.

[3] G. Forman, E. Kirshenbaum, and J. Suermondt. Pragmatic text mining: minimizing human effort to quantify many issues in call logs. In *KDD '06: Proc. of the 12th ACM SIGKDD int'l conf. on Knowledge discovery and data mining*, pages 852–861. ACM Press, 2006.

[4] E. Gabrilovich and S. Markovitch. Text categorization with many redundant features: using aggressive feature selection to make SVMs competitive with C4.5. In *ICML '04: Proc. of the 21st Int'l Conf. on Machine learning*, pages 321–328, 2004.

[5] E. Gabrilovich and S. Markovitch. Feature generation for text catego-

rization using world knowledge. In *Proc. of The 19th Int'l Joint Conf. for Artificial Intelligence*, pages 1048–1053, Edinburgh, Scotland, 2005.

[6] I. Guyon and E. Elisseef, A. Special issue on variable and feature selection. *J. of Machine Learning Research*, 3:1157–1461, 2003.

[7] I. Guyon, J. Weston, S. Barnhill, and V. Vapnik. Gene selection for cancer classification using support vector machines. *Machine Learning*, 46(1-3):389–422, 2002.

[8] T. Joachims. Text categorization with suport vector machines: Learning with many relevant features. In *ECML'98: Proc. of the European Conf. on Machine Learning*, pages 137–142. Springer-Verlag, New York, 1998.

[9] T. Joachims. Training linear SVMs in linear time. In *Proc. of the 12th ACM SIGKDD International Conference on Knowledge Discovery and Data Mining*, pages 217–226, 2006.

[10] C. D. Manning and H. Schütze. *Foundations of Statistical Natural Language Processing*. MIT Press, Cambridge, MA, 1999.

[11] A. McCallum and K. Nigam. A comparison of event models for naive bayes text classification. In *AAAI/ICML-98 Workshop on Learning for Text Categorization*, TR WS-98-05, pages 41–48. AAAI Press, 1998.

[12] D. Mladenic and M. Globelnik. Word sequences as features in text learning. In *Proceedings of the 17th Electrotechnical and Computer Science Conference (ERK98), Ljubljana, Slovenia*, pages 145–148, 1998.

[13] F. Sebastiani. Machine learning in automated text categorization. *ACM Comput. Surveys*, 34(1):1–47, 2002.

Chapter 14

A Bayesian Feature Selection Score Based on Naïve Bayes Models

Susana Eyheramendy

Ludwig-Maximilians Universität München

David Madigan

Rutgers University

14.1   Introduction

The past decade has seen the emergence of truly massive data analysis challenges across a range of human endeavors. Standard statistical algorithms came into being long before such challenges were even imagined, and spurred on by a myriad of important applications, much statistical research now focuses on the development of algorithms that scale well. Feature selection represents a central issue on this research.

Feature selection addresses scalability by removing irrelevant, redundant, or noisy features. Feature or variable selection has been applied to many different problems for many different purposes. For example, in text categorization problems, feature selection is often applied to select a subset of relevant words that appear in documents. This can help to elucidate the category or class of unobserved documents. Another area of application that is becoming popular is in the area of genetic association studies, where the aim is to try to find genes responsible for a particular disease (e.g., [13]). In those studies, hundreds of thousands or even a couple of million positions in the genome are genotyped in individuals who have the disease and individuals who do not have the disease. Feature selection in this context seeks to reduce the genotyping of correlated positions in order to decrease the genotyping cost while still being able to find the genes responsible for a given disease.

Feature selection is an important step in the preprocessing of the data. Removing irrelevant and noisy features helps generalization performance, and in addition reduces the computational cost and the memory demands. Reducing the number of variables can also aid in the interpretation of data and in the better distribution of resources.

In this chapter, we introduce a new feature selection method for classification problems. In particular, we apply our novel method to text categorization problems and compare its performance with other prominent feature selection methods popular in the field of text categorization.

Since many text classification applications involve large numbers of candidate features, feature selection algorithms play a fundamental role. The text classification literature tends to focus on feature selection algorithms that compute a score independently for each candidate feature. This is the so-called *filtering* approach. The scores typically contrast the counts of occurrences of words or other linguistic artifacts in training documents that belong to the target class with the same counts for documents that do not belong to the target class. Given a predefined number of words to be selected, d, one chooses the d words with the highest scores. Several score functions exist (Section 14.2 provides definitions). The authors of [14] show that Information Gain and χ^2 statistics performed best among five different scores. Reference [4] provides evidence that these two scores have correlated failures. Hence, when choosing optimal pairs of scores, these two scores work poorly together. Reference [4] introduced a new score, the Bi-Normal Separation, that yields the best performance on the greatest number of tasks among 12 feature selection scores. The authors of [12] compared 11 scores under a naïve Bayes classifier and found that the Odds Ratio score performs best in the highest number of tasks.

In regression and classification problems in statistics, popular feature selection strategies depend on the same algorithm that fits the models. This is the so-called *wrapper* approach. For example, *best subset regression* finds for each k the best subset of size k based on residual sum of squares. *Leaps and bounds* is an efficient algorithm that finds the best set of features when the number of predictors is no larger than about 40. An extensive discussion on subset selection on regression problems is provided in [11]. The recent paper [9] gives a detailed categorization of all existing feature selection methods.

In a Bayesian context and under certain assumptions, reference [1] shows that for selection among normal linear models, the best model contains those features that have overall posterior probability greater than or equal to $1/2$. Motivated by this study, we introduce a new feature selection score (PIP) that evaluates the posterior probability of inclusion of a given feature over all possible models, where the models correspond to a set of features. Unlike typical scores used for feature selection via filtering, the PIP score *does* depend on a specific model. In this sense, this new score straddles the filtering and wrapper approaches.

We present experiments that compare our new feature selection score with

TABLE 14.1:
Two-way contingency
table of word F and
category k.

	k	$\bar{k}$	
F	n_{kF}	$n_{\bar{k}F}$	n_F
$\bar{F}$	$n_{k\bar{F}}$	$n_{\bar{k}\bar{F}}$	$n_{\bar{F}}$
	n_k	$n_{\bar{k}}$	M

five other feature selection scores that have been prominent in the studies mentioned above. The feature selection scores that we consider are evaluated on two widely-used benchmark text classification datasets, Reuters-21578 and 20-Newsgroups, and implemented on four classification algorithms. Following previous studies, we measure the performance of the classification algorithms using the F_1 measure.

We have organized this chapter as follows. Section 14.2 describes the various feature selection scores we consider, both our new score and the various existing alternatives. In Section 14.3 we mention the classification algorithms that we use to compare the feature selection scores. The experimental settings and experimental results are presented in Section 14.4. We conclude in Section 14.5.

14.2 Feature Selection Scores

In this section we introduce a new methodology to define a feature score and review the definitions of other popular feature selection scores.

Before we list the feature selection scores that we study, we introduce some notation. In the context of our text categorization application, Table 14.1 show the basic statistics for a single word and a single category (or class).

n_{kF} : n° of documents in class k with word F.
$n_{k\bar{F}}$: n° of documents in class k without word F.
$n_{\bar{k}F}$: n° of documents not in class k with word F.
$n_{\bar{k}\bar{F}}$: n° of documents not in class k without word F.
n_k : total n° of documents in class k.
$n_{\bar{k}}$: total n° of documents that are not in class k.
n_F : total n° of documents with word F.
$n_{\bar{F}}$: total n° of documents without word F.
M : total n° of documents.

We refer to F as a word or feature occuring in documents and x as the value that depends on the number of times the word F appears in a document. For example, consider a document that consists of the phrase "curiosity begets curiosity". If F_1 represents the word "curiosity", then x_1 can take the value 1 if we consider the presence or absence of the words in the documents, or x_1 can take the value 2 if the actual frequency of appearance is considered.

14.2.1 Posterior Inclusion Probability (PIP)

Consider a classification problem in which one has M instances in training data, $Data = \{(y_1, \mathbf{x}_1), \ldots, (y_M, \mathbf{x}_M)\}$, where y_i denotes the class label of instance i that takes values in a finite set of C classes, and $\mathbf{x}_i$ is its corresponding vector of N features. We consider a naïve Bayes model where the probability of the training data instances can be expressed as the product of the individual conditional probablities of each feature given the class membership, times the probablities of the class memberships,

$$Pr((y_1, \mathbf{x}_1), \ldots, (y_M, \mathbf{x}_M)) = \Pi_{i=1}^M Pr(y_i) \Pi_{j=1}^N Pr(x_{ij}|y_i) \qquad (14.1)$$

We aim to select a subset of the features with which one can infer accurately the class label of new instances using a prediction function or rule that links the vector of features with the class label.

Given N features, one can consider 2^N different models, each one containing a different subset of features. We denote each model by a vector of length the number of features N, where each component is either 1 if the feature is present or 0 if the feature is absent. For two features, Figure 14.1 shows a graphical representation of the four possible models. For example, model $M_{(1,1)}$ contains both features, and the distribution of each feature depends on the class label of the document. This is represented in the graph with an arrow from the node y to each of the features x_1 and x_2.

Without assuming any distribution on the conditional probabilities in Equation (14.1), we propose as a feature score the Posterior Inclusion Probability (PIP) for feature F_j and class k, which is defined as

$$PIP(F_j, k) = \sum_{\mathbf{l}:l_j=1} Pr(M_{\mathbf{l}}|Data) \qquad (14.2)$$

where $\mathbf{l}$ is a vector of length the number of features and the jth component takes the value 1 if the jth feature F_j is included in model $M_{\mathbf{l}}$, otherwise it is 0. In other words, we define as the feature selection score the posterior probability that each feature is included in a model, for all features appearing in documents or instances of class k.

Each feature appears in 2^{N-1} models. For moderate values of N, the sum in Equation (14.2) can be extremely large. Fortunately, we show in the next section that it is not necessary to compute the sum in Equation (14.2) because it can be expressed in closed form.

Note that for each class, each feature is assigned a different score. The practitioner can either select a different set of features for each of the classes

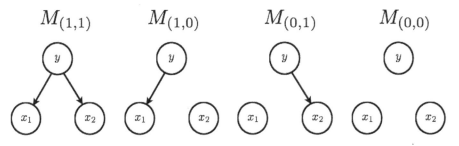

FIGURE 14.1: Graphical model representation of the four models with two features, x_1 and x_2.

or a single score can be obtained by weighting the scores over all the classes by the frequency of instances in each class. We follow the latter approach in all features selection scores considered in this study. The next two sections implement the PIP feature selection score. The next section assumes a Bernoulli distribution for the conditional probabilities in Equation (14.1) and the subsequent section assumes a Poisson distribution.

14.2.2 Posterior Inclusion Probability (PIP) under a Bernoulli distribution

Consider first that the conditional probabilities $P(x_{ij}|y_i,\theta)$ are Bernoulli distributed with parameter θ, and assume a Beta prior distribution on θ. This is the binary naïve Bayes model for the presence or absence of words in the documents. Section 14.2.3 considers a naïve Bayes model with Poisson distributions for word frequency. This score for feature F and class k can be expressed as

$$PIP(F,k) = \frac{l_{0Fk}}{l_{0Fk} + l_{Fk}} \tag{14.3}$$

where

$$l_{0Fk} = \frac{B(n_{kF} + \alpha_{kF}, n_{k\overline{F}}\beta_{kF})}{B(\alpha_{kF}, \beta_{kF})} \tag{14.4}$$

$$\times \frac{B(n_{\overline{k}F} + \alpha_{\overline{k}F}, n_{\overline{k}\overline{F}} + \beta_{\overline{k}F})}{B(\alpha_{\overline{k}F}, \beta_{\overline{k}F})} \tag{14.5}$$

$$l_{Fk} = \frac{B(n_F + \alpha_F, n_{\overline{F}} + \beta_F)}{B(\alpha_F, \beta_F)} \tag{14.6}$$

$B(a,b)$ is the *Beta* function, which is defined as $B(a,b) = \frac{\Gamma(a)\Gamma(b)}{\Gamma(a+b)}$, and α_{kF}, $\alpha_{k\overline{F}}$, α_F, β_{kF}, $\beta_{k\overline{F}}$, β_F are constants set by the practitioner. In our experiments we set them to be $\alpha_F = 0.2$, $\beta_F = 2/25$ for all words F, $\alpha_{kF} = 0.1$, $\alpha_{k\overline{F}} = 0.1$, $\beta_{kF} = 1/25$, and $\beta_{k\overline{F}} = 1/25$ for all categories k and feature F. These settings correspond to rather diffuse priors.

We explain this score in the context of a two-candidate-word model. The likelihoods for each model and category k are given by

$$M_{(1,1)} : \prod_{i=1}^{M} Pr(x_{i1}, x_{i2}, y_i | \theta_{1k}, \theta_{2k}) = \prod_{i=1}^{M} \mathcal{B}(x_{i1}, \theta_{k1})\mathcal{B}(x_{i1}, \theta_{\overline{k}1})\mathcal{B}(x_{i2}, \theta_{k2})$$
$$\times \mathcal{B}(x_{i2}, \theta_{\overline{k}2})Pr(y_i|\theta_k)$$

$$M_{(1,0)} : \prod_{i=1}^{M} Pr(x_{i1}, x_{i2}, y_i | \theta_{1k}, \theta_2) = \prod_{i=1}^{M} \mathcal{B}(x_{i1}, \theta_{k1})\mathcal{B}(x_{i1}, \theta_{\overline{k}1})\mathcal{B}(x_{i2}, \theta_2)$$
$$\times \mathcal{B}(x_{i2}, \theta_2)Pr(y_i|\theta_k)$$

$$M_{(0,1)} : \prod_{i=1}^{M} Pr(x_{i1}, x_{i2}, y_i | \theta_1, \theta_{2k}) = \prod_{i=1}^{M} \mathcal{B}(x_{i1}, \theta_1)\mathcal{B}(x_{i1}, \theta_1)\mathcal{B}(x_{i2}, \theta_{k2})$$
$$\times \mathcal{B}(x_{i2}, \theta_{\overline{k}2})Pr(y_i|\theta_k)$$

$$M_{(0,0)} : \prod_{i=1}^{M} Pr(x_{i1}, x_{i2}, y_i | \theta_1, \theta_2) = \prod_{i=1}^{M} \mathcal{B}(x_{i1}, \theta_1)\mathcal{B}(x_{i1}, \theta_1)\mathcal{B}(x_{i2}, \theta_2)$$
$$\times \mathcal{B}(x_{i2}, \theta_2)Pr(y_i|\theta_k)$$

where x_{ij} takes the value 1 if document i contains word F_j and 0 otherwise, y_i is 1 if document i is in category k, and otherwise is 0, $Pr(y_i|\theta_k) = \mathcal{B}(y_i, \theta_k)$, and $\mathcal{B}(x, \theta) = \theta^x(1 - \theta)^{1-x}$ denotes a Bernoulli probability distribution.

Therefore, in model $M_{(1,1)}$, the presence or absence of both words in a given document depends on the document class. θ_{k1} corresponds to the proportion of documents in category k with word F_1 and $\theta_{\overline{k}1}$ to the proportion of documents not in category k with word F_1. In model $M_{(1,0)}$, only word F_1 depends on the category of the document and θ_2 corresponds to the proportion of documents with word F_2 regardless of the category associated with them. θ_k is the proportion of documents in category k and $Pr(y_i|\theta_k)$ is the probability that document i is in category k.

We assume the following prior probability distributions for the parameters: $\theta_{kF} \sim Beta(\alpha_{kF}, \beta_{kF})$, $\theta_{\overline{k}F} \sim Beta(\alpha_{\overline{k}F}, \beta_{\overline{k}F})$, $\theta_F \sim Beta(\alpha_F, \beta_F)$, and $\theta_k \sim Beta(\alpha_k, \beta_k)$, where $Beta(\alpha, \beta)$ denotes a Beta distribution, i.e., $Pr(\theta|\alpha, \beta) = \frac{1}{B(\alpha,\beta)}\theta^{\alpha-1}(1 - \theta)^{\beta-1}$, $k \in \{1, ..., C\}$, and $F \in \{F_1, ..., F_N\}$.

Then the marginal likelihoods for each of the four models above can be expressed as the products of three terms,

$$Pr(data|M_{(1,1)}) = l_0 \times l_{0F_1k} \times l_{0F_2k}$$
$$Pr(data|M_{(1,0)}) = l_0 \times l_{0F_1k} \times l_{F_2k}$$
$$Pr(data|M_{(0,1)}) = l_0 \times l_{F_1k} \times l_{0F_2k}$$
$$Pr(data|M_{(0,0)}) = l_0 \times l_{F_1k} \times l_{F_2k}$$

where l_{0Fk} and l_{Fk} are defined in Equations (14.4) for $F \in \{F_1, F_2, ..., F_N\}$

and l_0 is defined as

$$l_0 = \int_0^1 \prod_i Pr(y_i|\theta_k)Pr(\theta_k|\alpha_k, \beta_k)d\theta_k \tag{14.7}$$

which is the marginal probability for the category of the documents.

It is straightforward to show that $PIP(F, k)$ in Equation (14.2) is equivalent to $PIP(F, k)$ in Equation (14.3) if we assume that the prior probability density for the models is uniform, e.g., $Pr(M_l) \propto 1$.

In the example above, the posterior inclusion probability for feature F_1 is given by

$$Pr(F_1|y_k) = Pr(M_{(1,1)}|data) + Pr(M_{(1,0)}|data)$$
$$= \frac{l_{0F_1k}}{l_{0F_1k} + l_{F_1k}}$$

To get a single "bag of words" for all categories we compute the weighted average of $PIP(F, k)$ over all categories:

$$PIP(F) = \sum_k Pr(y = k)PIP(F, k)$$

We note that the authors of [2] present similar manipulations of the naïve Bayes model but for model averaging purposes rather than finding the median probability model.

14.2.3   Posterior Inclusion Probability (PIPp) under Poisson distributions

A generalization of the binary naïve Bayes model assumes class-conditional Poisson distributions for the word frequencies in a document. As before, assume that the probability distribution for a word in a document might or might not depend on the category of the document. More precisely, if the distribution for feature x depends on the category k of the document, we have

$$Pr(x|y = k) = \frac{e^{-\lambda_{kF}}\lambda_{kF}^x}{x!}$$
$$Pr(x|y \neq k) = \frac{e^{-\lambda_{\bar{k}F}}\lambda_{\bar{k}F}^x}{x!}$$

where x is the number of times word F appears in the document and λ_{kF} ($\lambda_{\bar{k}F}$) represents the expected number of times word F appears in documents in category k ($\bar{k}$). If the distribution for x does not depend on the category of the document, we then have

$$Pr(x) = \frac{e^{-\lambda_F}\lambda_F^x}{x!}$$

where λ_F represents the expected number of times word F appears in a document regardless of the category of the document.

Assume the following conjugate prior probability densities for the parameters:

$$\lambda_{kF} \sim Gamma(\alpha_{kF}, \beta_{kF})$$
$$\lambda_{\bar{k}F} \sim Gamma(\alpha_{\bar{k}F}, \beta_{\bar{k}F})$$
$$\lambda_F \sim Gamma(\alpha_F, \beta_F)$$

where $\alpha_{kF}, \beta_{kF}, \alpha_{\bar{k}F}, \beta_{\bar{k}F}, \alpha_F$, and β_F are hyperparameters to be set by the practitioner.

Now, as before, the posterior inclusion probability for Poisson distributions (PIPp) is given by

$$PIPp(F, k) = \frac{l_{0Fk}}{l_{0Fk} + l_{Fk}}$$

where

$$l_{0Fk} = \frac{\Gamma(N_{kF} + \alpha_{kF})}{\Gamma(\alpha_{kF})\beta_{kF}^{\alpha_{kF}}} \frac{\Gamma(N_{\bar{k}F} + \alpha_{\bar{k}F})}{\Gamma(\alpha_{\bar{k}F})\beta_{\bar{k}F}^{\alpha_{\bar{k}F}}}$$
$$\times (\frac{\beta_{kF}}{n_k \beta_{kF} + 1})^{n_{kF} + \alpha_{kF}} (\frac{\beta_{\bar{k}F}}{n_{\bar{k}} \beta_{\bar{k}F} + 1})^{n_{\bar{k}F} + \alpha_{\bar{k}F}}$$
$$l_{Fk} = \frac{\Gamma(N_F + \alpha_F)}{\Gamma(\alpha_F)} (\frac{\beta_F}{\beta_F n + 1})^{n_F + \alpha_F} \frac{1}{\beta_F^{\alpha_F}}.$$

This time, $N_{kF}, N_{\bar{k}F}$, and N_F denote:
N_{kF}: $n°$ of times word F appears in documents in class k.
$N_{\bar{k}F}$: $n°$ of times word F appears in documents not in class k.
N_F: total $n°$ of times that word F appears in all documents.

As before, to get a single "bag of words" for all categories, we compute the weighted average of $PIPp(F, k)$ over all categories:

$$PIPp(F) = \sum_{k}^{C} Pr(y = k)PIPp(F, k)$$

14.2.4 Information Gain (IG)

Information gain is a popular score for feature selection in the field of machine learning. In particular, it is used in the C4.5 decision tree inductive algorithm. Reference [14] compared five different feature selection scores on two datasets and showed that Information Gain is among the two most effective. The information gain of word F is defined to be

$$IG(F) = -\sum_{k=1}^{C} Pr(y = k) \log Pr(y = k)$$

$$+ Pr(F) \sum_{k=1}^{C} Pr(y = k|F) \log Pr(y = k|F)$$

$$+ Pr(\overline{F}) \sum_{k=1}^{C} Pr(y = k|\overline{F}) \log Pr(y = k|\overline{F})$$

where $\{1, \ldots, C\}$ is the set of categories and $\overline{F}$ is the absence of word F. It measures the decrease in entropy when the feature is present versus when the feature is absent.

14.2.5   Bi-Normal Separation (BNS)

The Bi-Normal Separation score, introduced in [4], is defined as

$$BNS(F, k) = |\Phi^{-1}(\frac{n_{kF}}{n_k}) - \Phi^{-1}(\frac{n_{\overline{k}F}}{n_{\overline{k}}})|$$

where Φ is the standard normal distribution and Φ^{-1} is its corresponding inverse. $\Phi^{-1}(0)$ is set to be equal to 0.0005 to avoid numerical problems following [4]. By averaging over all categories, we get a score that selects a single set of words for all categories:

$$BNS(x) = \sum_{k=1}^{C} Pr(y = k)|\Phi^{-1}(\frac{n_{kF}}{n_k}) - \Phi^{-1}(\frac{n_{\overline{k}F}}{n_{\overline{k}}})|$$

To get an idea of what this score is measuring, assume that the probability that a word F is contained in a document is given by $\Phi(\delta_k)$ if the document belongs to class y_k and otherwise is given by $\Phi(\delta_{\overline{k}})$. A word will discriminate with high accuracy between a document that belongs to a category from one that does not, if the value of δ_k is small and the value of $\delta_{\overline{k}}$ is large, or vice versa, if δ_k is large and $\delta_{\overline{k}}$ is small. Now, if we set $\delta_k = \Phi^{-1}(\frac{n_{kF}}{n_k})$ and $\delta_{\overline{k}} = \Phi^{-1}(\frac{n_{\overline{k}F}}{n - n_k})$, the Bi-Normal Separation score is equivalent to the distance between these two quantities, $|\delta_{\overline{k}} - \delta_k|$.

14.2.6   Chi-Square

The chi-square feature selection score, $\chi^2(F, k)$, measures the dependence between word F and category k. If word F and category k are independent, $\chi^2(F, k)$ is equal to zero. When we select a different set of words for each category, we utilize the following score:

$$\chi^2(F, k) = \frac{n(n_{kF} n_{\overline{k}\overline{F}} - n_{\overline{k}F} n_{k\overline{F}})^2}{n_k n_F n_{\overline{k}} n_{\overline{F}}}$$

Again, by averaging over all categories, we get a score for selecting a single set of words for all categories:

$$\chi^2(F) = \sum_{k=1}^{C} Pr(y=k)\chi^2(F,k)$$

14.2.7 Odds Ratio

The Odds Ratio measures the odds of word F occuring in documents in category k divided by the odds of word F not occuring in documents in category k. Reference [12] found this to be the best score among eleven scores for a naïve Bayes classifier. For category k and word F, the oddsRatio is given by

$$OddsRatio(F,k) = \frac{\frac{n_{kF}+0.1}{n_k+0.1} / \frac{n_{k\overline{F}}+0.1}{n_k+0.1}}{\frac{n_{\overline{k}F}+0.1}{n_{\overline{k}}+0.1} / \frac{n_{\overline{k}\overline{F}}+0.1}{n_{\overline{k}}+0.1}}$$

where we add the constant 0.1 to avoid numerical problems. By averaging over all categories we get

$$OddsRatio(F) = \sum_{k=1}^{C} Pr(y=k)OddsRatio(F,k)$$

14.2.8 Word Frequency

This is the simplest of the feature selection scores. In the study in [14] they show that word frequency is the third best after information gain and χ^2. They also point out that there is a strong correlation between these two scores and word frequency. For each category k, word frequency (WF) for word F is the number of documents in category k that contain word F, i.e., $WF(F,k) = n_{kF}$.

Averaging over all categories we get a score for each F:

$$WF(F) = \sum_{k=1}^{C} Pr(y=k)WF(F,k) = \sum_{k=1}^{C} Pr(y=k)n_{kF}$$

14.3 Classification Algorithms

To determine the performance of the different feature selection scores, the classification algorithms that we consider are the Multinomial, Poisson, and Binary naïve Bayes classifiers (e.g., [10], [8], and [3]), and the hierarchical probit classifier of [5]. We choose these classifiers for our analysis for two reasons.

The first one is the different nature of the classifiers. The naïve Bayes models are generative models while the probit is a discriminative model. Generative classifiers learn a model of the joint probability $Pr(x, y)$, where x is the input and y the label. These classifiers make their predictions by using Bayes rule to calculate $Pr(y|x)$. In contrast, discriminative classifiers model the conditional probability of the label given the input ($Pr(y|x)$) directly. The second reason is the good performance that they achieve. In [3], the multinomial model, notwithstanding its simplicity, achieved the best performance among four naïve Bayes models. The hierarchical probit classifier of [5] achieves state-of-the-art performance, comparable to the performance of the best classifiers such as SVM ([7]). We decided to include the binary and Poisson naive Bayes models (see [3] for details) because they allow us to incorporate information of the probability model used to fit the categories of the documents into the feature selection score. For instance, in the Binary naïve Bayes classifiers, the features that one can select using the PIP score correspond exactly to the features with the highest posterior inclusion probability. We want to examine whether or not that offers an advantage over other feature selection scores.

14.4 Experimental Settings and Results

Before we start the analysis we remove common noninformative words taken from a standard *stopword* list of 571 words and we remove words that appear less than three times in the training documents. This eliminates 8, 752 words in the Reuters dataset (38% of all words in training documents) and 47, 118 words in the Newsgroups dataset (29% of all words in training documents). Words appear on average in 1.41 documents in the Reuters dataset and in 1.55 documents in the Newsgroups dataset.

We use F1 to measure the performance of the different classifiers and feature selection scores. F1 is the harmonic mean between recall and precision. We average the F1 scores across all categories to get a single value. The micro F1 is a weighted average, where the weights for each category are proportional to the frecuency of documents in the category. The macro F1 gives equal weight to all categories.

14.4.1 Datasets

The 20-Newsgroups dataset contains 19, 997 articles divided almost evenly into 20 disjoint categories. The category topics are related to computers, politics, religion, sports, and science. We split the dataset randomly into 75% for training and 25% for testing. We took this version of the dataset from http://www.ai.mit.edu/people/jrennie/20Newsgroups/. Another dataset that we used comes from the Reuters-21578 news story collection. We used a subset of the ModApte version of the Reuters$-21,578$ collection, where each document has assigned at least one topic label (or category) and this topic label belongs to any of the 10 most populous categories - earn, acq, grain, wheat, crude, trade, interest, corn, ship, money-fx. It contains $6,775$ documents in the training set and $2,258$ in the testing set.

14.4.2   Experimental Results

In these experiments we compare seven feature selection scores, on two benchmark datasets, Reuters-21578 and Newgroups (see Section 14.4.1), under four classification algorithms (see Section 14.3).

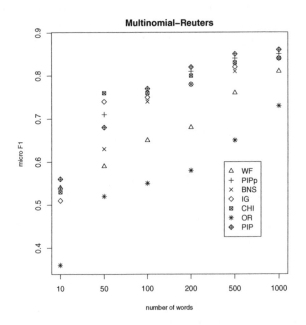

FIGURE 14.2: Performance (for the multinomial model) for different numbers of words measured by micro F1 for the Reuters dataset.

We compare the performance of the classifiers for different numbers of words and vary the number of words from 10 to 1000. For larger numbers of words, the classifiers tend to perform somewhat more similarly, and the effect of choosing the words using a different feature selection procedure is less noticeable.

Figures 14.2 - 14.5 show the micro average F1 measure for each of the feature selection scores as we vary the number of features selected for the

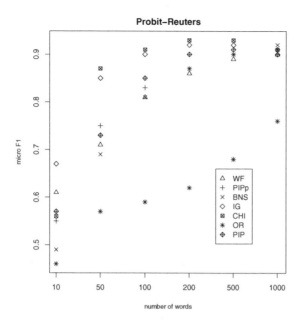

FIGURE 14.3: Performance (for the probit model) for different numbers of words measured by micro F1 for the Reuters dataset.

four classification algorithms we considered: multinomial, probit, Poisson, and binary respectively.

We notice that PIP gives, in general, high values to all very frequent words. To avoid that bias we remove words that appear more than 2000 times in the Reuters dataset (that accounts for 15 words) and more than 3000 times in the Newsgroups dataset (that accounts for 36 words). We now discuss the results for the two datasets:

Reuters. Like the results of [4], if for scalability reasons one is limited to a small number of features (< 50), the best available metrics are IG and χ^2, as Figures 14.2 – 14.5 show. For larger numbers of features (> 50), Figure 14.2 shows that PIPp and PIP are the best scores for the mutinomial classifier. Figures 14.4 and 14.5 show the performance for the Poisson and binary classifiers. PIPp and BNS achive the best performance in the Poisson classifier and PIPp achieves the best performance in the binary classifier. WF performs poorly compared to the other scores in all the classifiers, achieving the best performance with the Poisson one.

Newsgroups. χ^2 followed by BNS, IG, and PIP are the best-performing measures in the probit classifier. χ^2 is also the best one in the multinomial model, followed by BNS and the binary classifier with the macro F1 measure. OR performs best in the Poisson classifier. PIPp is best in the binary classifier under the micro F1 measure. WF performs poorly compared to the other

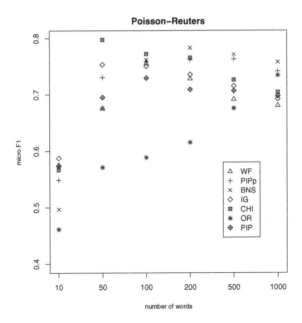

FIGURE 14.4: Performance (for the poisson model) for different numbers of words measured by micro F1 for the Reuters dataset.

scores in all classifiers. Because of lack of space we do not show a graphical display of the performance of the classifiers in the Newsgroups dataset, and only the micro F1 measure is displayed graphically for the Reuters dataset.

In Table 14.2 and Table 14.3 we summarize the overall performance of the feature selection scores considered by integrating the curves formed when the dots depicted in Figures 14.2 – 14.5 are joined. Each column corresponds to a given feature selection. For instance, the number 812 under the header "Multinomial model Reuters-21578" and the row "micro F1" corresponds to the area under the *IG* "curve" in Figure 14.2. In 7 out of 16 instances, χ^2 is the best-performing score and in 3 it is the second best. PIPp in 4 out of 16 is the best score and in 6 is the second best. BNS is the best in 2 and second best in 6. In bold are the best two performance scores.

14.5   Conclusion

Variable or feature selection has become the research focus of many researchers working in applications that contain a large number of potential features. The main goals of feature selection procedures in classification problems are to reduce data dimensionality in order to allow for a better interpre-

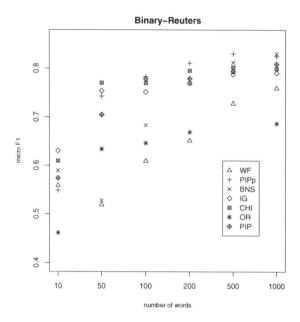

FIGURE 14.5: Performance (for the binary naïve Bayes model) for different numbers of words measured by micro F1 for the Reuters dataset.

tation of the underlying process generating the data, and to provide faster, computationally tractable, and accurate predictors.

In this chapter, we introduced a flexible feature selection score, PIP. Unlike other popular feature selection scores following the filtering approach, PIP is a model-dependent score, where the model can take several different forms. The advantage of this score is that the selected features are easy to interpret, the computation of the score is inexpensive, and, when the predicting model corresponds to a naïve Bayes model, the score depends also on the predicting model, which can lead to better classification performance. While feature selection scores following the filtering approach are computationally inexpensive as well, most do not provide a clear interpretation of the selected features.

The value that the PIP score assigns to each word has an appealing Bayesian interpretation, being the posterior probability of inclusion of the word in a naïve Bayes model. Such models assume a probability distribution on the words of the documents. We consider two probability distributions, Bernoulli and Poisson. The former takes into account the presence or absence of words in the documents, and the latter, the number of times each word appears in the documents. Future research could consider alternative PIP scores corresponding to different probabilistic models.

This score can be applied as a regular filtering score as part of the preprocessing of the data for dimensinality reduction, before fitting the predicting

TABLE 14.2:   Performance of the Binary and Poisson models.

	IG	χ^2	OR	BNS	WF	PIP	PIPp
	\multicolumn{7}{c}{Poisson model Reuters-21578 dataset}						
micro F_1	708	719	670	**763**	684	699	**755**
macro F_1	618	**628**	586	**667**	590	618	**667**
	\multicolumn{7}{c}{Poisson model 20-Newsgroups dataset}						
micro F_1	753	808	**928**	812	684	777	**854**
macro F_1	799	841	**936**	841	773	813	**880**
	\multicolumn{7}{c}{Berboulli model Reuters-21578 dataset}						
micro F_1	779	794	669	**804**	721	786	**822**
macro F_1	680	**698**	618	709	614	696	**746**
	\multicolumn{7}{c}{Bernoulli model 20-Newsgroups dataset}						
micro F_1	531	**566**	508	556	436	534	**650**
macro F_1	628	**673**	498	**652**	505	627	650

This table summarizes an overall performance of the feature selection scores considered by integrating the curves formed by joining the dots depicted in Figures 14.2 – 14.5. In bold are the best two performing score.

model. Alternatively, it can be applied in conjuction with a naïve Bayes model, in which case the score is built based on the predicting model, which bears a resemblance to the scores that follow the so-called wrapper approach for feature selection.

The wrapper approach attempts to identify the best feature subset to use with a particular algorithm and dataset, whereas the filtering approach attempts to assess the merits of features from the data alone. For some naïve Bayes models like the Binary naïve model or Poisson naïve model, the score computed by PIP Bernoulli and PIP Poisson depends on the classification algorithm. Our empirical results do not corroborate the benefit of using the same model in the feature selection score and in the classification algorithm. The strong assumption that naïve Bayes models make about the independence of the features given the label is well known not to be suitable for textual datasets, as words tend to be correlated. Despite the correlation structure of words, naïve Bayes classifiers have been shown to give highly accurate predictions. The reasons for that are clearly explained in [6]. The authors are currently exploring extensions of this method of feature selection to applications where the naïve Bayes assumption appears to be more suitable.

Our results regarding the performance of the different scores are consistent with [14] in that χ^2 and IG seem to be strong scores for feature selection in discriminative models, but disagree in that WF appears to be a weak score in most instances. Note that we do not use exactly the same WF score. Ours is a weighted average by the category proportion.

χ^2, PIPp, and BNS are the best-performing scores. Still, feature selection scores and classification algorithms seem to be highly data- and model-

TABLE 14.3: Performance of the Multinomial and Probit models.

	IG	χ^2	OR	BNS	WF	PIP	PIPp
	Multinomial model Reuters-21578 dataset						
micro F_1	812	822	644	802	753	**842**	**832**
macro F_1	723	733	555	713	644	**762**	**753**
	Multinomial model 20-Newsgroups dataset						
micro F_1	535	**614**	575	**584**	456	564	575
macro F_1	594	**644**	565	**634**	486	604	585
	Probit model Reuters-21578 dataset						
micro F_1	**911**	**921**	674	891	881	901	891
macro F_1	**861**	**861**	605	842	753	842	**851**
	Probit model 20-Newsgroups dataset						
micro F_1	703	**723**	575	**713**	565	693	644
macro F_1	693	**723**	565	**703**	565	683	624

This table summarizes an overall performance of the feature selection scores considered by integrating the curves formed by joining the dots depicted in Figures 14.2 – 14.5. In bold are the best two performing score.

dependent. The feature selection literature reports similarly mixed findings. For instance, the authors of [14] found that IG and χ^2 are the strongest feature selection scores. They performed their experiments on two datasets, Reuters-22173 and OHSUMED, and under two classifiers, kNN and a linear least square fit. The authors of [12] found that OR is the strongest feature selection score. They performed their experiments on a naïve Bayes model and used the Yahoo dataset. Reference [14] favors bi-normal separation.

A better understanding of the dependency between the correlation structure of textual datasets, potential feature selection procedures, and classification algorithms is still an important challenge to be further addressed that we intend to pursue in the future.

References

[1] M. Barbieri and J. Berger. Optimal predictive model selection. *Annals of Statistics*, 32:870–897, 2004.

[2] D. Dash and G. Cooper. Exact model averaging with naive bayesian classifiers. In *Proceedings of the Nineteenth International Conference on Machine Learning*, pages 91–98, 2002.

[3] L. D. Eyheramendy, S. and D. Madigan. On the naive bayes classifiers for text categorization. In *Proceedings of the ninth international workshop on Artificial Intelligence and Statistics*, 2003.

[4] G. Forman. An extensive empirical study of feature selection metrics for text classification. *Journal of Machine Learning Research*, 3:1289–1305, 2003.

[5] L. D. E. S. J. W. Genkin, A. and D. Madigan. Sparse bayesian classifiers for text categorization. *JICRD*, 2003.

[6] D. Hand and K. Yu. Idiot's bayes not so stupid after all? *International Statistical Review*, 69:385–398, 2001.

[7] T. Joachims. Text categorization with support vector machines: Learning with many relevant features. In *Proceedings of ECML-98*, pages 137–142, 1998.

[8] D. Lewis. Naive (bayes) at forty: The independence assumption in information retrieval. In *Proceedings of ECML-98*, pages 4–15, 1998.

[9] H. Liu and L. Yu. Towards integrating feature selection algorithms for classification and clustering. *IEEE Transactions in Knowledge and Data Ingeneering*, 17(3):1–12, 2005.

[10] A. McCallum and K. Nigam. A comparison of event models for naive bayes text classification. In *AAAI/ICML Workshop on Learning for Text Categorization*, pages 41–48, 1998.

[11] A. Miller. *Subset selection in regression.* second edition edition, 2002.

[12] D. Mladenic and M. Grobelnik. Feature selection for unbalanced class distribution and naive bayes. In *Proceedings ICML-99*, pages 258–267, 1999.

[13] L. C. Pardi, F. and J. Whittaker. Snp selection for association studies: maximizing power across snp choice and study size. *Ann Human Genetics*, 69:733–746, 2005.

[14] Y. Yang and J. Pedersen. A comparative study on feature selection in text categorization. In *Proceedings ICML-97*, pages 412–420, 1997.

Chapter 15

Pairwise Constraints-Guided Dimensionality Reduction

Wei Tang

Florida Atlantic University

Shi Zhong

Yahoo! Inc.

15.1 Introduction

High-dimensional data are commonly seen in many practical machine learning and data mining problems and present a challenge in both classification and clustering tasks. For example, document classification/clustering often deals with tens of thousands of input features based on bag-of-words representation (where each unique word is one feature dimension). In market basket data analysis, the input dimensionality is the same as the number of products seen in transactions, which can also be huge. Although there are already some algorithms that can handle high-dimensional data directly (e.g., support vector machines and naïve Bayes models), it is still a good practice to reduce the number of input features. There are several good reasons for this practice: a) Many features may be irrelevant to or uninformative about the target of our classification/clustering tasks; b) reduced dimensionality makes it possible to use more choices of classification/clustering algorithms; and c) lower dimensionality is more amenable to computational efficiency.

Common dimensionality reduction approaches include feature projection [11], feature selection [13], and feature clustering [9]. Feature projection methods project high-dimensional data to a lower-dimensional space, where each projected feature is a linear combination of the original features. The objective of feature projection is to learn a projection matrix $P_{N \times K}$ that maps the original

N-dimensional instances into a K-dimensional space. Feature selection refers to selecting a small set of input features based on some measure of feature usefulness. Most feature selection algorithms are used for classification problems since the usefulness metric for a feature can be more well-defined with class labels available. Feature clustering aims to cluster the original feature set into different groups and use cluster centroids to form the reduced feature set. Both feature selection and feature clustering can be considered as special cases of feature projection with specific projection matrices.

The widely-used principal component analysis (PCA) and linear discriminant analysis (LDA) are two representative feature projection techniques – the former is for unsupervised learning and the latter for classification. PCA [17] is an orthonormal transformation of data to a low-dimensional space such that maximum variance of the original data is preserved. latent semantic indexing (LSI) [7], essentially just a different name for PCA when applied to analyzing text data, has been used to map text documents to a low-dimensional "topic" space spanned by some underlying latent concept vectors. PCA works well for a lot of data analysis problems but does not fit well for classification purposes. LDA and other supervised feature selection techniques are better positioned for classification in that they reduce the input features in such a way that maximum separability between target classes is preserved.

This chapter focuses on dimensionality reduction for *semi-supervised clustering* problems. In this learning task, the exact class labels are not available. Instead, there is some "weak" supervision in the form of pairwise instance constraints. The goal of semi-supervised clustering is still to categorize data instances into a set of groups, but the groups are usually not pre-defined due to the lack of class labels. A pairwise instance constraint specifies whether a pair of instances must or must not be in the same group, naturally called *must-link* constraints and *cannot-link* constraints, respectively. Pairwise instance constraints are a common type of background knowledge that appears in many practical applications. For example, in text/image information retrieval, user feedback on which retrieved results are similar to a query and which are not can be used as pairwise constraints. These constraints help better organize the underlying text/image database for more efficient retrievals. For clustering GPS data for lane-finding [19], or grouping different actors in movie segmentation [1], the complete class information may not be available, but pairwise instance constraints can be extracted automatically with minimal effort. Also, a user who is not a domain expert is more willing to provide an answer to whether two objects are similar/dissimilar than to specify explicit group labels.

The techniques mentioned above (PCA or LDA) cannot easily exploit pairwise constraints for reducing the number of features. To the best of our knowledge, the most related work is the relevant component analysis (RCA) algorithm [1], which learns a Mahalanobis distance based on *must-link* constraints and using a whitening transform [12]. Since there are no class labels, such methods are usually evaluated on (semi-supervised) clustering problems.

Another related work is metric learning with pairwise constraints [20, 3], which learn the parameters associated with a parameterized distance metric function from pairwise data constraints. This kind of approach, however, does not reduce the input dimensionality directly.

In this chapter, we propose two methods of leveraging pairwise instance constraints for dimensionality reduction: *pairwise constraints-guided feature projection* and *pairwise constraints-guided co-clustering*. The first approach projects data onto a low-dimensional space such that the sum-squared distance between each group of *must-link* data instances (and their centroids) is minimized and that between *cannot-link* instance pairs is maximized in the projected space. The solution to this formulation reduces to an elegant eigenvalue decomposition problem similar in form to PCA and LDA. The second approach is a feature clustering approach and benefits from pairwise constraints via a constrained co-clustering mechanism [8]. Even though the constraints are imposed only on data instances (rows), the feature clusters (columns) are influenced since row clustering and column clustering are interleaved together and mutually reinforced in the co-clustering process.

We evaluate our proposed techniques in three sets of experiments on various real-world datasets. The evaluation metric is based on the improvement to the clustering performance through pairwise instance constraints. The experiments reported in this chapter were conducted separately at different times during the period we worked on this topic. The first set of experiments aim to show the effectiveness of pairwise constraints-guided feature projection in improving the clustering performance and the superiority of feature projection over adaptive metric learning [20]. In the second set of experiments, we compare our constraints-guided feature projection method with RCA. The last set of experiments are to demonstrate the superiority of the proposed pairwise constraints-guided co-clustering algorithm.

This chapter is organized as follows. Section 15.2 describes the proposed pairwise constraints-guided feature projection algorithm. Section 15.3 presents the pairwise constraints-guided co-clustering algorithm. Three experimental studies are presented in Section 15.4. Section 15.5 concludes this chapter with discussions and remarks on future work.

15.2 Pairwise Constraints-Guided Feature Projection

In this section, we first present the pairwise constraints-guided feature projection approach, and then describe how it can be used in conjunction with semi-supervised clustering algorithms.

15.2.1   Feature Projection

Given a set of pairwise data constraints, we aim to project the original data to a low-dimensional space, in which *must-link* instance pairs become close and *cannot-link* pairs far apart.

Let $X = \{x | x \in R^N\}$ be a set of N-dimensional column vectors (i.e., data instances) and $P_{N \times K} = \{P_1, \ldots, P_K\}$ a projection matrix containing K orthonormal N-dimensional vectors. Suppose the function $f : R^N \mapsto C$ maps each data instance to its target group. Then $C_{ml} = \{(x_1, x_2) | f(x_1) = f(x_2)\}$ is the set of all must-link instance pairs and $C_{cl} = \{(x_1, x_2) | f(x_1) \neq f(x_2)\}$ the set of all cannot-link instance pairs. We aim to find an optimal projection matrix P that maximizes the following objective function:

$$f(P) = \sum_{(x_1, x_2) \in C_{cl}} \|P^T x_1 - P^T x_2\|^2 - \sum_{(x_1, x_2) \in C_{ml}} \|P^T x_1 - P^T x_2\|^2 \quad (15.1)$$

subject to the constraints

$$P_i^T P_j = \begin{cases} 1 \text{ if } i = j \\ 0 \text{ if } i \neq j \end{cases} \quad (15.2)$$

where $\|\cdot\|$ denotes the L_2 norm.

There exists a direct solution to the above optimization problem. The following theorem shows that the optimal projection matrix $P_{N \times K}$ is given by the first K eigenvectors of matrix $Q = CDC^T$, where each column of matrix $C_{N \times M}$ is a difference vector $x_1 - x_2$ for a pair (x_1, x_2) in C_{ml} or C_{cl} and $D_{M \times M}$ is a diagonal matrix with each value on the diagonal corresponding to a constraint (1 for a cannot-link pair and -1 for a must-link pair).

THEOREM 15.1

Given the reduced dimensionality K, the set of must-link constraints C_{ml}, and cannot-link constraints C_{cl}, construct matrix $Q = CDC^T$, where C and D are defined above. Then the optimal projection matrix $P_{N \times K}$ is comprised of the first K eigenvectors of Q corresponding to the K largest eigenvalues.

PROOF   Consider the objective function

$$f(P) = \sum_{\substack{(x_1,x_2) \\ \in C_{cl}}} \|P^T(x_1 - x_2)\|^2 - \sum_{\substack{(x_1,x_2) \\ \in C_{ml}}} \|P^T(x_1 - x_2)\|^2$$

$$= \sum_{\substack{(x_1,x_2) \\ \in C_{cl}}} \sum_l P_l^T(x_1 - x_2)(x_1 - x_2)^T P_l$$

$$\quad - \sum_{\substack{(x_1,x_2) \\ \in C_{ml}}} \sum_l P_l^T(x_1 - x_2)(x_1 - x_2)^T P_l$$

$$= \sum_l P_l^T \left[\sum_{\substack{(x_1,x_2) \\ \in C_{cl}}} (x_1 - x_2)(x_1 - x_2)^T - \sum_{\substack{(x_1,x_2) \\ \in C_{ml}}} (x_1 - x_2)(x_1 - x_2)^T \right] P_l$$

$$= \sum_l P_l^T(CDC^T)P_l$$

$$= \sum_l P_l^T Q P_l \tag{15.3}$$

where the P_l's are subject to the constraints $P_l^T P_h = 1$ for $l = h$ and 0 otherwise.

Using the traditional Lagrange multiplier optimization technique, we write the Lagrangian to be

$$L_{P_1,\dots,P_k} = f(P_1,\dots,P_k) + \sum_{l=1}^{k} \xi_l(P_l^T P_l - 1) \tag{15.4}$$

By taking the partial derivative of $L_{P_1,\dots,P_k}$ with respect to each P_l and set it to zero, we get

$$\frac{\partial L}{\partial P_l} = 2QP_l + 2\xi_l P_l = 0 \qquad \forall l = 1,\dots,K \tag{15.5}$$

$$\Rightarrow QP_l = -\xi_l P_l \qquad \forall l = 1,\dots,K \tag{15.6}$$

Now it is obvious that the solution P_l is an eigenvector of Q and $-\xi_l$ the corresponding eigenvalue of Q. To maximize F, P must be the first K eigenvectors of Q that makes F the sum of the K largest eigenvalues of Q. ☐

When N is very large, $Q_{N \times N}$ is a huge matrix that can present difficulties to the associated eigenvalue decomposition task. In this case, we don't really need to compute Q since its rank is most likely much lower than N and we can use the Nystrom method [4] to calculate the top K eigenvectors more efficiently.

Algorithm: Projection-based semi-supervised spherical k-means using pairwise constraints (PCSKM+P)

Input: Set of unit-length documents $\mathcal{X} = \{x_i\}_1^N$, set of *must-links* C_{ml}, set of *cannot-links* C_{cl}, pre-specified reduced dimension K, and number of clusters C.

Output: C partitions of the documents.

Steps:

1. Use the method stated in Theorem 15.1 to project the original N-dimensional documents into K-dimensional vectors

2. Initialize the C unit-length cluster centroids $\{\mu_h\}_{h=1}^C$, set $t \leftarrow 1$

3. Repeat until *convergence*
   For $i = 1$ to N

   (a) For each document x_i which does not involve any *cannot-links*, find the closest centroid $y_n = \arg\max_k x_i^T \mu_k$;

   (b) For each pair of documents (x_i, x_j) involved in *must-link* constraints, find the closest centroid $y_n = \arg\max_k x_i^T \mu_k + x_j^T \mu_k$;

   (c) For each pair of documents (x_i, x_j) involved in *cannot-links*, find two different centroids μ_k and $\mu_{k'}$ that maximize $x_i^T \mu_k + x_j^T \mu_{x'}$;

   (d) For cluster k, let $\mathcal{X}_k = \{x_i | y_i = k\}$, the centroid is estimated as $\mu_k = \sum_{x \in \mathcal{X}_k} / \|\sum_{x \in \mathcal{X}_k}\|$;

4. $t \leftarrow t + 1$.

FIGURE 15.1: Projection-based semi-supervised spherical k-means using pairwise constraints.

15.2.2   Projection-Based Semi-supervised Clustering

The feature projection method enables us to represent the original instances in a low-dimensional subspace that conforms to *the pairwise instance constraints*. In this section, we will show how we can enhance the performance of semi-supervised clustering using feature projection. Since we shall apply this to text document clustering problems and text documents are often represented as unit-length vectors [10], we shall use the standard spherical k-means algorithm as the baseline method to construct our algorithm.

Given a set of documents $\mathcal{X} = \{x_1, \ldots, x_M\}$, a set of *must-link* constraints C_{ml}, a set of *cannot-link* constraints C_{cl}, and a predefined number of clusters C, we aim to find C disjoint partitions that conform to the given constraints. For the *must-link* constraints, since they represent an equivalence relation, we can easily put any pair of instances involved in C_{ml} into the same cluster. For the *cannot-link* constraints, finding a feasible solution for the *cannot-link* constraints is much harder than that for the *must-link* constraints (actually NP-complete) [6]. Therefore, we adopt a local greedy heuristic to update the cluster centroids. Given each *cannot-link* constraint $C_{cl}(x_i, x_j)$, we find two different cluster centroids μ_{x_i} and μ_{x_j} such that

$$x_i^T \mu_{x_i} + x_j^T \mu_{x_j} \qquad (15.7)$$

is maximized and assign x_i and x_j to these two different centroids to avoid violating the *cannot-link* constraints. Note that all x's and μ's need to be L_2-normalized vectors for (15.7) to work. Our algorithm for semi-supervised clustering using feature projection is shown in Figure 15.1.

It is worth noting that the constrained spherical k-means incorporated in our algorithm does not distinguish the different priorities among the *cannot-link* constraints as shown in [3]. For fair comparison in our experiment, we adopt this version of constrained k-means algorithm to implement the typical *distance-based* method in [20]. We will hold off the detailed discussion until Section 15.4.

15.3 Pairwise Constraints-Guided Co-clustering

As we mentioned in Section 15.1, feature clustering in general can be regarded as a special case of feature projection, with each cell in the projection matrix restricted to be a binary value. Usually dimensionality reduction method only acts as a pre-processing step in data analysis. The transformed data in the reduced low-dimensional space will be used for subsequent classification or clustering. However, the co-clustering method discussed in this section cannot be simply regarded as a special case of feature projection since it involves clustering the instances and features at the same time. The co-clustering algorithm used here is proposed in [8] and aims to minimize the following objective function:

$$I(X;Y) - I(\hat{X};\hat{Y}) \qquad (15.8)$$

subject to the constraints on the number of row and column clusters. $I(X;Y)$ is the mutual information between the row random variable X, which governs the distribution of rows, and the column random variable Y, which governs the distribution of columns. $\hat{X}$ and $\hat{Y}$ are variables governing the distribution of clustered rows and clustered columns, respectively. An iterative algorithm was used in [8] to alternate between clustering rows and clustering columns to reach a local minimum of the above objective function.

Due to space limits, we omit a detailed discussion of the co-clustering algorithm and readers are referred to [8]. Also, here we just concisely describe how we involve constraints in the co-clustering process: The constraints only affect the row/data clustering step algorithmically and the impact on column/feature clustering is implicit. For must-link data pairs, we merge the rows and replace each instance by the average; for cannot-link data pairs, we

separate a pair if they are in the same cluster after an iteration of row clustering, by moving the instance that is farther away from the cluster centroid to a
different cluster. Essentially, the idea of handling constraints is similar to the
existing work [19, 2], but we get features clustered through the co-clustering
process. This combination of pairwise constraints and co-clustering seems to
have not appeared before in the literature.

15.4   Experimental Studies

To measure clustering performance, we adopt *normalized mutual information* (NMI) as the evaluation criterion. NMI is an external validation metric
and estimates the quality of clustering with respect to the given true labels of
the datasets [18]. Let $\hat{Z}$ be the random variable denoting the cluster assignments of the instances and Z the random variable denoting the underlying
class labels. Then, NMI is defined as

$$NMI = \frac{I(\hat{Z}; Z)}{(H(\hat{Z}) + H(Z))/2} \tag{15.9}$$

where $I(\hat{Z}; Z) = H(Z) - H(Z|\hat{Z})$ is the mutual information between the
random variables $\hat{Z}$ and Z, $H(Z)$ the Shannon entropy of Z, and $H(Z|\hat{Z})$ is
the conditional entropy of Z given $\hat{Z}$ [5].

15.4.1   Experimental Study – I

In our first set of experiments, we used four subsets from the 20-newsgroup
data [14] for comparison. The 20-newsgroup dataset consists of approximately 20,000 newsgroup articles collected evenly from 20 different Usenet
newsgroups. Many of the newsgroups share similar topics and about 4.5% of
the documents are cross-posted over different newsgroups making the class
boundary rather fuzzy. We applied the same pre-processing steps as in [8],
i.e., removing stopwords, ignoring file headers and subject lines, and selecting
the top 2000 words by mutual information. Specific details of the datasets
are given in Table 15.1. The Bow [15] library was used for generating these
four subsets from the 20-newsgroup corpus.

The algorithms we compared are listed below:

- SPKM: the standard spherical k-means algorithm that does not make
  use of any pairwise constraints [10];

- PCSKM: the pairwise constrained spherical k-means algorithm described
  in Section 15.2.2;

TABLE 15.1: Datasets from 20-newsgroup corpus for Experiment I.

Dataset	Newsgroup	No. doc. per group	Tot. doc.
Binary	talk.politics.mideast talk.politics.misc	250	500
Multi5	comp.graphics rec.motorcycles rec.sports.baseball sci.space talk.politics.mideast	100	500
Multi10	alt.atheism comp.sys.mac.hardware misc.forsale rec.autos, sci.crypt rec.sport.hockey sci.electronics sci.med, sci.space talk.politics.gun	50	500
Science	sci.crypt, sci.med sci.electronics sci.space	500	2000

- PCSKM+M: the *distance-based* pairwise constrained spherical k-means algorithm introduced in [20];

- PCSKM+P: the *projection-based* pairwise constrained spherical k-means algorithm described in Figure 15.1 that reduces dimensionality first using our proposed feature projection algorithm.

We implemented all the algorithms in MATLAB and conducted our experiments on a machine running Linux with 4 Intel Xeon 2.8GHz CPUs and 2G main memory. For each dataset, we randomly repeated each experiment for 20 trials. In each trial, we randomly generated 500 pairwise constraints from half of the dataset and tested the clustering performance on the whole dataset. The average results over 20 trials are presented.

We performed an extensive comparative study on the algorithms listed above (SPKM, PCSKM, PCSKM+M, and PCSKM+P). The clustering performance of different algorithms are compared at different numbers of pairwise constraints.

The results are shown in Figure 15.2, where the x-axis denotes the number of pairwise constraints, and the y-axis denotes the clustering performance in terms of NMI. The number of reduced dimensionality K is set to 30 in the SPKM+P algorithm for all the datasets. It is worth noting that we did not utilize the pairwise constraints to initialize the cluster centroids. Although it has been demonstrated that seeding the initial centroids by constraint information can give further improvement to the clustering result [2], here we decided not to do so because we want to measure the improvement contributed from the feature projection or metric learning only.

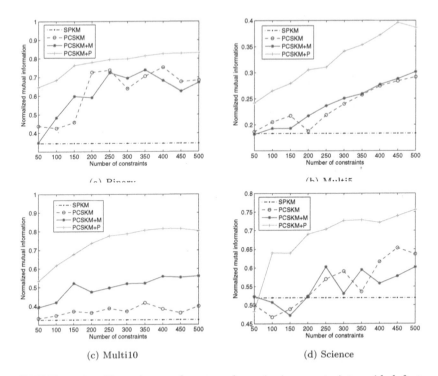

(a) Binary

(b) Multi5

(c) Multi10   (d) Science

FIGURE 15.2: Clustering performance by pairwise constraints-guided feature projection on four datasets from the 20-newsgroup collection.

It is clear that, on most datasets, clustering performance is constantly improved along with the increasing number of pairwise constraints. Specifically, the PCSKM+P algorithm (i.e., PCSKM + feature projection) is more stable compared to all the other methods and almost always outperforms all the other methods. This may be due to the fact that constraint-guided feature projection can produce more condensed and more meaningful representations for each instance. On the other hand, PCSKM+M is not significantly better than the PCSKM algorithm except for the Multi10 datasets. This indicates that, for the original high-dimensional sparse data, it is difficult for metric learning to get a meaningful distance measure.

We also compared the impact of *must-link* and *cannot-link* constraints on the performance of the PCSKM+P algorithm on the first three datasets. To compare the different impacts of these two types of constraints on improving clustering performance, we incorporated a parameter β into the objective function in Equation (15.8) to adjust the relative importance between the *must-link* and *cannot-link* constraints:

$$f = (1 - \beta) \cdot \sum_{(x_1,x_2) \in C_{CL}} \|F^T(x_1 - x_2)\|^2$$
$$-\beta \cdot \sum_{(x_1,x_2) \in C_{ML}} \|F^T(x_1 - x_2)\|^2 \qquad (15.10)$$

It is clear that $\beta = 0$ is equivalent to using only *cannot-link* constraints and $\beta = 1$ is equivalent to using only *must-link* constraints. In our experiments, we varied the value of β from 0 to 1 with a stepsize of 0.1. The performances of the clustering results, measured by NMI, are plotted in Figure 15.3. In the figure, the x-axis denotes the different values of parameter β and the y-axis the clustering performance measured by NMI.

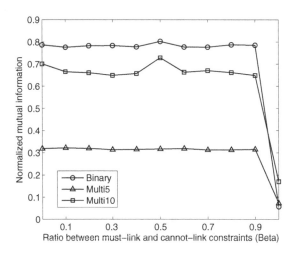

FIGURE 15.3: Relative importance of must-link vs. cannot-link constraints.

As can be seen, there is no significant difference when β is in the range of 0.1–0.9. However, when using only *must-link* constraints ($\beta = 1$) the clustering performance deteriorates sharply, indicating that *must-link* constraints have little value compared to *cannot-link* constraints in guiding the feature projection process to find a good low-dimensional representation.

15.4.2   Experimental Study – II

Our second set of experiments were performed on a couple of real-world datasets from rather different application domains. For the low-dimensional datasets, we used six UCI datasets [16]: *balance-scale, ionosphere, iris, soybean, vehicle,* and *wine.* These datasets have been used for distance metric learning [20] and for constrained feature projection via RCA [1]. For the high-dimensional datasets, we chose six subsets derived from some TREC collections (available at http://trec.nist.gov).

For each test dataset, we repeated each experiment for 20 trials. For the UCI datasets, we randomly generated 100 pairwise constraints in each trial. For the TREC datasets collection, we randomly generated 500 pairwise constraints from half of the dataset and tested the clustering performance on the whole dataset. Also, the final result is the average of the results from the 20 trials.

We evaluated the performance of our feature projection method relative to other dimensionality reduction methods such as PCA and RCA (source code available at http://www.cs.huji.ac.il/~tomboy/code/RCA.zip). For a thorough comparison, we used both relatively low-dimensional datasets from the UCI repository and high-dimensional data from the TREC corpus. As for the low-dimensional UCI datasets, we used the standard k-means algorithm to serve as the baseline algorithm. As for the high-dimensional TREC datasets, we again chose the spherical k-means algorithm [10] as the baseline.

Figure 15.4 shows the clustering performance of standard k-means applied to the original and projected data from different algorithms on six UCI datasets with different numbers of pairwise constraints. Note that N represents the size of the dataset, C the number of clusters, D the dimensionality of original data, and d the reduced dimensionality after projection. As shown in Figure 15.4, RCA performs extremely well on the low-dimensional datasets and the performance improves significantly when the number of available constraints increases. However, for some datasets such as *vehicle* and *wine,* when only providing limited constraints, the performance of RCA is even worse than PCA, which is unsupervised and does not use any pairwise constraints. Our method, Projection, on the other hand, is always comparable to or better than PCA. In some cases, such as for the *soybean* and *iris* datasets, Projection has comparable performance to RCA.

Although the performance of RCA is good for low-dimensional datasets, it is computationally prohibitive to directly apply RCA to high-dimensional datasets, such as the TREC datasets. Our projection-based methods, on the other hand, have no problem handling very high-dimensional text data. For the purpose of getting some comparison between Projection and RCA for high-dimensional text data, we first applied PCA to project the original data into a 100-dimensional subspace, and then applied different algorithms to further reduce the dimensionality to 30. Note that, even without the PCA step, our method is applicable and generates similar performance. Figure 15.5

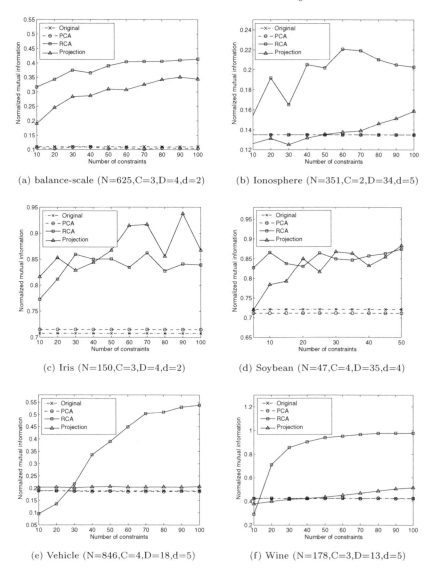

FIGURE 15.4: Clustering performance on UCI datasets with different feature projection methods (N: size of dataset; C: number of clusters; D: dimensionality of original data; d: reduced dimensionality after projection).

gives the clustering performance of spherical k-means applied to the original and projected data from different algorithms on six TREC datasets with different numbers of pairwise constraints. In Figure 15.5, we also included the clustering performance by directly applying PCA without any other additional dimensionality reduction algorithms. As can be seen in Figure 15.5,

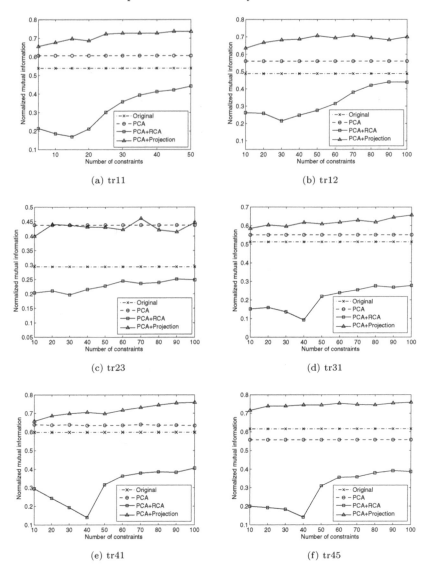

FIGURE 15.5: Clustering performance on TREC datasets with different feature projection methods, where "PCA+RCA" denotes the method with PCA being applied first, followed by RCA, and "PCA+Projection" means the method with PCA being applied first, followed by our Projection method.

PCA+Projection almost always achieves the best performance of dimensionality reduction on all test datasets. In contrast, RCA performs the worst for the text datasets, indicating that RCA is not a desirable method for high-dimensional data.

TABLE 15.2: Datasets from 20-newsgroup corpus for Experiment III.

Dataset	Newsgroups	Instances	Dims	Classes
News-Similar-3	comp.graphics comp.os.ms-windows comp.windows.x	295	1864	3
News-Related-3	talk.politics.misc talk.politics.guns talk.politics.mideast	288	3225	3
News-Different-3	alt.atheism rec.sport.baseball sci.space	300	3251	3

15.4.3 Experimental Study – III

In our third set of experiments, we compared our pairwise constraints-guided co-clustering algorithm with the standard information theoretic co-clustering [8]. We constructed three datasets from the 20-newsgroup collection [14]. From the original dataset, three datasets were created by selecting some particular group categories. *News-Similar-3* consists of three newsgroups on similar topics: comp.graphics, comp.os.ms-windows, and comp.windows.x, with significant overlap between clusters due to cross-posting. *News-Related-3* consists of three newsgroups on related topics: talk.politics.misc, talk.politics.guns, and talk.politics.mideast. *News-Different-3* consists of three well-separat-ed newsgroups that cover quite different topics: alt.atheism, rec.sport.baseball and sci.space. All the datasets were converted to the vector-space representation following several steps—tokenization, stop-word removal, and removing words with very high frequency and low frequency [10]. The semi-supervised co-clustering algorithm directly clusters the normalized document-term matrix (treated as a probability distribution) without any TF-IDF weighting. Table 15.4.3 summarizes the properties of the datasets.

We denoted our method as `co-clustering+pc` and the algorithm proposed in [8] as `co-clustering`. The results are shown in Figure 15.6, from which we can see that as the number of pairwise constraints increases, the performance of the constrained co-clustering algorithm improves significantly compared to the unguided version. As the co-clustering algorithm does simultaneous instance (row) clustering and feature (column) clustering, imposing pairwise constraints on instances indirectly contributes to the feature clustering part as well.

It is interesting to see that, when the number of constraints is small (e.g., smaller than 100), the constraints-guided co-clustering algorithm actually performs worse than the regular co-clustering algorithm. We suspect the reason is that the constraints-guided co-clustering algorithm runs into worse local optima more frequently when the guidance is too limited, but this needs to be further investigated in future research.

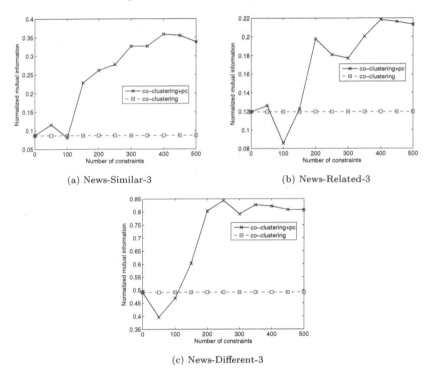

(a) News-Similar-3

(b) News-Related-3

(c) News-Different-3

FIGURE 15.6: Performance of pairwise constraints-guided co-clustering on dataset from the 20-newsgroup collection.

15.5   Conclusion and Future Work

In this chapter, we have introduced two different pairwise constraints-guided dimensionality reduction techniques, and investigated how they can be used to improve semi-supervised clustering performance, especially for very high-dimensional data. The proposed pairwise constraints-guided dimensionality reduction techniques seem to be a promising new way of leveraging "weak" supervision to improve the quality of clustering, as demonstrated by the experimental results on the selected text datasets.

Although the feature projection via pairwise constraints can make certain achievements, the number of projected features is currently chosen in an ad hoc way in our experiments. How to find out the "best" number for the feature projection is an interesting problem for future research.

References

[1] A. Bar-Hillel, T. Hertz, N. Shental, and D. Weinshall. Learning distance functions using equivalence relations. In *Proc. of the Twentieth International Conference on Machine Learning*, 2003.

[2] S. Basu, A. Banerjee, and R. J. Mooney. Semi-supervised clustering by seeding. In *Proc. of Ninteenth International Conference on Machine Learning*, pages 19–26, 2002.

[3] S. Basu, M. Bilenko, and R. Mooney. A probabilistic framework for semi-supervised clustering. In *Proc. of Tenth ACM SIGKDD International Conference on Knowledge Discovery and Data Mining*, pages 59–68, Seattle, WA, August 2004.

[4] C. Burges. Geometric methods for feature extraction and dimensionality reduction: a guided tour. Technical Report MSR-TR-2004-55, Microsoft, 2004.

[5] T. Cover and J. Thomas. *Elements of Information Theory*. Wiley-Interscience, 1991.

[6] I. Davidson and S. S. Ravi. Clustering with constraints: Feasibility issues and the *k*-means agorithm. In *Proc. of Fifth SIAM International Conference on Data Mining*, Newport Beach, CA, April 2005.

[7] S. Deerwester, S. Dumais, T. Landauer, G. Furnas, and R. Harshman. Indexing by latent semantic analysis. *Journal of the American Society of Information Science*, 41(6):391, 1990.

[8] I. Dhillon, S. Mallela, and D. Modha. Information-theoretic co-clustering. In *Proc. of Ninth ACM SIGKDD International Conference on Knowledge Discovery and Data Mining*, pages 89–98, August 2003.

[9] I. S. Dhillon, S. Mallela, and R. Kumar. Enhanced word clustering for hierarchical text classification. In *Proc. of Eighth ACM SIGKDD International Conference on Knowledge Discovery and Data Mining*, pages 446–455, July 2002.

[10] I. S. Dhillon and D. S. Modha. Concept decompositions for large sparse text data using clustering. *Machine Learning*, 42(1):143–175, 2001.

[11] R. Duda, P. Hart, and D. Stork. *Pattern Classification*. Wiley-Interscience, 2nd edition, 2000.

[12] K. Fukunaga. *Statistical pattern recognition*. Academic Press, San Diego, CA, 2nd edition, 1990.

[13] I. Guyon and A. Elisseeff. An introduction to variable and feature selec-

tion. *Journal of Machine Learning Research*, 3:1157–1182, 2003.

[14] K. Lang. News weeder: learning to filter netnews. In *Proc. of Twelfth International Conference on Machine Learning*, pages 331–339, 1995.

[15] A. K. McCallum. Bow: A toolkit for statistical language modeling, text retrieval, classification and clustering, 1996. http://www.cs.cmu.edu/~mccallum/bow.

[16] D. J. Newman, S. Hettich, C. L. Blake, and C. J. Merz. UCI repository of machine learning databases, 1998. http://www.ics.uci.edu/~mlearn/MLRepository.html.

[17] K. Pearson. On lines and planes of closest fit to systems of points in space. *Philosophical magazine*, 2(6):559, 1901.

[18] A. Strehl, J. Ghosh, and R. J. Mooney. Impact of similarity measures on web-page clustering. In *AAAI Workshop on AI for Web Search*, pages 58–64, July 2000.

[19] K. Wagstaff, C. Cardie, S. Rogers, and S. Schroedl. Constrained k-means clustering with background knowledge. In *Proc. of Eighteenth International Conference on Machine Learning*, pages 577–584, 2001.

[20] E. P. Xing, A. Y. Ng, M. I. Jordan, and S. Russell. Distance metric learning with application to clustering with side-information. In S. T. S. Becker and K. Obermayer, editors, *Proc. of Advances in Neural Information Processing Systems 13*, pages 505–512, MIT Press, Cambridge, MA, 2003.

Chapter 16

Aggressive Feature Selection by Feature Ranking

Masoud Makrehchi

University of Waterloo

Mohamed S. Kamel

University of Waterloo

16.1 Introduction

Recently, text classification has become one of the fastest growing applications of machine learning and data mining [15]. There are many applications that use text classification techniques, such as natural language processing and information retrieval [9]. All of these applications use text classification techniques in dealing with natural language documents. Since text classification is a supervised learning process, a good many learning methods such as K-nearest neighbor (KNN), regression models, naïve Bayes classifier (NBC), decision trees, inductive rule learning, neural networks, and support vector machines (SVM) can be employed [1].

Most text classification algorithms use vector space model, and bag-of-words representation, as proposed by Salton [22], to model textual documents. Some extensions of the vector space model have also been proposed that utilize the semantic and syntactic relationships between terms [14]. In the vector space model, every word or group of words (depending on whether one is working with a single word or a phrase) is called a term, which represents one dimension of the feature space. A positive number, reflecting the relevancy and significance, is assigned to each term. This number can be the frequency of the term in the document [19].

The major problem of text classification with vector space modeling is its high dimensionality. A high-dimensional feature space addresses a very large vocabulary that consists of all terms occurring at least once in the collection of documents. High-dimensional feature space has a destructive influence on the performance of most text classifiers. Additionally, it increases the complexity of the system. To deal with high dimensionality and avoid its consequences, dimensionality reduction is strongly advised [12, 21].

One well-known approach for excluding a large number of irrelevant features is feature ranking [12, 6]. In this method, each feature is scored by a feature quality measure such as information gain, χ^2, or odds ratio. All features are sorted based on their scores. For feature selection, a small number of best features are kept and the rest are removed. This method has a serious disadvantage, however, in that it ignores the redundancies among terms. This is because the ranking measures consider the terms individually. An experiment, detailed in the next section, shows that the impact of term redundancy is as destructive as noise.

Due to the high dimensionality of text classification problems, computational efficiency and complexity reduction are very important issues. One strategy in dimensionality reduction is aggressive feature selection, in which the classification task is performed by very few features with minimum loss of performance and maximum reduction of complexity. In aggressive feature selection, more than 90% of non-discriminant, irrelevant, and non-informative features are removed. In [12], the number of selected features is as low as 3% of features. More aggressive feature selection, including only 1% of all features, has also been reported in [6].

In this chapter, a new approach for feature selection is proposed, with a more than 98% reduction in features. The method is based on a multi-stage feature selection including: *(i)* pre-processing tasks to remove stopwords, infrequent words, noise, and errors; *(ii)* a feature ranking, such as information gain, to identify the most informative terms; and *(iii)* removing redundant terms among those that have been already selected by the feature ranking measure.

This chapter consists of five sections. After the introduction, feature selection by feature ranking is briefly reviewed in Section 16.2. In Section 16.3, the proposed approach to reducing redundancy is detailed. Experimental results and the summary are presented in Sections 16.4 and 16.5, respectively.

16.2   Feature Selection by Feature Ranking

A class of filter approach of feature selection algorithms is feature ranking methods. Feature ranking aims to retain a certain number of features, specified by ranking threshold, with scores determined according to a measure of term relevance, discriminating capability, information content, or quality index. Simply defined, feature ranking is sorting the features according to a "feature quality index," which reflects the relevance, information, and discriminating power of the feature.

Feature ranking requires a decision about the following three factors:

- **The Feature Ranking Scope:** The process of feature ranking is either local or global. In the local case, feature ranking is performed for each class individually, which implies employing a local vocabulary. In global feature ranking, we are dealing with one unified vocabulary associated with the training dataset.

- **The Type of Feature Ranking Measure:** Selecting a ranking measure must be performed by considering the classification model and the characteristics of the dataset. There is a link between feature ranking measures and classifiers. It means some classifiers work better with a particular set of feature ranking measures. For example, the NBC classifier works better with odds ratio, one of the feature ranking methods, such that features with a higher ranking in odds ratio are more influential in NBC [2, 11]. It can be also shown that the performance of feature ranking methods vary from one dataset to the other. For instance, in [11], it has been shown that odds ratio feature ranking performs more successfully with moderately sparse datasets, for example, 10 to 20 terms per document vector, while the classifiers are NBC or SVM. Due to this correlation, one challenging problem is selecting the appropriate ranking method for a particular dataset.

- **Feature Ranking Threshold:** One crucial problem in feature ranking is to determine the appropriate threshold at which to filter out noise and stopwords. This threshold represents the number of desired features and reflects the complexity of the classifier. The ranking threshold can be applied either to the value of the scoring metrics or to the number of features.

Feature ranking methods, despite their scalability and lower cost algorithms, suffer from lower performance as compared to the search-based feature selection such as wrappers. The low performance of feature ranking techniques arises from two major issues: *(i)* ignoring the correlation between terms and implementing an univariate scheme while the nature of text classification problems is multivariate; and *(ii)* failing in rejecting redundant terms. These two issues are investigated, but the focus in this chapter will be on improving the feature ranking performance by extracting and removing the redundant terms.

16.2.1   Multivariate Characteristic of Text Classifiers

As a major disadvantage, feature ranking methods ignore the correlation and dependency between terms because of their univariate function nature. Feature selection based on ranking is a univariate approach, in which only one feature is considered to be retained or removed. In other words, feature ranking measures such as information gain simply ignore the dependency and correlation between terms. The consequences can be low discriminating capacity and increased redundancy. Neglecting term correlation causes two problems. Let t_1 and t_2 be two relevant, and t_3 and t_4 be two irrelevant terms:

1. Most feature ranking measures rank t_1 and t_2 higher than t_3 and t_4, while in textual data, especially in natural language texts, sometimes two individually irrelevant terms, such as t_3 and t_4, are jointly relevant. A well-known example is the phrase "to be or not to be," in which all terms are individually noise but are meaningful as part of a phrase.

2. By any feature ranking, t_1 and t_2 will be kept, while in textual data these two terms can be redundant as well as relevant, such as synonym terms.

In spite of feature ranking, text classifiers behave based on the combination of features. Adopted from [5], a simple test is provided here to explain the impact of the multivariate characteristic of text classifiers on their performance. Two scoring metrics including information gain and random feature ranking (RND) are applied to the "20 newsgroups" dataset. In RND, the rank of each feature is randomly assigned. In addition to these two ranking methods, the third ranking measure called single term prediction (STP), which is defined based on the predicting capacity of every feature, is introduced. Let $h(f_i)$, $1 \leq i \leq m$ be a classifier using the feature set including only feature f_i. Here h is a Rocchio classifier. $STP(f_i)$ is defined as the performance (for example, the accuracy or macro-averaged F-measure) of the $h(f_i)$ when it is applied to the dataset. After estimating STP for all features, the terms are sorted based on their corresponding STPs.

The classifier performance of the three ranking methods is estimated across all levels of filtering (ranking threshold) for the dataset. Figure 16.1 depicts the classifier performance vs. filter levels for 50% of the best features for the three ranking methods. It shows that STP ranking always performs very poorly as compared to the other methods, including random ranking. It means ignoring the correlation and dependency between terms is as destructive as noise in feature ranking.

16.2.2   Term Redundancy

All terms of the vocabulary with respect to their contribution to the categorization and retrieval processes can be grouped into four classes: *(i)* non-redundant and relevant; *(ii)* non-redundant and irrelevant; *(iii)* redundant

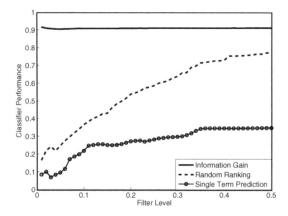

FIGURE 16.1: The impact of ignoring term correlation on classifier performance.

and relevant; and *(iv)* redundant and irrelevant. In feature selection for text classifiers, we are only interested in the first group, which is non-redundant and relevant terms. Measuring the relevancy of the terms, by employing strong feature ranking methods, such as information gain, is quite feasible. The difficulty is to extract term redundancies.

Redundancy is a kind of data dependency and correlation that can be estimated by different measures, such as the Jaccard, cosine, co-occurrence, and correlation coefficients [4, 20, 16]. In this chapter, redundancy between two terms is measured by mutual information. If two terms have similar probability distributions on class labels, one of the terms might be considered as a redundant term such that removing it does not hurt the classifier performance. The problem is to find the possible redundancies and identify the redundant terms to be removed.

In this section, the result of an experiment illustrating the influence of redundancy on the classifier performance is presented. Two different text classifiers are employed: a Rocchio classifier, which is a weak classifier and sensitive to noise, and an SVM classifier with a linear kernel, as an optimum classifier that is commonly used as a text classifier. The data collection is the well-known 20-Newsgroups (20NG) dataset. Macro-averaged F-measure is employed to evaluate the classifiers.

We show that adding redundancy, in the case of a very low number of features, can degrade the accuracy. The testing process is as follows: Let T be the set of N terms of the vocabulary, $T = \{t_1, t_2, \ldots, t_N\}$. The terms are ranked by a feature ranking method, for instance, information gain, such that t_1 is the best term and t_N the worst. A smaller set V, called the set of selected features, is a subset of T with n terms such that $V = \{v_1, v_2, \ldots, v_n\}, V \subset T, n \ll N$. Three different versions of V are generated by the following setups:

- n best terms: The n first terms of the set T are selected such that $v_i = t_i, 1 \leq i \leq n$.

- $n/2$ best terms + $n/2$ redundant terms: The vector V has two parts. For the first part, $n/2$ best terms of T are selected. The $n/2$ terms of the second part are artificially generated by adding a very small amount of noise to each term of the first part. The result is a set of redundant terms. Using this setup, the rate of redundancy is at least 50%.

- $n/2$ best terms + $n/2$ noise: It is the same as the previous setup, except that the second part consists of noise and stopwords. Due to the use of feature ranking measures, $n/2$ last (worst) terms should be noisy and less informative. Therefore, we do not have to generate artificial noise.

$$T = \{\overbrace{t_1, t_2, \ldots, t_{n/2}}^{P_1}, \ldots, \overbrace{t_{N+1-n/2}, \ldots, t_{N-1}, t_N}^{P_2}\}, V = P_1 \cup P_2 \quad (16.1)$$

where P_1 is the set of the most informative terms and P_2 includes noise.

We use five-fold cross validation for estimating the performance of classifiers. In this process, the collection (whole dataset) is divided into five subsets. The experiment is repeated five times. Each time we train the classifier with four subsets and leave the fifth one for the test phase. The average of the five measures is the estimated classifier performance, which is the macro-average F-measure. Since the main objective is to select a very small number of features, all three feature vectors with different and very small values for n, $n = \{5, 10, \ldots, 40\}$, are submitted to the SVM and Rocchio classifiers and the average of the performance of eight classifications is calculated. Figure 16.2 illustrates the results. It clearly shows that redundancy and noise reduce accuracy. Comparing the performance of the two last feature vectors, including redundant and noisy terms, they have a similar impact on both classifiers. Precisely speaking, Table 16.1 shows that Rocchio, as the weaker classifier with less generalization capability in comparison with SVM, is more sensitive to redundancy. This fact is clearly seen in Figure 16.2(a) and 16.2(b). In the figures, relative performance measures vs. the number of features have been depicted. Let F be the performance measure of a classifier, for example, accuracy, using the original feature vector (100% best features) with no added extra noise or redundancy; F_r the measure using artificially added redundancy; and F_n the same measure using added noise. The relative measure is calculated using F_r/F and F_n/F. Figure 16.2(b) shows that Rocchio's deviation from the original performance by adding redundancy is worse than the case of adding noise to the original feature vector.

In a small feature vector, the risk of having redundant terms is quite high. For example, in a five-term feature vector, if there is only one redundant term, we are actually using four terms instead of five because one of the terms is useless. By removing the redundant term, we make room for other terms (or any next non-redundant term), which can improve the discriminating power of the feature vector.

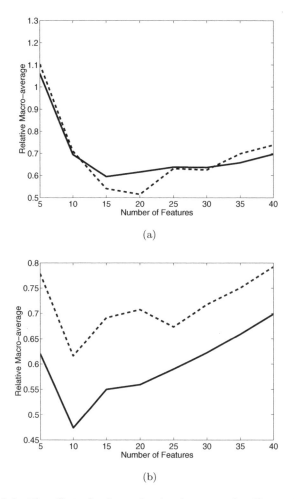

(a)

(b)

FIGURE 16.2: The effect of noise and redundancy on classifier performance. The solid lines represent added redundant terms, while the dashed lines represent added noise terms. (a) SVM, and (b) Rocchio classifier.

In conclusion, redundant terms not only have no discriminating benefits for the classifier, but also reduce the chance that other less informative but non-redundant terms can contribute to the classification process.

TABLE 16.1: The impact of redundancy and noise on the accuracy of the SVM and Rocchio text classifiers with feature selection using information gain ranking.

feature vector scheme	SVM Classifier	Rocchio Classifier
100% best terms (original)	0.6190	0.5946
50% redundancy	0.4067	0.3642
50% noise	0.4040	0.4289

16.3   Proposed Approach to Reducing Term Redundancy

In this chapter, a technique is proposed for feature selection with a high rate of reduction, by which the number of selected features V is much less than those in the original vocabulary T. We propose a three-stage feature selection strategy including pre-processing tasks, feature ranking, and removing redundant terms.

16.3.1   Stemming, Stopwords, and Low-DF Terms Elimination

In most information retrieval and text classification problems, stopwords are removed and whole terms reduced to their root by a stemming algorithm such as a Porter stemmer. Unlike stopword removal, which removes only a few hundred terms from the vocabulary, stemming can reduce by up to 40% the vocabulary size [7].

In most text classification researches, *low document frequency terms* (low-DF terms) are also removed from the vocabulary. Low-DF terms include very rare terms or phrases, spelling errors, and those having no significant contribution to classification. Although these words from an information retrieval point of view may play a critical role for indexing and retrieval, in the classification process, they have no information content and can be treated as noise. Another reason can be explained as follows: Since a class of feature ranking methods, in particular χ^2, behave unreliably and are not robust in the case of low frequent features, eliminating low-DF terms can prevent this drawback.

Although major low-DF terms are considered as noise, misspellings, or non-informative terms, in the case of difficult classes having less sparse vocabulary, or classes with very few samples, they may have a more discriminating role. In conclusion, in eliminating low-DF terms, one should be aware of class difficulty and class imbalance.

16.3.2   Feature Ranking

In the second stage, a feature ranking measure is employed to select the most informative and relevant terms. Adopted from [21], information gain is one of the most effective ranking measures and calculated as follows:

$$IG(t_j) = -\sum_{k=1}^{C} P(c_k).\log P(c_k) +$$

$$P(t_j)\sum_{k=1}^{C} P(c_k|t_j).\log P(c_k|t_j) + \quad (16.2)$$

$$P(\bar{t_j})\sum_{k=1}^{C} P(c_k|\bar{t_j}).\log P(c_k|\bar{t_j})$$

where $P(c_k)$ is the probability of a document belonging to the class c_k, $P(t_j)$ is the probability of a document containing the term t_j, and $P(c_k|t_j)$ is the conditional probability of c_k given term t_j. The number of classes is denoted by C. In practice, information gain is estimated as follows:

$$IG(t_j) = -\sum_{k=1}^{C} \frac{n(c_k)}{n}\log\frac{n(c_k)}{n} + \frac{n(t_j)}{n}\sum_{k=1}^{C}\frac{n(t_j;c_k)}{n(t_j)}\log\frac{n(t_j;c_k)}{n(t_j)} + \quad (16.3)$$

$$(1 - \frac{n(t_j)}{n})\sum_{k=1}^{C}\frac{n(c_k) - n(t_j;c_k)}{n - n(t_j)}\log\frac{n(c_k) - n(t_j;c_k)}{n - n(t_j)}$$

where n is the total number of documents in the training data, $n(c_k)$ depicts the number of documents in the kth class, and $n(t_i)$ is the number of documents, which contain term t_i. The number of documents, which belongs to the kth class and includes the term t_i, is expressed by $n(t_i;c_k)$.

Using the entropy of t_j, information gain can be normalized as follows:

$$NIG(t_j) = \frac{IG(t_j)}{-\frac{n(t_j)}{n}\log\frac{n(t_j)}{n}} \quad (16.4)$$

Information gain is one of the most efficient measures of feature ranking in classification problems [6]. Yang and Pedersen [21] have shown that with various classifiers and different initial corpora, sophisticated techniques such as information gain or χ^2 can reduce the dimensionality of the vocabulary by a factor of 100 with no loss (or even with a small increase) of effectiveness. In our application, the original vocabulary after pre-processing, including $28,983$ terms, is ranked by information gain. Next, the best 10% of terms are chosen for this stage. Similar to other ranking methods, information gain has serious drawbacks such as ignoring the redundancy among higher ranked features.

16.3.3 Redundancy Reduction

It has been previously explained that by employing a small number of features, any term redundancy can influence the classifier performance. It has also been reported that redundancy reduction can improve the performance of feature selection algorithms [12]. In the third stage, by reducing term redundancies, about 80% to 95% of ranked features are removed.

The problem of redundancy reduction is to find an efficient redundancy extraction algorithm in terms of low computational complexities. The major difficulty in redundancy extraction, in addition to choosing proper correlation measure, is calculating pairwise correlation between features. This last calculation can be expensive. Although few simplified term redundancy reductions such as the μ-occurrence measure proposed by [16] have been reported, they propose special cases such as binary class problems or assessing only pairwise term redundancy without considering the class labels of the terms, which can increase the complexity of the problem.

The proposed approach has two core elements, mutual information and inclusion index. These are detailed in the following subsections:

16.3.3.1 Mutual Information

Mutual information is a measure of statistical information shared between two probability distributions. Based on the definition in [10], mutual information $I(x; y)$ is computed by the relative entropy of a joint probability distribution, such as $P(x, y)$ and the product of the marginal probability distributions $P(x)$ and $P(y)$:

$$I(x; y) = D(P(x, y) || P(x)P(y)) = \sum_x \sum_y P(x, y) log \frac{P(x, y)}{P(x)P(y)} \qquad (16.5)$$

which is called the Kullback-Leibler divergence. Mutual information, such as other information theoretic measures, widely used in language modeling, has been applied in text mining and information retrieval for applications such as word association [3] and feature selection [18]. Mutual information is viewed as the entropy of co-occurrence of two terms when observing a class. We practically compute mutual information between two other mutual information measures. Each measure represents shared information between a term such as t_i and a class such as c_k. Since we are interested in the distribution of a pair of terms given a specific class, the joint distribution is considered as the probability of occurrence of the two terms t_i and t_j in those documents belonging to the class c_k. Equation 16.5 can be rewritten as follows:

$$I(t_i; c_k) = P(t_i, c_k) log \frac{P(t_i, c_k)}{P(t_i)P(c_k)} \qquad (16.6)$$

where $I(t_i; c_k)$ is the mutual information of the distribution of term t_i and class c_k. Equation 16.6 might be written for term t_j in exactly the same way. In other words, $I(t_i; c_k)$ is the entropy of $P(t_i, c_k)$, which is the joint probability distribution of the term t_i and the class c_k. The total mutual information (φ) of the distribution of two terms when observing the class c_k is calculated as follows:

$$\varphi\{I(t_i; c_k); I(t_j; c_k)\} = \varphi(t_i \cap c_k, t_j \cap c_k) \tag{16.7}$$

$$\varphi(t_i \cap c_k, t_j \cap c_k) = P(t_i \cap c_k, t_j \cap c_k) log \frac{P(t_i \cap c_k, t_j \cap c_k)}{P(t_i \cap c_k).P(t_j \cap c_k)} \tag{16.8}$$

$\varphi\{I(t_i; c_k); I(t_j; c_k)\}$ is a pointwise mutual information. The total mutual information of two terms when observing whole class information is the average of the mutual information over $c_k, 1 \leq k \leq C$. This measure is simply represented by the summarized form $\varphi(t_i; t_j)$:

$$\varphi(t_i; t_j) = \sum_{k=1}^{C} \varphi(t_i \cap c_k, t_j \cap c_k) \tag{16.9}$$

Although the Venn diagram is widely used to illustrate information theoretic concepts, Mackay [10] showed that it is sometimes misleading, especially in the case of three or more joint probability ensembles such as (t_i, t_j, c_k). Adopted from [10], Figure 16.3 depicts the concept of φ more precisely. Since φ has no upper bound, normalized mutual information Φ, which has an upper bound and is a good measure to compare two pieces of shared information, is proposed as follows [17]:

$$\Phi(t_i; t_j) = \frac{\varphi(t_i; t_j)}{\sqrt{I(t_i; c).I(t_j; c)}}, 0 \leq \Phi(t_1; t_2) \leq 1 \tag{16.10}$$

From [17], φ and $I(t_i; c)$ can be estimated by

$$I(t_i; c) = \sum_{k=1}^{C} \frac{n(t_i; c_k)}{n} log \frac{\frac{n(t_i; c_k)}{n}}{\frac{n(t_i)}{n}.\frac{n(c_k)}{n}} = \frac{1}{n} \sum_{k=1}^{C} n(t_i; c_k) log \frac{n.n(t_i; c_k)}{n(t_i).n(c_k)} \tag{16.11}$$

$$\varphi(t_i; t_j) = \sum_{k=1}^{C} \frac{n(t_i, t_j; c_k)}{n} log \frac{\frac{n(t_i, t_j; c_k)}{n}}{\frac{n(t_i, t_j)}{n}.\frac{n(c_k)}{n}} = \frac{1}{n} \sum_{k=1}^{C} n(t_i, t_j; c_k) log \frac{n.n(t_i, t_j; c_k)}{n(t_i, t_j).n(c_k)} \tag{16.12}$$

where $n(t_i, t_j)$ is the number of documents that contain both terms t_i and t_j. The number of documents that belong to the kth class and include t_i and t_j

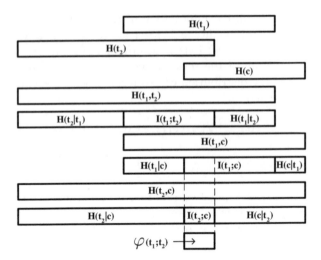

FIGURE 16.3: $\varphi(t_1; t_2)$ is the mutual information between two other mutual information measures $I(t_1; c)$ and $I(t_2; c)$.

is expressed by $n(t_i, t_j; c_k)$. Equation 16.10 is estimated as follows:

$$\Phi(t_i; t_j) = \frac{\sum_{k=1}^{C} n(t_i, t_j; c_k) log \frac{n.n(t_i, t_j; c_k)}{n(t_i, t_j).n(c_k)}}{\sqrt{\sum_{k=1}^{C} n(t_i; c_k) \log \frac{n.n(t_i; c_k)}{n(t_i).n(c_k)} \cdot \sum_{k=1}^{C} n(t_j; c_k) \log \frac{n.n(t_j; c_k)}{n(t_j).n(c_k)}}} \tag{16.13}$$

If the two terms are completely identical and correlated when observing a class, then $\Phi = 1$, and $\Phi = 0$ if the two terms are completely uncorrelated. It should be noted that, although pointwise mutual information $\varphi \{I(t_i; c_k); I(t_j; c_k)\}$ can be negative [10], the average mutual information $\varphi(t_i; t_j)$ is always positive and its normalized version is less than or equal to one.

The Φ measure is calculated for all possible pairs of terms in the vocabulary. The result is a matrix such as $\Phi \in \mathrm{R}^{M \times M}$, where M is the size of the vocabulary or the number of terms. We know that Φ is a symmetric measure and $\Phi(t_i; t_i) = 1$. Then, to construct the matrix Φ, we need to calculate $\frac{M(M-1)}{2}$ mutual information values. One approach to reduce this number is to calculate the matrix Φ for a very small subset of the most relevant terms V of the vocabulary T. This means that, instead of the full Φ matrix, a submatrix of Φ is provided. In other words, we need to calculate Φ measures for the most likely correlated terms. Let us suppose that there are n_s groups of correlated terms in the vocabulary. The problem is identifying these groups and calculating Φ for each of them. We propose an *inclusion matrix* for this purpose.

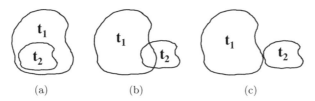

(a)   (b)   (c)

FIGURE 16.4: Inclusion relations between terms t_1 and t_2: (a) t_1 includes t_2, (b) t_1 partially includes t_2, (c) no inclusion relation between t_1 and t_2.

16.3.3.2 Inclusion Index

Let $D = \{d_1, d_2, \ldots, d_n\}$ be the collection of documents. Every document is represented by a vector of words, called the document vector. For example,

$$d_i = \{w_{i,1}.t_1, w_{i,2}.t_2, \ldots, w_{i,M}.t_M\} \tag{16.14}$$

where $w_{i,j}$ is the weight of the jth term in the ith document. Here we use binary weighting, which reflects if the term is in the document or not. As a consequence, D can be represented by an $n \times M$ matrix in which every row (d_i) shows a document and every column (t_j) represents the occurrence of the term in every document. Based on this notation, inclusion, which is a term-term relation, is defined [13]. The inclusion index $Inc(t_i, t_j)$, measuring how much t_i includes t_j, is calculated by:

$$Inc(t_i, t_j) = \frac{||t_i \cap t_j||}{||t_j||} = \frac{n(t_i, t_j)}{n(t_j)}, \quad Inc(t_i, t_j) \neq Inc(t_j, t_i) \tag{16.15}$$

where $||.||$ is the cardinal number of the set. $Inc(t_i, t_j) = 1$ when t_i is completely covering t_j and is full inclusive. $Inc(t_i, t_j) = 0$ means there is no overlap between the two terms. There is also partial inclusion when $0 < Inc(t_i, t_j) < 1$. t_j is called more inclusive than t_i if $Inc(t_i, t_j) < Inc(t_j, t_i)$ (see Figure 16.4). The inclusion matrix **Inc** is an $M \times M$ matrix in which each entry is an inclusion index between two terms.

16.3.4 Redundancy Removal Algorithm

The main idea in identifying redundant terms is finding the sets of correlated terms. For example, $\{$"rec", "hockei", "motorcycl", "bike", "nhl", "playoff"$\}$ shows one of these sets including six correlated terms. The sets are extracted using the inclusion matrix **Inc**.

Let S_q be the qth set of correlated terms. Instead of calculating the full matrix of Φ, it is only obtained for the terms in S_q. The resulting matrix is represented by Φ_q, which is a submatrix of Φ. We do the same for $\mathbf{Inc_q}$. Matrix $\mathbf{R_q}$, which is called a redundancy matrix, is calculated by entry-entry multiplication of $\mathbf{\Phi_q}$ and $\mathbf{Inc_q}$ as follows:

$$R_q(i,j) = \Phi_q(i,j).Inc_q(i,j), \quad 1 \leq i,j \leq n_q \qquad (16.16)$$

where n_q is the number of terms in S_q. The ith row of $\mathbf{R_q}$, which is an $n_q \times n_q$ matrix, shows the ith term (in S_q) with which terms are being covered. In each row the maximum entry is kept and the others are set to zero. Finally, every term and its corresponding column in $\mathbf{R_q}$ is full zero (all elements are zero) is assigned as a redundant term because it does not include any other term. Table 16.2 shows the resulting matrices for a set of correlated terms.

16.3.5 Term Redundancy Tree

A tree representation is also proposed for visualizing the redundant terms. A term redundancy tree is a directed and incomplete graph in which both initial and terminal nodes are assigned to terms such as $t_1 = $ "hockei" and $t_2 = $ "nhl". An edge, connecting t_1 to t_2, states that t_1 includes t_2 and can effectively cover most of its occurrences. Figure 16.5 shows four examples. The direction of each edge depends on the value of $R_q(i,j)$ and $R_q(j,i)$ (see Table 16.2(d)). If $R_q(i,j) < R_q(j,i)$, then the direction is from the jth to i^{th} node, otherwise the direction is reversed. Finally, each node that is the terminal node (and not the initial node for another edge) is assigned as the redundant term (Figure 16.5).

16.4 Experimental Results

The proposed approach has been applied to the 20-Newsgroups (20NG) dataset using the SVM (with linear kernel) and Rocchio text classifiers. Recently, SVM has outperformed most classifiers in text categorization [8, 6]. Although there are some reports showing feature selection for an SVM classifier is not only unnecessary but also can reduce its performance [12, 8], in this chapter we show that for a very small size of feature vector, SVM performance can be improved by feature selection through redundancy reduction [6].

The proposed approach has been evaluated by comparing its results with those of stand-alone information gain ranking. A five-fold cross validation is used for better estimation of classifier performance. Each method has been applied to the SVM and Rocchio classifiers with eight levels of aggressive feature selections. The detailed results of both classifiers for eight different lengths of feature vector are presented in Figure 16.6(a) and 16.6(b).

The results show that in both classifiers, the proposed method outperforms the stand-alone information gain ranking. From the findings, the following conclusions can be made:

TABLE 16.2: An example of the process of extracting redundant terms: (a) Normalized mutual information matrix $\mathbf{\Phi_q}$ for q^{th} set of correlated terms, (b) inclusion sub-matrix $\mathbf{Inc_q}$ for q^{th} set of correlated terms, (c) multiplication of the two matrices ($\mathbf{\Phi_q}$ and $\mathbf{Inc_q}$), (d) term redundancy matrix $\mathbf{R_q}$ for q^{th} set of correlated terms. Based on $\mathbf{R_q}$, all terms whose corresponding columns are entirely zero are redundant and should be removed.

a						
	rec	hockei	motorcycl	bike	nhl	playoff
rec	1	0.4448	0.4415	0.2866	0.2078	0.2059
hockei	0.4448	1	0	0	0.4555	0.4300
motorcycl	0.4415	0	1	0.5886	0	0
bike	0.2866	0	0.5886	1	0	0
nhl	0.2078	0.4555	0	0	1	0.1754
playoff	0.2059	0.4300	0	0	0.1754	1

b						
	rec	hockei	motorcycl	bike	nhl	playoff
rec	1	0.2221	0.2255	0.1162	0.0669	0.0680
hockei	0.9951	1	0	0	0.2998	0.2883
motorcycl	0.9903	0	1	0.4911	0	0
bike	0.9906	0	0.9530	1	0	0
nhl	0.9945	0.9945	0	0	1	0.2623
playoff	1	0.9459	0	0	0.2595	1

c						
	rec	hockei	motorcycl	bike	nhl	playoff
rec	0	0.0988	0.0995	0.0333	0.0139	0.0140
hockei	0.4426	0	0	0	0.1366	0.1240
motorcycl	0.4372	0	0	0.2891	0	0
bike	0.2839	0	0.5609	0	0	0
nhl	0.2067	0.4530	0	0	0	0.0460
playoff	0.2059	0.4067	0	0	0.0455	0

d						
	rec	hockei	motorcycl	bike	nhl	playoff
rec	0	0	0.0995	0	0	0
hockei	0.4426	0	0	0	0	0
motorcycl	0.4372	0	0	0	0	0
bike	0	0	0.5609	0	0	0
nhl	0	0.4530	0	0	0	0
playoff	0	0.4067	0	0	0	0

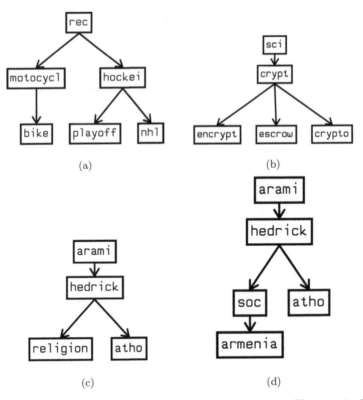

FIGURE 16.5: Four examples of term redundancy tree. The terminal nodes are representing the redundant terms: (a) "bike", "nhl", and "playoff"; (b) "escrow", "crypto", "encrypt", and "sci"; (c) "religion", and "atho"; (d) "armenia", and "atho" are redundant.

- Let n_s be the number of selected features in an aggressive approach before removing redundancies. In both classifiers, with high and low values of n_s (less than 10 and more than 30), information gain performs better than the proposed method. The main reason can be understood intuitively as follows: Referring to Figure 16.7, illustrating the sorted information gain for the first 100 best terms, when n_s is less than 10, term redundancy reduction is being held in the sharp slope region of the curve (between points "A" and "B"). It means with removing a redundant term from the feature vector, most likely a much less informative term will be substituted, but in the case of working in a smooth region of the curve (between points "B" and "C"), the proposed method may outperform information gain. It is referring to the cost of redundancy reduction, which might be high if the set of features to be substituted is unexpectedly poor in information content and less discriminant in comparison with the redundant term to be removed. These results confirm the findings in [6].

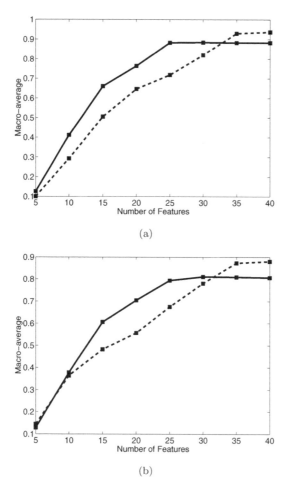

FIGURE 16.6: Text classifier performance vs. the number of features for two aggressive feature selection methods. Solid lines represent the proposed method and dashed lines represent information gain feature ranking: (a) SVM, and (b) Rocchio classifier.

- The SVM classifier result, according to Table 16.3, shows better overall performance than that of Rocchio. The fact is, although an SVM classifier rarely needs feature selection, and by employing the complete feature vector in the classifier we usually achieve good results, it can perform more efficiently if redundancy is reduced. Informally, let V_1 and V_2 be two feature vectors including the best features according to the information gain ranking. Unlike V_1, which includes some redundant terms, there is no redundancy in V_2. If the removed redundant term is from the smoothly sloped region of the sorted information gain

TABLE 16.3: Comparing the results of two aggressive feature selections using information gain ranking and the proposed method on SVM and Rocchio text classifiers.

Feature Selection	SVM Classifier	Rocchio Classifier
information gain	0.6190	0.5946
Information gain + redundancy reduction	0.6868	0.6298

curve (Figure 16.7), most likely the SVM classifier performance with V_2 will be superior to that of the V_1 feature vector. Otherwise, redundancy reduction can be risky.

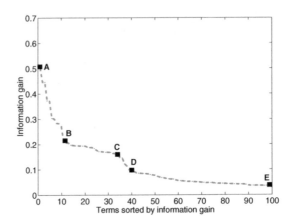

FIGURE 16.7: Sorted information gain for first 100 best terms. Redundancy reduction may hurt the performance if the redundant terms are located in the sharp slope of the curve (such as "A" to "B" and "C" to "D"). On the other hand, it can improve the performance in regions whose slope is smooth (such as "B" to "C" and "D" to "E").

16.5 Summary

Aggressive feature selection with higher than 95% feature reduction was discussed. The proposed approach is applicable to text classifiers while having a large vocabulary. Since the length of the feature vector in this strategy is quite short, the text classifiers, working with very small feature vectors, are very sensitive to noise, outliers, and redundancies. Because of these restrictions, improving any classical feature selection method such as feature ranking for aggressive reduction is strongly necessary.

Term redundancies in text classifiers cause a serious complication in most feature rankings, such as information gain, because they always ignore correlation between terms. The results of an experiment in the chapter showed that the effect of term redundancies can be worse than noise. To deal with redundancy, a method for improving aggressive feature selection by information gain ranking for text classifiers was proposed. The method was based on identifying and removing term redundancy using a mutual information measure and inclusion index. Terms were grouped in a few sets of correlated terms using an inclusion matrix. In the next step, each set was modeled by the term redundancy matrix. Using this matrix, term redundancies were recognized. In addition to the matrix representation, term redundancies were visualized by a graph called a term redundancy tree.

Aggressive feature selection approaches by stand-alone information gain ranking and the proposed method (removing the redundant terms from the ranked feature vector by information gain) were compared in SVM and Rocchio text classifier frameworks. Results showed that the proposed approach outperformed the aggressive feature selection by the stand-alone information gain. The proposed method improved information gain results 9.5% in macro-average F-measure. Better results are expected for other feature ranking methods such as χ^2 and odds ratio, since information gain is obviously more effective than other feature ranking methods and it has already been outperformed by the proposed method.

References

[1] M. W. Berry. *Survey of Text Mining: Clustering, Classification, and Retrieval.* Springer, New York, 2004.

[2] J. Brank, M. Groblenik, N. Milic-Frayling, and D. Mladenic. Interaction of feature selection methods and linear classification models. In *ICML-2002 Workshop on Text Learning of International Conference on Machine Learning, Sydney, Australia,* 2002.

[3] K. W. Church and P. Hanks. Word association norms, mutual information, and lexicography. *Comput. Linguist.,* 16(1):22–29, 1990.

[4] I. Dagan, L. Lee, and F. C. N. Pereira. Similarity-based models of word cooccurrence probabilities. *Machine Learning,* 34(1-3):43–69, 1999.

[5] E. R. Dougherty. Feature-selection overfitting with small-sample classifier design. *IEEE Intelligent Systems,* 20(6):64–66, 2005.

[6] E. Gabrilovich and S. Markovitch. Text categorization with many redundant features: Using aggressive feature selection to make SVMs competitive with C4.5. In *Proceedings of the Twenty-First International Conference on Machine learning,* pages 321–328, Banff, Alberta, Canada, 2004.

Morgan Kaufmann.

[7] S. Haynes. Stemming and stopwording effects on word frequency. In *Proceedings of the Thirteenth Midwest Artificial Intelligence and Cognitive Science Conference: MAICS 2002, S. Conlon, ed., Chicago, IL*, pages 71–75, 2002.

[8] T. Joachims. Text categorization with support vector machines: learning with many relevant features. In C. Nédellec and C. Rouveirol, editors, *Proceedings of ECML-98, 10th European Conference on Machine Learning*, number 1398, pages 137–142, Chemnitz, DE, Springer-Verlag, Heidelberg, 1998.

[9] W. Lam, M. E. Ruiz, and P. Srinivasan. Automatic text categorization and its applications to text retrieval. *IEEE Transactions on Knowledge and Data Engineering*, 11(6):865–879, 1999.

[10] D. Mackay. *Information Theory, Inference and Learning Algorithms*. Cambridge University Press, New York, 2003.

[11] D. Mladenic, J. Brank, M. Grobelnik, and N. Milic-Frayling. Feature selection using linear classifier weights: interaction with classification models. In *Proceedings of the 27th Annual International Conference on Research and Development in Information Retrieval*, pages 234–241, 2004.

[12] M. Rogati and Y. Yang. High-performing feature selection for text classification. In *Proceedings of the 11th International Conference on Information and Knowledge Management*, pages 659–661, 2002.

[13] G. Salton. Recent trends in automatic information retrieval. In *SIGIR '86: Proceedings of the 9th Annual International ACM SIGIR Conference on Research and Development in Information Retrieval*, pages 1–10, 1986.

[14] S. Scott and S. Matwin. Feature engineering for text classification. In *Proceedings of the 16th International Conference on Machine Learning*, pages 379–388. Morgan Kaufmann, 1999.

[15] F. Sebastiani. Machine learning in automated text categorization. *ACM Computing Surveys*, 34(1):1–47, 2002.

[16] P. Soucy and G. W. Mineau. A simple feature selection method for text classification. In B. Nebel, editor, *Proceeding of IJCAI-01, 17th International Joint Conference on Artificial Intelligence*, pages 897–902, Seattle, US, 2001.

[17] A. Strehl and J. Ghosh. Cluster ensembles – a knowledge reuse framework for combining partitionings. In *Proceedings of AAAI 2002, Edmonton, Canada*, pages 93–98. AAAI, July 2002.

[18] G. Wang and F. H. Lochovsky. Feature selection with conditional mutual

information maximin in text categorization. In *CIKM '04: Proceedings of the Thirteenth ACM conference on Information and Knowledge Management*, pages 342–349, 2004.

[19] S. K. M. Wong and V. V. Raghavan. Vector space model of information retrieval: a reevaluation. In *Proceedings of the 7th Annual International ACM SIGIR Conference on Research and Development in Information Retrieval*, pages 167–185, 1984.

[20] J. Xu and W. B. Croft. Corpus-based stemming using cooccurrence of word variants. *ACM Trans. Inf. Syst.*, 16(1):61–81, 1998.

[21] Y. Yang and J. O. Pedersen. A comparative study on feature selection in text categorization. In D. H. Fisher, editor, *Proceedings of ICML-97, 14th International Conference on Machine Learning*, pages 412–420, Nashville, TN, Morgan Kaufmann, San Francisco, 1997.

[22] C. T. Yu, K. Lam, and G. Salton. Term weighting in information retrieval using the term precision model. *J. ACM*, 29(1):152–170, 1982.

Part V

Feature Selection in Bioinformatics

Chapter 17

Feature Selection for Genomic Data Analysis

Lei Yu

Binghamton University

17.1   Introduction

The rapid advances of gene expression microarray technology have provided scientists, for the first time, the opportunity of observing complex relationships between various genes in a genome by simultaneously measuring the expression levels of the tens of thousands of genes in massive experiments. Analysis of large-scale genomic data in order to extract biologically meaningful insights presents unprecedented opportunities and challenges for data mining in areas such as gene clustering [3], sample class discovery, and classification [4]. In this chapter, we first introduce the challenges of microarray data analysis and some traditional solutions of feature selection, and then present a redundancy-based feature selection solution and demonstrate its effectiveness and efficiency on some benchmark microarray datasets.

17.1.1   Microarray Data and Challenges

The description of microarray technologies is beyond the scope of this chapter. In a nutshell, gene expression microarrays are silicon chips that simultaneously measure the mRNA expression levels of tens of thousands of genes. The expression levels of the same sets of genes under study are normally measured from different samples or under different conditions, and eventually recorded in a data matrix. In a typical microarray data matrix as shown in Table 17.1, each column represents a gene and each row represents a sample (or a condition). Each value f_{ij} is the measurement of the expression level of the jth gene for the ith sample, where $i = 1, ..., M$ and $j = 1, ..., N$. In a classification task, a microarray dataset is provided as a training set of samples with class labels c_M. The task is to build a classifier that accurately predicts the classes (diseases or phenotypes) of unlabeled samples.

TABLE 17.1: A typical gene expression matrix.

	Gene 1	Gene 2	. . .	Gene N	Class
Sample 1	f_{11}	f_{12}	. .	f_{1N}	c_1
Sample 2	f_{21}	f_{22}	. .	f_{2N}	c_2
.	.	.	.	.	.
.	.	.	.	.	.
.	.	.	.	.	.
Sample M	f_{M1}	f_{M2}	. . .	f_{MN}	c_M

A typical microarray dataset has the following three characteristics: (1) high dimensionality due to tens of thousands of genes; (2) severely limited amount of samples - usually in tens or at most a couple of hundreds due to the expense of obtaining microarray samples; and (3) abundance of redundancy among genes - if the change of expression of one gene is correlated to the change of the phenotypes, a great many of other genes can be co-regulated in the same or opposite way. Such data characteristics pose severe challenges to classification tasks. Computational learning theory suggests that the search space is exponentially large in terms of N and the required number of samples for reliable learning about given phenotypes is on the scale of $O(2^N)$ [13]. However, even the minimum requirement ($M = 10 * N$) as a statistical "rule of thumb" is patently impractical for a microarray dataset [7]. With very limited samples, a large set of genes leads to too many statistically relevant hypotheses that are equally valid in interpreting the same dataset. Therefore, selecting a small number of discriminative genes is essential for successful classification. From a practical point of view, the selection of a small subset of discriminative genes often helps identify genes that are relevant to the cause or consequences of disease or can be used as biomarkers for diagnostic of diseases, measuring drug toxicology, and efficacy [20]. A compact gene set is desirable to biologists because of the heavy expenses associated with follow-up biological or clinical validation of selected genes.

17.1.2 Feature Selection for Microarray Data

Feature selection methods can broadly fall into the wrapper model and the filter model [9]. The wrapper model uses the predictive accuracy of a predetermined learning algorithm to determine the goodness of a selected subset. It is computationally very expensive for data with a large number of features, and the selected subset is dependent on the learning algorithm used [9]. The filter model separates feature selection from classifier learning and relies on the general characteristics of the training data to select features.

Traditional methods in gene selection are within the filter model, and often evaluate genes in isolation without considering the gene-to-gene correlation. They rank genes according to their individual relevance or discriminative power to the target class and select top-ranked genes. Some methods based on statistical tests or information gain have been used in [4, 12]. These methods are computationally efficient due to linear time complexity $O(N)$ in terms of dimensionality N. However, they cannot remove redundant genes. The issue of redundancy among genes has been studied in recent literature. It is pointed out in a number of studies [2, 20] that simply combining a highly ranked gene with another highly ranked gene often does not form a better gene set because these two genes could be highly correlated. The effect of redundancy among selected genes is two-fold. On one hand, the selected gene set can have a less comprehensive representation of the target class than one of the same size but without redundant genes; on the other hand, in order to include all representative genes, redundant genes will unnecessarily increase the size of the selected gene set, which will in turn affect the mining performance on the small sample. Besides incapability of handling redundant genes, most gene ranking methods require certain domain knowledge or trial-and-error to determine a threshold for the number of genes to be selected (e.g., a threshold of the top 50 genes was arbitrarily determined in the work of Golub et al. [4]).

Subset search methods have also been applied to select discriminative genes while taking into account gene redundancy [2, 19, 20]. In Xiong's work [20], a method in the wrapper model was developed that searches through possible subsets of genes using the classification accuracy as a measure of goodness for a particular subset. A sequential forward floating search was applied to generate subsets. Different subsets of genes were selected based on three learning algorithms: linear discriminant analysis, logistic regression, and support vector machines. Because a classifier has to be built for every subset of genes visited in the search procedure, these methods are very expensive to run. In [2, 19], subset search methods within the filter model were proposed that employ correlation measures to evaluate the relevance and redundancy of various gene sets of the same size during the search. In order to determine a threshold for the size of the finally selected gene set, different learning algorithms were applied to evaluate the classification accuracy of subsets of different sizes. These "hybrid" methods are more efficient than wrapper methods, but they are dependent on the classifiers used and computationally more costly than filter methods. In addition, expertise in various classifiers is needed to empirically tune these methods in determining an optimal size of the final subset.

A key challenge in gene selection from microarray data is to provide biologists an efficient filter method that effectively identifies and removes both irrelevant and redundant genes in an automatic manner. In the rest of this chapter, we tackle this challenge by providing a general framework for redundancy analysis and an efficient algorithm developed under this framework.

17.2   Redundancy-Based Feature Selection

In this section, we first introduce definitions on feature relevance and redundancy, we next propose a framework of efficient feature selection based on explicit redundancy analysis, and we then present and evaluate a specific algorithm developed under this framework.

17.2.1   Feature Relevance and Redundancy

In general, the goal of feature selection can be formalized as selecting a minimum subset G such that $p(C \mid G)$ is equal or as close as possible to $p(C \mid F)$, where $p(C \mid F)$ or $p(C \mid G)$ is the probability distribution of the class values C given the feature values in F or G, respectively. Such a minimum subset is also called an *optimal* feature subset in feature selection [10] and can be illustrated by the example below.

Let features $F_1, ..., F_5$ be Boolean. The target concept is $C = g(F_1, F_2)$, where g is a Boolean function. With $F_2 = \overline{F_3}$ and $F_4 = \overline{F_5}$, there are only eight possible instances. In order to determine the target concept, F_1 is indispensable; one of F_2 and F_3 can be disposed of (note that C can also be determined by $g(F_1, \overline{F_3})$), but we must have one of them; both F_4 and F_5 can be discarded. Either $\{F_1, F_2\}$ or $\{F_1, F_3\}$ is an optimal subset. The goal of feature selection is to find either of them.

Based on a review of previous definitions of feature relevance, John, Kohavi, and Pfleger classified features into three disjoint categories, namely, strongly relevant, weakly relevant, and irrelevant features [9]. Let F be a full set of features, F_i a feature, and $S_i = F - \{F_i\}$. These categories of relevance can be formalized as follows.

DEFINITION 17.1   **(Strong relevance)** *A feature F_i is strongly relevant iff*

$$P(C \mid F_i, S_i) \neq P(C \mid S_i)$$

DEFINITION 17.2   **(Weak relevance)** *A feature F_i is weakly relevant iff*

$$P(C \mid F_i, S_i) = P(C \mid S_i), \textbf{ and}$$

$$\exists \, S_i' \subset S_i, \text{ such that } P(C \mid F_i, S_i') \neq P(C \mid S_i')$$

DEFINITION 17.3 **(Irrelevance)** *A feature F_i is irrelevant iff*

$$\forall\ S_i' \subseteq S_i,\ P(C \mid F_i,\ S_i') = P(C \mid S_i')$$

Strong relevance of a feature indicates that the feature is always necessary for an optimal subset; it cannot be removed without loss of discriminative power. Weak relevance suggests that the feature is not always necessary but may become necessary to the discrimination of the class under certain conditions. Irrelevance indicates that the feature can never contribute to the discrimination of the class. According to these definitions, it is clear that in our illustrative example, feature F_1 is strongly relevant; F_2; F_3 are weakly relevant; and F_4, F_5 are irrelevant. An optimal subset should only include all strongly relevant features and a subset of weakly relevant features. However, the definition of weak relevance only reveals feature redundancy (i.e., the information a feature has about the class is subsumed by other features) but cannot help identify which features among the weakly relevant ones should be selected while others eliminated. Therefore, it is necessary to define feature redundancy among relevant features.

Notions of feature redundancy are normally in terms of feature correlation. It is widely accepted that two features are redundant to each other if their values are completely correlated (for example, features F_2 and F_3). In reality, it may not be so straightforward to determine feature redundancy when a feature is correlated (perhaps partially) with a set of features. In order to devise an approach to explicitly identify and eliminate redundant features, in our previous work [21], we formally defined feature redundancy based on the definition of a feature's Markov blanket [14].

DEFINITION 17.4 **(Markov blanket)** *Given a feature F_i, let $M_i \subset F$ ($F_i \notin M_i$). M_i is said to be a Markov blanket for F_i iff*

$$P(F - M_i - \{F_i\},\ C \mid F_i,\ M_i) = P(F - M_i - \{F_i\},\ C \mid M_i)$$

It is easy to see that if M_i is a Markov blanket of F_i, the class C is conditionally independent of F_i given M_i. Moreover, the Markov blanket condition is stronger than conditional independence. It requires that M_i subsume not only the information that F_i has about C, but also about all of the other features. While it might be difficult to find such a set M_i, it is pointed out in [10] that an optimal subset can be obtained by using Markov blankets as the basis for feature elimination. Let G be the current set of features ($G = F$ in the beginning), at any phase, if there exists a Markov blanket for F_i within G, F_i is judged as unnecessary for an optimal subset and thus removed from G. It is proved that this process guarantees a feature removed in an earlier phase will still find a Markov blanket in any later phase, that is, removing a feature in a later phase will not render the previously removed features necessary to be included in an optimal subset [10]. Thus, the Markov blanket criterion only removes features that are really unnecessary: features that are irrelevant to the target class, and features that are weakly relevant but redundant given other features. According to previous definitions of feature relevance, we can prove that strongly relevant features cannot find any Markov blankets. Since irrelevant features should be removed anyway, we exclude them from our definition of redundant features.

DEFINITION 17.5   (**Redundant feature**) *Let G be the current set of features. A feature is redundant and hence should be removed from G iff it is weakly relevant and has a Markov blanket M_i within G.*

From the property of a Markov blanket, it is easy to see that a redundant feature removed earlier remains redundant when other features are removed. Figure 17.1 depicts the relationships between definitions of feature relevance and redundancy introduced so far. It shows that an entire feature set can be conceptually divided into four basic disjoint parts: irrelevant features (I), redundant features (II, part of weakly relevant features), weakly relevant but non-redundant features (III), and strongly relevant features (IV). An optimal subset essentially contains all the features in parts III and IV. It is worthy to point out that although parts II and III are disjoint, different partitions of them can result from the process of Markov blanket filtering. In our illustrative example, either of F_2 or F_3, but not both, should be removed as a redundant feature.

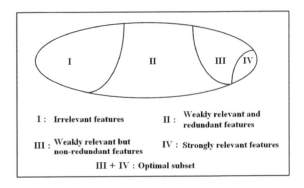

FIGURE 17.1: A view of feature relevance and redundancy.

In terms of gene selection, an optimal subset of genes is a minimum subset of genes of maximum discriminative power; it should only include all strongly relevant genes and a subset of weakly relevant but mutually non-redundant genes. In search of an optimal subset for gene expression microarray data, efficient methods are needed for two reasons. First, the Markov blanket criterion given in Definition 17.4 is combinatorial in nature. It is obvious that an exhaustive or complete search is prohibitive with thousands of genes. Second, an optimal subset is defined based on the full population where the true data distribution is known. It is generally assumed that a training dataset is only a small portion of the full population, especially in a high-dimensional space as in microarray data.

In search of a suboptimal subset of genes, our goal is to efficiently find for a gene F_i an approximate Markov blanket M_i that subsumes the information content of F_i. As mentioned previously, if M_i is a true Markov blanket for F_i, the class C is conditionally independent of F_i given M_i, i.e., $p(C \mid F_i, M_i) = p(C \mid M_i)$. However, finding a subset M_i for every gene is still combinatorial in nature. We present an efficient framework in the next section.

17.2.2   An Efficient Framework for Redundancy Analysis

In our framework, we first differentiate two types of correlations between genes and the class: individual C-correlation and combined C-correlation.

DEFINITION 17.6   (Individual C-correlation) *The correlation between any gene F_i and the class C is called individual C-correlation, denoted by $r(F_i, C)$.*

DEFINITION 17.7   (Combined C-correlation) *The correlation between any pair of genes F_i and F_j $(i \neq j)$ and the class C is called combined C-correlation, denoted by $r(F_{i,j}, C)$.*

In combined C-correlation, we treat genes F_i and F_j as one single feature $F_{i,j}$. An immediate issue is how to decide the feature values of a virtual gene represented by the vector $F_{i,j}$. If the expression values of genes F_i and F_j are numerical values, $F_{i,j}$ can be some linear combination of F_i and F_j. If the expression values have been discretized into nominal states, we can use the cartesian product of the domains of F_i and F_j as the domain of $F_{i,j}$. For example, if both F_i and F_j assume binary values (0 or 1), the combined C-correlation aims to measure the correlation between the joint states (0,0), (0,1), (1,0) (1,1) and the class label C.

Our method determines whether a single gene F_i can be an approximate Markov blanket for another gene F_j based on both individual C-correlations and the combined C-correlation. It assumes that a gene with a larger individual C-correlation value contains by itself more information about the class than a gene with a smaller individual C-correlation value. For two genes F_i

and F_j with $r(F_i, C) \geq r(F_j, C)$, it chooses to evaluate whether gene F_j can be approximately redundant to gene F_i (instead of F_i to F_j) in order to maintain more information about the class. In addition, if combining F_j with F_i does not provide more discriminative power than F_i alone, it heuristically decides that F_i forms an approximate Markov blanket for F_j. Thus, an approximate Markov blanket is defined as follows.

DEFINITION 17.8   (**Approximate Markov blanket**) *For two genes F_i and F_j, F_i forms an approximate Markov blanket for F_j iff $r(F_i, C) \geq r(F_j, C)$ and $r(F_i, C) \geq r(F_{i,j}, C)$.*

Recall that Markov blanket filtering, a backward elimination procedure based on a feature's Markov blanket in the current set, guarantees that a feature removed in an earlier phase will still find a Markov blanket in any later phase when another feature is removed. It is easy to verify that this is not the case for backward elimination based on a feature's approximate Markov blanket in the current set. For instance, if F_j is the only gene that forms an approximate Markov blanket for F_k, and F_i forms an approximate Markov blanket for F_j, after removing F_k based on F_j, further removing F_j based on F_i will result in no approximate Markov blanket for F_k in the current set. However, we can avoid this situation by removing a gene only when it can find an approximate Markov blanket formed by a predominant gene, defined as follows.

DEFINITION 17.9   (**Predominant gene**) *A gene is predominant iff it does not have any approximate Markov blanket in the current set.*

Predominant genes will not be removed at any stage. If a gene F_j is removed based on a predominant gene F_i in an earlier phase, it is guaranteed that it will still find an approximate Markov blanket (the same F_i) in any later phase when another gene is removed. To determine predominant genes, all genes can be ranked according to their individual C-correlation values. Since the gene with the highest individual C-correlation value does not have any approximate Markov blanket, it must be one of the predominant genes and can be used as the starting point to eliminate other unnecessary genes.

In summary, our framework for redundancy analysis is to find all predominant genes and eliminate the rest. Comparing with traditional gene selection methods that evaluate the relevance of each gene individually, our framework has the following distinct characteristics: (1) It efficiently handles redundancy among relevant genes; (2) it is able to consider gene-to-gene interactions to some extent by evaluating combined C-correlation; and (3) it removes irrelevant genes as well as relevant but redundant genes based on the same criterion. The last characteristic makes it unnecessary to determine a threshold for selecting relevant genes. In search of approximate Markov blankets, we can extend the algorithm to consider more complex combinations of genes other than a combined C-correlation of two genes. However, this will not only increase the time complexity of the search, but also cause an over-searching problem [8] due to the data characteristics of limited samples in a high-dimensional space.

17.2.3 RBF Algorithm

Under our search framework, different correlation measures can be applied to calculate individual C-correlations and combined C-correlations. For data with continuous gene expression values, linear correlation measures are widely used. Of linear correlation, the most well-known measure is *linear correlation coefficient*. For a pair of variables (X, Y), the linear correlation coefficient ρ is given by

$$\rho\left(X,Y\right) = \frac{\sum_i (x_i - \overline{x_i})(y_i - \overline{y_i})}{\sqrt{\sum_i (x_i - \overline{x_i})^2} \sqrt{\sum_i (y_i - \overline{y_i})^2}}$$

where $\overline{x_i}$ is the mean of X, and $\overline{y_i}$ is the mean of Y. The value of ρ lies between -1 and 1, inclusive. If X and Y are completely correlated, ρ takes the value of 1 or -1; if X and Y are independent, ρ is zero. Other measures in this category are basically variations of the above formula [4]. Linear correlation measures may not be able to capture relationships that are not linear in nature and are limited to numerical values.

To reduce the variance and noise of the original data, continuous expression values are often discretized into discrete values [2, 11]. For discrete data, information-theoretic measures are widely adopted [15]. They are based on the well-known concept of *entropy*, a measure of the uncertainty of a random variable. For nominal variables, the entropy of a variable X is defined as

$$H(X) = -\sum_i P(x_i) \log_2(P(x_i))$$

and the entropy of X after observing values of another variable Y is defined as

$$H(X|Y) = -\sum_j P(y_j) \sum_i P(x_i \mid y_j) \log_2(P(x_i \mid y_j))$$

where $P(x_i)$ is the prior probability for all values of X, and $P(x_i \mid y_i)$ is the posterior probability of X given the values of Y. The amount by which the entropy of X decreases reflects additional information about X provided by Y and is called *information gain*, given by

$$IG(X \mid Y) = H(X) - H(X \mid Y)$$

Information gain tends to favor variables with more values and can be normalized by their corresponding entropy. One way to normalize information gain is by *symmetrical uncertainty* (SU), defined as

$$SU(X,\ Y) = 2 \left[\frac{IG(X \mid Y)}{H(X) + H(Y)} \right]$$

which compensates for information gain's bias toward features with more values and restricts its values to the range $[0, 1]$. A value of 1 indicates that knowing the values of either feature completely predicts the values of the other; a value of 0 indicates that X and Y are independent.

We experimented with both linear correlation coefficient and symmetrical uncertainty under our general search framework and found that symmetrical uncertainty works more effectively than linear correlation coefficient. Therefore, we chose symmetrical uncertainty as the correlation measure in our algorithm RBF (redundancy-based filter). Individual C-correlation and combined C-correlation are thus measured by $SU(F_i, C)$ and $SU(F_{i,j}, C)$, respectively. For simplicity we refer to $SU(F_i, C)$ as ISU_i and $SU(F_{i,j}, C)$ as $CSU_{i,j}$.

---

input: $D(F_1, F_2, ..., F_N, C)$ // a training dataset
output: S_{best} // a selected subset
1   **begin**
2   for $i = 1$ to N do **begin**
3   calculate ISU_i for F_i;
4   append F_i to S_{list};
5   **end**;
6   order S_{list} in descending ISU_i value;
7        $F_i = getFirstElement(S_{list})$;
8   while $(F_i \neq$ NULL$)$ do **begin**
9            $F_j = getNextElement(S_{list}, F_i)$;
10   while $(F_j \neq$ NULL$)$ do **begin**
11   if $(ISU_i \geq CSU_{i,j})$ remove F_j from S_{list};
12               $F_j = getNextElement(S_{list}, F_j)$;
13   **end**;
14           $F_i = getNextElement(S_{list}, F_i)$;
15   **end**;
16       $S_{best} = S_{list}$;
17   **end**;

FIGURE 17.2: RBF algorithm.

As shown in Figure 17.2, the RBF algorithm first calculates the ISU value for each gene and orders all genes in a descending order according to their

ISU values (lines 2–6). It then further processes the ordered list S_{list} to select predominant genes (lines 7–15). Recall that a gene that has already been determined to be a predominant gene can always be used to filter out other genes for which it forms an approximate Markov blanket. Since the gene with the highest *ISU* value does not have any approximate Markov blanket, it must be one of the predominant genes. So the iteration starts from the first element in S_{list} (line 7) and continues as follows. For all the remaining genes, if F_i happens to form an approximate Markov blanket for F_j, F_j will be removed from S_{list} (line 11). After one round of filtering genes based on F_i, the algorithm will take the remaining gene right next to F_i as the new reference (line 14) to repeat the filtering process. The algorithm stops when no more predominant genes can be selected.

The majority of computation time of RBF involves calculation of *ISU* and *CSU* values, which has a linear time complexity in terms of the number of instances in a dataset. In terms of dimensionality N, to determine and rank the discriminative power of each gene, the algorithm has a linear time complexity $O(N)$; to determine predominant genes, it has a best-case complexity $O(N)$ when only one gene is selected and all of the remaining genes are removed, and a worse-case complexity $O(N^2)$ when all genes are selected. Such best-case and worse-case time complexities are comparable to gene selection methods based on greedy sequential search through possible gene sets, in which genes are, one at a time, added to the current subset (i.e., sequential forward selection) or removed from the current subset (i.e., sequential backward elimination). However, in general cases when k $(1 < k < N)$ genes are selected, the number of evaluations performed by RBF will typically be much less (and certainly never more) than the number of evaluations performed by a greedy sequential search because genes removed in each round are not considered in the next round and RBF typically removes a large number of genes in each round. This makes RBF substantially faster than gene selection methods based on greedy subset searches.

17.3 Empirical Study

In this section, we empirically evaluate the effectiveness and efficiency of our method on public gene expression microarray data sets.

17.3.1 Datasets

To evaluate our proposed framework of redundancy analysis and the RBF algorithm, we conducted experiments on four publicly available microarray datasets: colon cancer, leukemia, lung cancer, and breast cancer. We next

briefly describe these datasets and previously published results on them. A
summary of these datasets are presented in Table 17.2.

TABLE 17.2:   Summary of microarray datasets used in experiments.

Dataset	# Genes	# Samples	# Samples per Class			
Colon cancer	2000	62	Tumor	40	Normal	22
Leukemia	7129	72	ALL	47	AML	25
Lung cancer	12533	181	MPM	31	ADCA	150
Breast cancer	24481	97	Relapse	46	Non-relapse	51

Colon cancer data [1] has been frequently used in previous studies in cancer classification. It consists of gene expression profiles of 2000 genes for 62
tissue samples among which 40 are colon cancer tissues and 22 are normal
tissues. In [1], a hierarchical clustering method was applied to separate tumor and normal samples into two distinct clusters. Based on 20 genes with
the most statistically significant difference between tumors and normal tissues
according to t-test, the resulting dendrogram from hierarchical clustering misclassified 5 tumor samples and 3 normal samples into the opposite clusters.

Leukemia data [4] is another widely used benchmark dataset in cancer classification. It consists of gene expression profiles of two classes of leukemia:
acute lymphoblastic leukemia (ALL) and acute myeloblastic leukemia (AML).
The dataset consists of 7129 genes and 72 samples (47 ALL and 25 AML).
In [4], in order to distinguish between AML and ALL, a set of 50 genes mostly
correlated with AML-ALL distinction were selected from 38 training samples.
These genes were used to build a linear class predictor for the remaining 34
testing samples and achieved 85% predictive accuracy.

Lung cancer data [5] consists of gene expression profiles of 12533 genes
for 181 tissue samples (31 MPM and 150 ADCA). The problem is to distinguish between malignant pleural mesothelioma (MPM) and adenocarcinoma
(ADCA) of the lung. In [5], 8 genes were selected according to the most statistically significant difference in average expression levels between both tumor
types in the training set of 16 MPM and 16 ADCA samples. Based on these
genes, a ratio-based classifier was built on the training set and achieved 95%
accuracy in predicting diagnoses for the remaining 149 samples.

Breast cancer data [17] consists of gene expression profiles of 24481 genes for
97 samples (46 relapse breast cancer and 51 non-relapse breast cancer). In [17],
the correlation coefficient of the expression for each gene with disease outcome
was calculated, and 231 genes were found to be significantly associated with
disease outcome. These 231 genes were ranked according to the magnitude of
the correlation coefficient. A wrapper approach was then applied to determine
the optimal number of genes for the classifier by sequentially adding subsets
of 5 genes from the top of the ranking list and evaluating its power for correct
classification using 'leave-one-out' cross validation. The best accuracy (83%)
was achieved at a number of 70 genes.

17.3.2 Experimental Settings

We limit our comparisons to gene selection algorithms in the filter model as RBF is a filter algorithm designed for high-dimensional data. We choose two representative algorithms. One algorithm is ReliefF [16], which evaluates the discriminative power of individual genes without handling gene redundancy. ReliefF searches for the nearest neighbors of samples of different classes and weights genes according to how well they differentiate samples of different classes. The other algorithm is CFS [6], which exploits heuristic subset search based on some correlation measure. It evaluates the goodness of a subset by considering the discriminative power of each individual gene and the degree of correlation between them. Sequential forward selection is employed in CFS. For each of the four datasets, we use two classification algorithms, K-Nearest Neighbors (KNN) and Naive Bayes (NB) [18], to evaluate the predictive performance of subsets of genes selected by various gene selection algorithms. All these selected algorithms are from Weka's collection [18]. RBF is also implemented in the Weka environment.

The four original datasets contain continuous gene expression values. In order to reduce the noise, various discretization methods [11] can be used to transform continuous expression values of each gene into several nominal states. In this work, continuous values of each gene were discretized into three values -1, 0, and 1, representing the over-expression, baseline, and under-expression of genes, which correspond to $(-\infty, \mu - \sigma/2)$, $[\mu - \sigma/2, \mu + \sigma/2]$, and $(\mu + \sigma/2, +\infty)$, respectively. μ and σ respectively refer to the mean and standard deviation of all expression values for a given gene.

For each dataset, we apply KNN and NBC classifiers on the full set of genes in the original data and each subset of genes selected by RBF, ReliefF, and CFS, respectively. Since researchers who previously worked on these datasets either divided data into training and testing parts or employed "leave-one-out" cross validation (LOOCV) in assessing predictive performance of various gene sets, we adopt LOOCV in our experiments.

17.3.3 Results and Discussion

TABLE 17.3: Accuracy of KNN on selected genes for microarray data: Acc records leave-one-out cross validation accuracy rate (%).

	RBF		Full Set		ReliefF		CFS	
	# Genes	Acc	# Genes	Acc	# Genes	Acc	# Genes	Acc
Colon cancer	4	93.55	2000	70.97	4	87.10	26	85.48
Leukemia	16	94.44	7129	86.11	60	81.94	54	97.22
Lung cancer	7	99.45	12533	93.92	64	98.34	N/A	N/A
Breast cancer	34	94.85	24481	59.79	70	81.44	N/A	N/A

Table 17.3 reports the number of genes and associated predictive accuracy rates of KNN ($K = 3$) classifier for various gene sets across the four microarray datasets. From Table 17.3 we can clearly observe the degree of dimensionality reduction and the improvement on predictive accuracy due to RBF gene selection comparing with the full set. For example, based on the original colon cancer data (2000 genes), 18 out of 62 samples were incorrectly classified in LOOCV, resulting in an overall accuracy of 70.97%. RBF selected only 4 genes and helped to reduce the number of misclassified samples to 4 (increasing the overall accuracy to 93.55%). A similar trend of accuracy improvement with only a few genes selected by RBF can also be observed from other datasets. It is worth mentioning that accuracy improvement is not the sole purpose for gene selection. The selection of a small subset of discriminative genes often helps identify genes that are relevant to the cause or consequences of disease or can be used as biomarkers for the diagnosis of diseases, measuring drug toxicology, and efficacy [20]. Comparing RBF with ReliefF, RBF selected much smaller sets of genes than ReliefF for all the four datasets (except colon cancer data) and resulted in higher predictive accuracy. A similar trend can be observed when comparing RBF with CFS on colon cancer and leukemia data, except that CFS resulted in slightly higher accuracy than RBF on leukemia data. For lung cancer and breast cancer data, CFS did not produce any results because the program ran out of memory after a period of time due to its quadratic space complexity.

TABLE 17.4:   Accuracy of NBC on selected genes for microarray data: Acc records leave-one-out cross validation accuracy rate (%).

	RBF		Full Set		ReliefF		CFS-FS	
	# Genes	Acc	# Genes	Acc	# Genes	Acc	# Genes	Acc
Colon cancer	4	88.71	2000	58.06	4	85.48	26	85.48
Leukemia	16	98.61	7129	97.22	60	97.22	54	100.00
Lung cancer	7	97.79	12533	97.79	64	96.13	N/A	N/A
Breast cancer	34	93.81	24481	51.55	70	79.38	N/A	N/A

Table 17.4 reports the predictive accuracy rates of the NBC classifier on the same set of gene sets across the four microarray datasets. From Table 17.4 we can observe the same trend of dimensionality reduction and accuracy improvement due to RBF gene selection comparing with the full set as well as ReliefF and CFS. It is worth mentioning that, to our knowledge, the best reported result on breast cancer data was the LOOCV accuracy of 83% with 70 selected genes produced by the wrapper approach introduced in [17]. Our method, RBF, achieved an LOOCV accuracy of 94.85% (by KNN) with only 34 selected genes. Overall, the above results suggest that RBF is an effective method for gene selection in microarray sample classification.

We further evaluate the efficiency of RBF by examining its running time on different datasets. Table 17.5 records the running time of RBF, ReliefF,

TABLE 17.5: Comparison of running times (seconds) between RBF, ReliefF, and CFS.

	RBF	ReliefF	CFS
Colon cancer	0.1	2.5	16.4
Leukemia	0.4	12.1	702.4
Lung cancer	1.4	130.6	N/A
Breast cancer	3.5	75.1	N/A

and CFS on a Pentium IV PC for the four datasets used. It is clear that RBF is significantly faster (in degrees of magnitude) than ReliefF and CFS. The high efficiency of RBF allows us to exploit different variations of RBF. In the beginning of the search for approximate Markov blankets, all genes are ordered according to their individual C-correlation measure (ISU value in RBF). Different measures used to rank genes will result in different subsets of selected genes through the filtering process. Because of its efficiency, RBF can be easily repeated with different ranking strategies to get different gene selection results.

17.4   Summary

In this chapter, we have introduced the concept of an optimal gene set based on a Markov blanket, and proposed an efficient filter method to approximate the selection of discriminative and non-redundant genes. RBF has two desirable properties: First, it combines sequential forward selection with redundancy elimination and thus substantially reduces the number of gene pairs evaluated for redundancy; second, it removes both irrelevant and redundant genes in the filtering process and thus does not require a threshold for the number of selected genes. Experiments on microarray data have demonstrated RBF's effectiveness and efficiency.

Current research in gene selection mainly focuses on the selection of statistically significant predictors. One future research direction is to take advantage of available domain knowledge in finding both statistically significant and biologically relevant genes. The high efficiency of the RBF algorithm allows it to be used to search for biologically relevant genes by incorporating prior biological knowledge into the gene selection process. For example, a few seed genes of particular biological relevance can be appointed as predominant genes and placed on the very top of the ranking list, and the selection of additional predominant genes can then follow the filtering process of the RBF algorithm. By changing the seed genes, one can also exploit prior biological knowledge during gene selection.

References

[1] U. Alon, N. Barkai, and D. A. Notterman. Broad patterns of gene expression revealed by clustering analysis of tumor and normal colon tissues probed by oligonucleotide arrays. *Proc. Natl Acad. Sci. USA*, 96:6745–6750, 1999.

[2] C. Ding and H. Peng. Minimum redundancy feature selection from microarray gene expression data. In *Proceedings of the Computational Systems Bioinformatics conference (CSB'03)*, pages 523–529, 2003.

[3] M. Eisen, P. Spellman, P. Brown, and D. Botstein. Cluster analysis and display of genome-wide expression patterns. *Proc. Natl Acad. Sci. USA*, 95:14863–14868, 1998.

[4] T. R. Golub, D. K. Slonim, and P. Tamayo. Molecular classification of cancer: class discovery and class prediction by gene expression monitoring. *Science*, 286:531–537, 1999.

[5] G. J. Gordon, R. V. Jensen, and L. Hsiaoand. Translation of microarray data into clinically relevant cancer diagnostic tests using gene expression ratios in lung cancer and mesothelioma. *Cancer Research*, 62:4963–4967, 2002.

[6] M. A. Hall. Correlation-based feature selection for discrete and numeric class machine learning. In *Proceedings of the Seventeenth International Conference on Machine Learning*, pages 359–366, 2000.

[7] T. Hastie, R. Tibshirani, and J. Friedman. *The Elements of Statistical Learning*. Springer-Vergag, New York, 2001.

[8] D. D. Jensen and P. R. Cohen. Multiple comparisions in induction algorithms. *Machine Learning*, 38(3):309–338, 2000.

[9] R. Kohavi and G. H. John. Wrappers for feature subset selection. *Artificial Intelligence*, 97(1-2):273–324, 1997.

[10] D. Koller and M. Sahami. Toward optimal feature selection. In *Proceedings of the Thirteenth International Conference on Machine Learning*, pages 284–292, 1996.

[11] H. Liu, F. Hussain, C. L. Tan, and M. Dash. Discretization: An enabling technique. *Data Mining and Knowledge Discovery*, 6(4):393–423, 2002.

[12] H. Liu, J. Li, and L. Wong. A comparative study on feature selection

and classification methods using gene expression profiles and proteomic patterns. *Genome Informatics*, 13:51–60, 2002.

[13] T. M. Mitchell. *Machine Learning*. McGraw-Hill, New York, 1997.

[14] J. Pearl, editor. *Probabilistic reasoning in intelligent systems*. Morgan Kaufmann, 1988.

[15] W. H. Press, S. A. Teukolsky, W. T. Vetterling, and B. P. Flannery. *Numerical Recipes in C*. Cambridge University Press, Cambridge, UK, 1988.

[16] M. Robnik-Sikonja and I. Kononenko. Theoretical and empirical analysis of Relief and ReliefF. *Machine Learning*, 53:23–69, 2003.

[17] L. J. van't Veer, H. Dai, and M. J. van de Vijver. Gene expression profiling predicts clinical outcome of breast cancer. *Nature*, 415:530–536, 2002.

[18] I. H. Witten and E. Frank. *Data Mining - Pracitcal Machine Learning Tools and Techniques with JAVA Implementations*. Morgan Kaufmann, 2000.

[19] E. Xing, M. Jordan, and R. Karp. Feature selection for high-dimensional genomic microarray data. In *Proceedings of the Eighteenth International Conference on Machine Learning*, pages 601–608, 2001.

[20] M. Xiong, Z. Fang, and J. Zhao. Biomarker identification by feature wrappers. *Genome Research*, 11:1878–1887, 2001.

[21] L. Yu and H. Liu. Efficient feature selection via analysis of relevance and redundancy. *Journal of Machine Learning Research*, 5:1205–1224, 2004.

Chapter 18

A Feature Generation Algorithm with Applications to Biological Sequence Classification

Rezarta Islamaj Dogan

University of Maryland at College Park and National Center for Biotechnology Information

Lise Getoor

University of Maryland at College Park

W. John Wilbur

National Center for Biotechnology Information

18.1   Introduction

Many real-world data mining problems involve data best represented as sequences. Sequence data comes in many forms, including: 1) human communication such as speech, handwriting, and printed text; 2) time series such as stock market prices, temperature readings and web-click streams; and 3) biological sequences such as DNA, RNA and proteins. Sequence data in all domains contains useful "signals," or features, that enable the construction of classification algorithms.

In handwriting recognition, features may include horizontal and vertical profiles, internal holes, strokes, and other characteristics of the handwritten characters. In speech recognition, features may include the recognized phonemes, noise ratios, length of sounds, and many others. In the spam detection domain, features may include whether certain email headers are present or absent, whether the headers are well formed, the grammatical correctness of the text, n-gram frequency analysis, and many others. In biological sequence classification problems, gene and protein sequence features may be nucleotide or amino-acid blocks, their respective positions in the sequence, as well as many possible combinations.

In all these cases, extracting and interpreting the most useful features is known to be a hard problem and hand selection of good features forms the basis of almost all classification algorithms. Automatic methods usually take a brute force approach in which the sequence classification models are provided with a huge number of features in the hope that the important features are not overlooked. The large number of features introduces a dimensionality problem that has several disadvantages. First, enumerating all possible features is impractical. Second, many features are irrelevant to the classification task and have an adverse effect on accuracy. And third, knowledge discovery becomes difficult because of the large number of parameters involved.

As a result, feature selection methods are employed to select a representative feature set from the available features for application to classification algorithms. Here we present a scalable method for automatic feature generation for sequences. The algorithm uses sequence components and domain knowledge to construct features, explores the space of possible features, and identifies the most useful ones. This focused **feature generation algorithm (FGA)** integrates feature construction and feature selection in a systematic way. The method is scalable because it incrementally generates more complex features from currently selected ones.

18.2   Splice-Site Prediction

We validate the FGA method in the biological domain on the task of splice-site prediction for pre-mRNA sequences, which is an increasingly important task in bioinformatics. In the context of bioinformatics, automatic sequence classification can also be employed in a multitude of applications ranging from fast database search to the identification of patterns for some specific physical properties.

18.2.1   The Splice-Site Prediction Problem

Splice sites are the locations in the DNA sequence that are boundaries between protein coding and non-coding regions. Accurate location of splice sites is an important component in the gene finding problem. Gene finding is one of the first and most important steps in understanding the genome of a species once it has been sequenced. In eukaryotic organisms, especially complex organisms like humans, gene finding is challenging because of the splicing mechanism. A protein coding sequence in these genomes is divided into several parts known as exons, separated by intervening non-coding sequences known as introns. Typically, a protein-coding human gene sequence might be divided into a dozen exons, each often less than 200 nucleotides in length, some as short as 10 or 20. It may also include an exceptionally long exon that may extend more than 1,000 nucleotides. Notably, sequence characteristics like pre-mRNA sequence length, coding sequence length, number of exons and their lengths, and interrupting intron sequence lengths do not follow any known pattern, making it hard to design a highly effective computational approach.

Splice sites belong to two different categories: the *acceptor splice site*, which marks the start of an exon, and the *donor splice site*, which marks the end of an exon, as shown in Figure 18.1. The splice-site signals are short sequences of nucleotides that are preferred in the immediate splice-site neighborhood. These signals are probably the most critical signals used in the detection of splice sites. They can be compiled from thousands of sequences aligned at the annotated splice-site location. However, the resulting consensus sequence alone is not enough for an accurate prediction. A linear search along any genome sequence for splice-site signals produces false locations matching the consensus at a very high frequency [6]. To eliminate the false positives, and find missing true splice sites, other information is needed.

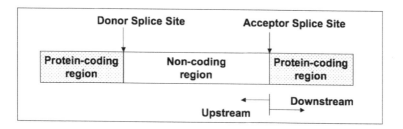

FIGURE 18.1: Depiction of a portion of a DNA gene sequence. The protein coding regions are called exons and the non-coding regions are called introns. Donor and acceptor sites mark the intron boundaries.

18.2.2 Current Approaches

Splice-site detection algorithms use statistical methods that are designed to capture the consensus signal. The weight matrix model (WMM) [15] computes the probabilities of nucleotides in each position in the splice-site sequence assuming independence between positions. The weight array model (WAM) [22] extends WMM by taking into account the dependencies between the adjacent nucleotides in the sequence. The maximal dependency decomposition (MDD) [3] is a decision tree model that improves on the previous models by capturing dependencies between non-adjacent as well as adjacent nucleotides in the splice-site sequence.

GeneSplicer, proposed by Pertea et al. [13], is a state-of-the-art computational tool for detecting splice sites that employs a combination of MDD and Markov modeling techniques. GeneSplicer views a splice site as a complex entity and is based on the following premise: Since a splice site (by definition) is surrounded by a coding region and a non-coding region, a splice-site model should take into consideration the coding difference between the two regions. Unlike the previous splice models, GeneSplicer models not only the splice-site signal but also the coding content in the upstream and the downstream sequence regions.

The GeneSplicer algorithm combines three different models for splice-site prediction. The first one is a statistical model of the immediate neighborhood of the site. Essentially, this is an MDD tree with the modification that a first order Markov chain, instead of an WMM, is built for each leaf of the decision tree. The next two models are second order Markov chains trained on coding and non-coding sequences. The final prediction for a given sequence is given by a combined score that adds the contribution of the three models. GeneSplicer is an accurate splice-site predictor, and has successfully combined the signal statistical models (WAM and MDD to capture the consensus signal) with the content sensor methods (Markov chains to capture coding/non-coding compositional differences).

In order to analyze a genomic sequence for the recognition of a target signal such as the splice site, it is important to employ all the information that can be extracted from the sequence. Specific candidate features can be generated and evaluated according to their relevance. The problem of how to select the relevant features has been the focus of intensive research. Recently, feature selection techniques have been receiving more attention for applications to biological data. The following is a non-comprehensive list. Liu and Wong [12] give a good introduction for filtering methods in the prediction of translation initiation sites. Degrovers et al. [4] describe a wrapper approach that uses both SVMs and naïve Bayes to select the relevant features for splice sites. Yeo et al. [18] use a model based on maximum entropy, in which only a small neighborhood around the splice site is considered. Zhang et al. [24] propose a recursive feature elimination approach using SVM.

Splice-site prediction has been the focus of other works, such as [1, 5, 20], that report promising results when compared with GeneSplicer. But, for a biologist, it is very difficult to interpret the features employed in these models. Especially, it is very difficult to relate them to actual biological signals. SpliceMachine [5] is similar to the approach we describe in this chapter because both methods employ sequence-based features. The SpliceMachine application performs a series of feature subset selection steps to find the best combination for an accurate splice-site prediction model. It details an extensive search for the best set of features, which is different from the guided feature generation algorithm that we discuss here.

18.2.3   Our Approach

In this chapter we describe a new approach to the biological sequence classification problem in general and a new solution to the splice-site prediction problem in particular. The feature generation algorithm uses the sequence properties to automatically construct useful features. These features are composed of two different components: the sequence alphabet and the relative position. The feature construction procedures produce complex features, including features containing elements that are not directly adjacent, and features that may be associated with a range of relative positions in the sequence. When the new features are constructed, feature selection techniques are employed to assess the constructed features and identify those that are most promising. Then, in an iterative fashion, feature construction procedures are called again. When building the features, this algorithm follows the GeneSplicer lead to consider a long subsequence window for splice-site prediction. The larger neighborhood provides information for other less-prominent but nevertheless important signals that are not usually considered in the gene-finding models. Then, a classification algorithm uses the identified features to predict splice sites.

Features constructed using the sequence domain knowledge are very important for knowledge discovery. Given a set of search and browsing procedures, a molecular biologist can explore the collection of these computationally identified signals. Such an exploration enables the researchers to discover new motifs and may guide them for experimental testing and validation.

We discuss the feature generation algorithm in the next section. We follow with an experimental evaluation of the algorithm for the splice-site prediction problem. Finally, we conclude with a discussion and offer some possible future directions.

18.3   Feature Generation Algorithm

In this section we describe the feature generation algorithm [7, 8]. This method generates features for splice-site prediction combining domain-specific feature construction methods and off-the-shelf feature selection methods. We start by defining general sequence feature types and the corresponding automatic construction methods. Generally, sequence feature construction methods use a sequence alphabet to construct words and sequence position information to construct position-specific words. Logical Boolean operators may be used to construct more complex features. These features have a generic definition and may be suitable for any sequence data. Once we have described feature construction, we discuss feature selection methods, including the different approaches and their characteristics, and explain how they can be adapted to the feature generation algorithm for different feature types. Then, we present the complete algorithm.

18.3.1 Feature Type Analysis

Sequence data has compositional and positional properties. Here we define a set of feature types that capture these properties and for each feature type we describe an incremental feature construction procedure. The feature construction starts with an initial set of features and produces an expanded set of features. Incrementally, it produces richer, more complex features in each iteration. We give examples for each feature type using DNA sequence data from the biological domain, but these rules and definitions apply to any sequence data defined over some fixed alphabet.

Compositional features A *general k-mer* is a string of k-characters. Given the alphabet for DNA sequences, $\{a, c, g, t\}$, the number of distinct k-mers is 4^k for each value of k. We consider the general k-mer composition of sequences for k ranging from 2 to 6. For each general k-mer, we count the number of times that the feature is present in the neighborhood of the splice site. There are a total of $5,456$ features for the values of k we consider. This feature type is useful for capturing information like coding potential and composition in the sequence.

Construction Method. Given an initial set of k-mer features, this construction method expands them to a set of $(k + 1)$-mers by appending the letters of the alphabet to each k-mer feature. As an example, suppose we begin with an initial set of 2-mers $F_{initial} = \{ac, cg\}$. From this set, we construct the extended set of 3-mers $F_{constructed} = \{aca, acc, acg, act, cga, cgc, cgg, cgt\}$. Incrementally, in this manner we construct level $k + 1$ from level k.

Region-specific compositional features Splice-site sequences characteristically have a coding region and a non-coding region, as shown in Figure 18.1. For donor splice-site sequences, the region of the sequence on the left of the splice-site position (upstream) is the coding region, and the region of the sequence on the right of the splice-site position (downstream) is the non-coding region. The opposite is true for acceptor splice sites. The upstream region is part of the intron and the downstream region is part of the exon. These regions may exhibit distinct compositional properties. In order to capture these differences, we introduce *region-specific k-mers*. A region-specific k-mer is a string of k-characters found in the specified region. In this work we have considered the upstream and the downstream regions. Other regions and interval specifications are also possible. Similar to general k-mers, we consider k values from 2 to 6 for these features. For each upstream (downstream) k-mer we count the number of times that feature is present in the upstream (downstream) neighborhood of the splice site. This results in a total of $10,912$ potential features.

Construction Method. The construction procedure of upstream and downstream k-mer features is the same as the general k-mer method, with the addition of a region indicator.

Positional features *Position-specific k-mers* are the most common features used for finding signals in the DNA stream data [21]. These features capture the correlations between different nucleotides and their relative positions. The nucleotides bordering the splice site are of primary importance as they may capture binding information. The simplest features of this type are position-specific 1-mers, which describe the occurrence of a specific nucleotide in a particular location in the sequence. These features also define the consensus sequence. We consider sequences of length 160, so there are 4×160 or 640 possible position-specific 1-mers. We use this basic feature set to construct position-specific k-mer features. Position-specific k-mers capture the correlations between k-adjacent nucleotides. At each position i in the sequence, these features represent the substrings appearing at positions $i, i+1, .., i+k-1$. This feature type is useful for discovering species-specific functional signals, as well as evolutionary conserved functional signals. For each position-specific k-mer we record the presence or absence of that feature in the neighborhood of the splice site. This results in a set of $(n-k+1) \times 4^k$ potential features for each value of k and sequence of length n.

Construction Method. This construction method starts with an initial set of position-specific k-mer features and extends them to a set of position-specific $(k+1)$-mers by appending the letters of the alphabet to each position-specific k-mer feature. As an example, suppose an initial set of 2-mers $F_{initial} = \{ac_2, cg_5\}$, where the subscript denotes the starting position. $F_{constructed} = \{aca_2, acc_2, acg_2, act_2, cga_5, cgc_5, cgg_5, cgt_5\}$ is the extended set of position-specific 3-mers. Incrementally, in this manner, we can construct level $k+1$ from level k.

Conjunctive positional features To capture the correlations between different nucleotides in non-consecutive positions in the sequence, we describe *conjunctive position-specific features*. We construct these complex features from conjunctions of basic position-specific features. This feature type is useful for discovering interacting functional signals in the sequence. The dimensionality of this kind of feature is inherently high. For each conjunctive positional feature, we record the presence or absence of that feature in the neighborhood of the splice site. For each iteration, if the number of conjuncts is k, we have a total of $\binom{n}{k} \times 4^k$ such features for a sequence of length n.

Construction Method. We construct conjunctions of basic features by starting with an initial conjunction of basic features and adding another conjunct basic feature in an unconstrained position. Let our basic set be $F_{basic} = \{a_1, c_1, \ldots, g_n, t_n\}$, where a_1 denotes nucleotide a at the first sequence position, and so on. If our initial set is $F_{initial} = \{a_1, g_2\}$, we can extend it to the level 2 set of position-specific base combinations $F_{constructed} =$

$\{a_1 \wedge a_2, a_1 \wedge c_2, \ldots, g_2 \wedge t_n\}$. A duplication check ensures that each feature in the $F_{constructed}$ set is unique. Incrementally, in this manner, we can construct higher levels. Given an initial set of k-conjuncts, this construction method selects from the set of basic position-specific features to add another conjunct in an unconstrained position, thereby constructing the set of $(k+1)$-conjuncts.

Other positional features The conjunctive positional features, as defined above, are constructed using position-specific nucleotides that can be adjacent to any position in the sequence of length n. Other variations are also possible, such as conjunctive positional features, which are region-specific or interval-specific. The difference between these other feature types and conjunctive positional features is the basic position-specific features set. The region-specific conjunctive features are constructed using position-specific nucleotides defined in the upstream or downstream sequence region as their basic feature set. This definition can be extended to other sequence regions or "user-defined intervals." In this case, each additive conjunct is selected from the basic feature set of position-specific nucleotides in a previously specified interval, i.e., the branch site interval.

The positional features that we have discussed so far define patterns of nucleotides in sequence positions that belong to a specific sequence interval or region. However, a biologist may also be interested to discover patterns of nucleotides in relative sequence positions. Motivated by this, we define another feature type, which we call the *floating conjunctive features* set. These features consist of basic conjuncts that belong to a short sequence window of length n_1, and the start of the first conjunct may be anywhere in the given sequence of length n, where $n_1 \leq n$. For each floating conjunctive feature we record the number of times that feature is present in the neighborhood of the splice site. As an example, consider the feature $a * *c$, or $a_i \wedge c_{i+3}$, and the sequence $aaccaggc$. This feature is constructed from two conjuncts in the window of length four, and occurs two times in the given sequence of length eight. The floating conjunctive feature set may have up to n_1 conjuncts. If all the conjuncts are used, then this feature set becomes a subset of the general n_1-mers.

18.3.2   Feature Selection

Feature selection methods prune the constructed feature set by reducing its size, keeping only the useful features for the task at hand. The problem of selecting useful features has been the focus of extensive research and many approaches have been proposed [2, 9, 10, 11, 17, 19].

Generally these approaches are divided into three major categories [2]. Filter approaches use the intrinsic properties of the dataset such as feature-class entropy to compute a feature relevance score. The low-scoring features are thus removed independent of the classifier algorithm. These approaches are usually very fast and are primarily used for high-dimensional datasets. Wrapper approaches are a second class of feature selection methods. These approaches perform a heuristic search through all the subsets using the classification algorithm as a guide to find promising subsets of features. These approaches have the disadvantage of being computationally intensive. This limits the wrapper approaches to datasets of low dimensionality. In the third group, embedded approaches, the feature selection method makes direct use of the parameters of the induction model to include or reject features.

In the experiments in the next section, we consider different feature selection methods to reduce the size of our constructed feature sets. We use several filter approaches including *Information Gain (IG)*, χ^2 *(CHI)*, *Mutual Information (MI)* and *KL-distance (KL)* for initial pruning of feature type sets during the generation stage. Here we give the definitions of these values as provided by Yang and Pedersen in [17]. IG is frequently employed as a feature-goodness criterion in the field of machine learning. It measures the number of bits of information obtained for category prediction by knowing the presence or absence of a feature. If the number of categories in the given dataset is m, the categories are $c_1, \ldots, c_m$, and P_r denotes probability, the information gain of feature f is defined to be

$$IG(f) = -\sum_{i=1}^{m} P_r(c_i) log P_r(c_i) + P_r(f) \sum_{i=1}^{m} P_r(c_i|f) log P_r(c_i|f)$$

$$+ P_r(\overline{f}) \sum_{i=1}^{m} P_r(c_i|\overline{f}) log P_r(c_i|\overline{f})$$

MI is a criterion commonly used in statistical language modeling of word associations. The MI between a feature f and the class c_i is defined to be

$$MI(f, c_i) = log \frac{P_r(f, c_i)}{P_r(f) \times P_r(c_i)}$$

We combine the category-specific scores to find the average mutual information value as $MI_{avg}(f) = \sum_{i=1}^{m} P_r(c_i) MI(f, c_i)$.

The χ statistic measures the lack of independence between feature f and the category c_i. The contingency table of a feature f and class c_i produces the following numbers: N_{fc_i}, the number of data points containing feature f and belonging to class c_i; N_{fn}, the number of times f occurs without c_i; N_{nc_i}, the number of times c_i occurs without f; and N_{nn}, and the number of times neither f nor c_i occurs. Assuming the size of dataset is N, the χ measure is defined as

$$CHI(f, c_i) = \frac{N \times (N_{fc_i} N_{nn} - N_{nc_i} N_{fn})^2}{(N_{fc_i} + N_{nc_i}) \times (N_{fn} + N_{nn}) \times (N_{fc_i} + N_{fn}) \times (N_{nc_i} + N_{nn})}$$

The KL criterion measures the divergence between the distribution of features present in a training sequence and the categories that sequence may belong to [14]. KL is defined as follows:

$$KL(f) = \sum_{i=1}^{m} P_r(c_i|f) log \frac{P_r(c_i|f)}{P_r(c_i)}$$

In the experiments we discuss in the next section, we found that MI performed best for selecting compositional features, *CHI* for positional features, and IG for conjunctive features.

Once we have performed feature generation for each feature type individually, we collect all the selected features into a mixed set. Starting with the mixed set, we use recursive feature elimination [24] to obtain the final set of features. A max-margin classifier similar to linear support vector machines (SVM) produces a decision boundary to discriminate between two different categories. The weights w_i of this decision boundary can be used as feature weights, assigned to each feature f_i, to derive feature ranking. We use the C-modified least squares (CMLS) classifier [23] to learn the decision boundary and the individual feature weights. We recursively train the classifier and remove low scoring features.

18.3.3 Feature Generation Algorithm (FGA)

The traditional feature selection approaches consider a single brute force selection over a large set of all features of all different types. By categorizing the features into different feature types it is possible to apply appropriate construction and selection methods suitable to the different types. Thus we can extract relevant features from each feature type set more efficiently than if a single selection method had been applied to the whole set.

The type-oriented feature selection approach allows the use of different feature selection models for each type set; i.e., for a feature set whose dimensionality is not too high one may use a wrapper approach in the selection step, while for a large feature type set one may use filter approaches. Also, this allows features of different types to be generated in a parallel fashion.

In order to employ the information embedded in the selected features for sequence prediction, we use the following algorithm:

- *Feature Generation.* The first stage generates feature sets for each feature type. For each defined feature type, we tightly couple the corresponding feature construction step with a specified feature selection step. We iterate through these steps to generate richer and more complex features. During each iteration, we eliminate features that are assigned a low selection score by the feature selection method.

- *Feature Collection and Selection.* We collect the features of different types and apply another selection step. The selection method we apply is recursive feature elimination. We recursively train the CMLS classifier and remove the low scoring features. We produce a final set of

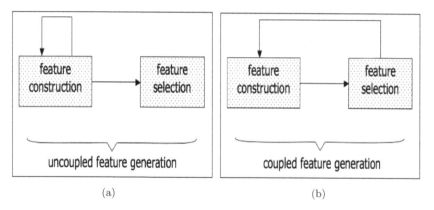

FIGURE 18.2: Feature generation component operating in (a) uncoupled and (b) coupled modes. When the feature generation operates in the coupled mode, the features scoring below the decided threshold, after the feature selection step, are not allowed to expand in the next iteration.

features originating from different feature types and different selection procedures.

- *Classification.* The last stage of our algorithm builds a classifier over the final set of features. The CMLS algorithm, described by Zhang and Oles in [23], is a max-margin method similar to SVM. Relative to SVM, CMLS has a smoother penalty function that allows calculation of gradients that provide faster convergence.

While feature generation remains a computationally intensive process, the organization of the generation process according to the different types allows us to search a much larger space efficiently. In addition, this feature generation approach has other advantages such as the flexibility to adapt with respect to the feature type and the possibility to incorporate the module in a generic learning algorithm. To deal with the large number of features, we use CMLS, which is very efficient.

The feature generation stage is also very generic and offers the flexibility to accomodate several different scenarios. This component may operate in the *coupled* or *uncoupled* mode, as shown in Figure 18.2.

When this component is in the uncoupled mode (see Figure 18.2(a)), the feature construction and selection steps are independent of each other. All the features constructed in iteration step i, regardless of the scores they are assigned by the feature selection method, are used in the next feature construction step. This mode allows even the low scoring features to expand in the next iteration. In the experiments described in the following section, we allow this component to operate in the uncoupled mode during compositional features generation.

When this component is in the coupled mode (see Figure 18.2(b)), the quality of the features produced by the feature construction method in the next iteration depends on the ability of the feature selection method to detect the useful features in the current iteration. The features scoring below the decided threshold are not allowed to expand in the next iteration. This mode of operation is useful when the dimensionality of the feature set is very high, as is the case in our experiments with conjunctive positional features.

18.4 Experiments and Discussion

We conducted a wide range of experiments and here we present a summary of them. We discuss our results based on these performance evaluation criteria: *11-point average precision, false positive rate*, and *Receiver Operating Characteristic* analysis. For any recall ratio, one can calculate the precision at the threshold, which achieves that recall ratio. The average precision of 11 recall points (11ptAvg Precision) [16] is calculated as follows. The 11ptAvg Precision is the average of precisions estimated at recall values 0%, 10%, 20%, ..., 100%. The ability of our algorithm to discriminate true splice-site sequences from normal sequences is also evaluated using Receiver Operating Characteristic (ROC) curve analysis. Another performance measure commonly used for biological data is the false positive rate (FPr) , defined as $FPr = \left(\frac{FP}{FP+TN}\right)$, where FP and TN are the number of false positives and true negatives, respectively. FPr can be computed for all recall values by varying the decision threshold of the classifier. We also present results using this measure. In all our experiments, the results reported use three-fold cross-validation.

18.4.1 Data Description

The dataset used for feature generation is a collection of $4,000$ human Ref-Seq pre-mRNA sequences. All the splice sites in these pre-mRNA sequences contain the consensus di-nucleotides AG for acceptors and GT for donors. Following the GeneSplicer format, we marked the splice sites and formed subsequences consisting of 80 nucleotides upstream and 80 nucleotides downstream from the sites. We constructed negative examples for the acceptor or donor datasets by choosing random AG-pair or GT-pair locations that were not annotated splice sites and selecting subsequences as we did for the true sites. The acceptor site data contains 20,996 positive instances and 200,000 negative instances. The donor site data contains 20,761 true positive instances and 200,000 negative instances. For further evaluation we tested the classification model of the final set of features on the B2hum dataset, provided by the

GeneSplicer team. This dataset contains 1,115 human pre-mRNA sequences. There is no overlap between the set of these sequences and the set the FGA algorithm is trained on.

Next, we discuss the prediction of acceptor and donor splice sites using the feature generation algorithm. Acceptor splice-site prediction is considered to be a harder problem than donor, which is characterized by a better conserved sequence.

18.4.2   Feature Generation

We begin with a brief evaluation of the effectiveness of the different feature types used in isolation.

Compositional features and region-specific compositional features
K-mer composition plays an important role in distinguishing sites and functional regions. In this analysis we aim to identify those k-mer features that can help recognize the splice sites. We examine each k-mer feature set independently for each value of k from 2 to 6. Figure 18.3 shows the process of feature generation for general and region-specific feature sets for the donor and acceptor dataset. We show the accuracy results for each general k-mer and region-specific k-mer feature sets after each iteration. In these experiments, after ranking the features according to each feature selection score, we selected the top 50% for each value of k. The *MI* selection method worked the best for compositional features. The results show that k-mer features carry more information when they are associated with a specific region (upstream or downstream), and this is shown by the significant increase in their 11ptAvg precisions.

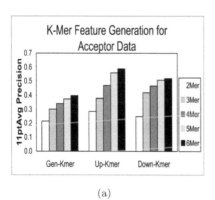   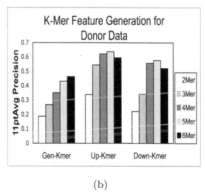

(a)   (b)

FIGURE 18.3: Feature generation comparison for performances of different feature type sets, general k-mers, upstream k-mers, and downstream k-mers, shown for different values of k for (a) acceptor splice-site prediction and (b) donor splice-site prediction

Positional features Next, we examine each *position-specific k-mer* feature set. *K*-mer compositional features adjacent to a particular site position may be used to discriminate such a site. In this analysis we explore *k*-values from 1 to 6. The prediction results for this feature type are shown in Table 18.1(a) for the acceptor splice-site prediction, and in Table 18.1(b) for the donor site. After each generation step we observe a gradual increase in performance until level 3, followed by a gradual decrease. This can be explained with the exponential increase in the number of features after each level; i.e., the feature set of position-specific 6-mers contains more than $600,000$ features. The statistics generated from the donor and acceptor datasets are not enough, so we experience this form of overfitting. In Table 18.1 we also list 11ptAvg precision results for the position-specific *k*-mer feature sets on acceptor and donor data when we use the *IG, MI, CHI,* and *KL* feature selection methods to select the best $1,000$ scoring features. The *IG* and *CHI* feature selection methods have a similar behavior. Our paired *t*-tests for statistical significance on the difference between their results reveal that the differences in these values are not statistically significant. The results on the position-specific 6-mer features on both datasets and position-specific 4-mer features for the acceptor data were statistically significant. The *KL* distance shows a good performance initially, but does not work well for more aggressive feature selections. This is most relevant for the set of position-specific 6-mers, where we have the largest reduction in feature set size. The *MI* method seems to be unreliable for the set of position-specific 3-mers for the donor data, but works well for the other cases. We choose *CHI* to work with this feature type, but *IG* would also be a good choice.

Conjunctive positional features Finally, we examine conjunctive positional features. Small groups of nucleotides adjacent to particular site positions, not necessarily adjacent to each other, may show a tendency to co-occur, therefore they may be used to discriminate the site. These feature sets are extremely large; for example, even for just three conjuncts there are 40 million unique combinations. We explored sets of 2 to 6 conjuncts denoted as $P2, P3, P4, P5, P6$. At each level, we used the *IG* selection method to select the top scoring $1,000$ features. We repeated the generation using this selected set to produce the next level of features.

Figure 18.4 depicts the performances of the conjunctive feature sets for acceptor and donor data. For comparison, we introduce a baseline method, which is the average of 10 trials of randomly picking $1,000$ conjunctive features from each level. We can see from the graphs in Figure 18.4 that the feature generation algorithm is picking up informative features that help distinguish the true splice-site locations. The 11ptAvg precision of these feature sets gradually drops as we generate more complex features. This happens because the feature set that is explored grows exponentially with each addition of another conjunct. The difference in precision values, however, between FGA and the baseline method is highly significant on every value of *k* (alpha =

TABLE 18.1:   Feature generation comparison for position-specific k-mer features for k from 1 to 6 for (a) acceptor and (b) donor splice-site predictions. We give the 11ptAvg precision for each set when all the features are used and for each selected set with different selection methods.

(a)

Pspec-Kmer	11ptAvg (Acc)	IG-1,000	MI-1,000	CHI-1,000	KL-1,000
1	79.85	-	-	-	-
2	85.96	84.91	76.49	84.68	84.84
3	86.54	82.43	74.36	82.46	79.54
4	84.92	73.94	72.59	75.96	70.09
5	80.60	72.59	71.94	72.65	60.94
6	68.64	58.84	58.58	59.31	30.27

(b)

Pspec-Kmer	11ptAvg (Don)	IG-1,000	MI-1,000	CHI-1,000	KL-1,000
1	82.11	-	-	-	-
2	86.47	85.61	82.75	85.02	85.20
3	87.46	84.58	65.42	84.45	84.06
4	87.31	80.80	79.15	80.77	77.18
5	86.31	80.34	80.93	80.48	77.77
6	84.93	68.94	70.16	70.35	47.21

0.005). Moreover, the generated features of this type can capture important functional biological signals.

18.4.3   Prediction Results for Individual Feature Types

Next, we compared collections of different levels of the feature sets of different types. The results are summarized in Figure 18.5.

Compositional features and region-specific compositional features The first three bars in Figure 18.5(a) show the results for the best k-mer features for k ranging from 2 to 6 on acceptor data. The general k-mer feature set contains 700 features and the 11ptAvg precision is 39.84%. The upstream and downstream k-mer feature sets sizes are $1,500$ features and $1,800$ features, and their results are respectively 58.77% and 52.01%. Similarly in Figure 18.5(b), the first three bars summarize the results for the general and region-specific k-mer features on donor data. The general k-mer feature set contains $1,000$ features and its 11ptAvg precision is 47.82%. The upstream and downstream k-mer feature sets sizes are $1,200$ features each, and their results are respectively 62.52% and 60.65%.

Position-specific k-mers The fourth bar shows the results for position specific 1-mers. The respective precision results are 80.27% for acceptor data and 82.11% for donor data. The next bar in Figure 18.5(a) shows $5,000$

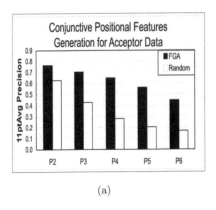
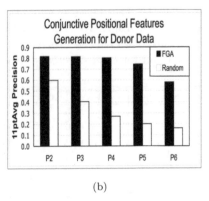

(a)   (b)

FIGURE 18.4: 11ptAvg results for the position-specific feature sets generated with the FGA algorithm vs. randomly generated features for (a) acceptor splice site data and (b) donor splice-site data

position-specific k-mer features selected using the *CHI* selection method. The 11ptAvg precision of this set is 85.94%. The result for $5,000$ position-specific k-mer features on donor data is 86.67%, represented by the fifth bar in Figure 18.5(b).

Conjunctive positional features The sixth bars on both graphs in Figure 18.5 show the results for conjunctive positional features. For acceptor data we have a collection of $3,000$ conjunctive positional features for k ranging from 2 to 6 selected using *IG*. The 11ptAvg precision that this collection set gives is 82.67%. The collection of conjunctive positional features for donor data results in an 11ptAvg precision of 83.95%. These results clearly show that using complex position-specific features is beneficial. Interestingly, these features typically are not considered by existing splice-site prediction algorithms.

Figure 18.5 also shows the performance of GeneSplicer on the same datasets as the last bar in the graph. We see that even in isolation, our positional features and our conjunctive positional features perform better than GeneSplicer. These results are also statistically significant.

18.4.4   Splice-Site Prediction with FGA Features

Once we collect all the features that we presented in Figure 18.5, general k-mers, upstream/downstream k-mers, position-specific k-mers, and conjunctive position-specific features, we run the CMLS classification algorithm for both acceptor and donor. We achieve 11ptAvg precision performances of 92.08% and 89.70%, respectively, in the acceptor and donor datasets that we built from the initial $4,000$ RefSeq pre-mRNA sequences. These improvements are highly statistically significant ($\alpha = 0.005$ for both acceptor and

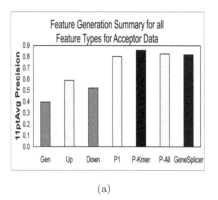
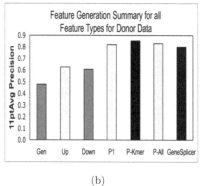

(a)   (b)

FIGURE 18.5: Performance results of the FGA method for different feature types as well as the GeneSplicer program in (a) acceptor splice data and (b) donor splice data. The depicted feature sets are as follows: Gen - selected general k-mers; Up - selected upstream k-mers; Down - selected downstream k-mers; P1 - position-specific nucleotides; P-Kmer - selected position-specific k-mers, comprising features from all considered values of k; P-All - conjunctive positional features comprising selected features for P2, P3, P4, P5, and P6.

donor) over one of the leading programs in splice-site prediction, GeneSplicer, which yields 11ptAvg precisions of 81.89% and 80.10% on the same datasets. The precision results of FGA-generated features at all individual recall points, shown in Figure 18.6, are consistently higher than those of GeneSplicer for both acceptor and donor site prediction. The break-even points for acceptor splice-site prediction for FGA and GeneSplicer are 67.8% and 54.9%, respectively. Donor splice-site prediction produced break-even values of 66.7% and 58.7%, respectively, for FGA and GeneSplicer.

In Figures 18.7(a) and 18.7(b) we explore more aggressive feature selection options using the more expensive recursive feature elimination method in order to select a smaller working feature set. The recursive feature elimination shows that the generated features using this algorithm are very robust. For donor splice-site prediction, even the feature set of size 500 yields an 11ptAvg precision of 89.66%. This is an improvement of 9.56% over GeneSplicer on the same dataset. For acceptor splice-site prediction, even the feature set of size 1,000 yields an 11ptAvg precision of 91.01%. This is an improvement of 9.12% over GeneSplicer on the same dataset.

Next, for further evaluation, we test both algorithms on the B2hum dataset provided by the GeneSplicer team, which contains 1,115 human pre-mRNA sequences. The FGA final feature sets for acceptor and donor splice-site prediction contain 3,000 and 1,500 features, respectively. In Figures 18.8(a) and (b) we present the false positive rates for a range of recall values from 5% to 95%. Figures 18.8(a) and (b) are actually ROC curves with the false

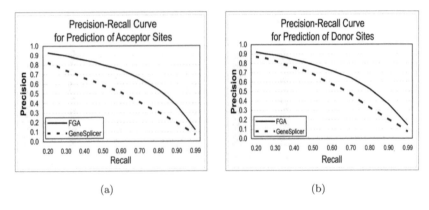

FIGURE 18.6: Precision results for each recall value for FGA with the complete set of features compared to GeneSplicer for (a) acceptor and (b) donor data.

positive rate shown on the y-axis. If we compare the AUC values for FGA and GeneSplicer, we get the following results. In the task of acceptor splice-site prediction, the FGA algorithm and GeneSplicer respectively score 99.37% and 98.71%. In the task of donor splice-site prediction, the AUC scores are 99.25% and 98.58% for FGA and GeneSplicer, respectively. The feature generation algorithm, with its rich set of features, consistently performs better than GeneSplicer in the B2hum dataset as well, which is the dataset the latter algorithm is trained on. FGA false positive rates, as depicted in Figure 18.8, are favorably lower at all recall values. At a 95% sensitivity rate, the FPr decreased from 6.2% to 2.5% for acceptor and from 6.7% to 3.3% for donor splice-site predictions. This significant reduction in false positive predictions can have a great impact when splice-site prediction is incorporated into a gene-finding program.

It should also be noted that there is no significant difference in the running time of FGA compared to GeneSplicer. Once the final set of features is determined, FGA performs a linear search (in terms of sequence length) along the given sequence in search of high scoring sites.

18.5   Conclusions

We have presented a general feature generation framework that integrates feature construction and feature selection in a flexible manner. We showed how this method can be used to build accurate sequence classifiers. We presented experimental results for the problem of splice-site prediction. Using the feature generation approach, we were able to search over an extremely large

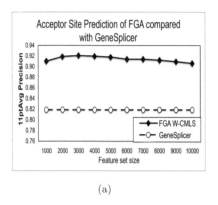 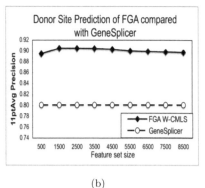

(a) (b)

FIGURE 18.7: 11ptAvg precision results for FGA compared to GeneSplicer for (a) acceptor and (b) donor data. We start with the complete set of features and recursively train the algorithm, eliminating 1,000 features at a time.

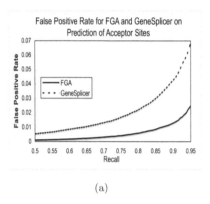 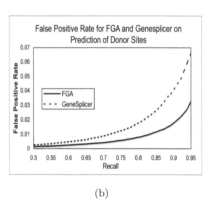

(a) (b)

FIGURE 18.8: The false positive rate results for FGA with the final feature set compared to GeneSplicer, varying the recall threshold for (a) acceptor and (b) donor data.

space of feature sets effectively, and we were able to identify the most useful set of features of each type. By using this mix of feature types, and searching over their combinations, we were able to build classifiers that achieved accuracy improvements of 10.6% and 9.5% over an existing state-of-the-art splice-site prediction algorithm. The specificity values are consistently higher for all sensitivity thresholds and the false positive rate has favorably decreased. These features have also shown a propensity to describe biologically significant functional elements. They are freely available to all interested researchers, and can be viewed at http://www.cs.umd.edu/projects/SplicePort/. This algorithm, with its systematic feature generation basis, can be applied to more complex feature types and other sequence prediction tasks, such as translation start-

Computational Methods of Feature Selection

site prediction, protein sequence classification problems, etc. Moreover, it can easily be extended to genomic data of other organisms.

[1] A. K. M. A. Baten, B. C. H. Chang, S. K. Halgamuge, and J. Li. Splice site identification using probabilistic parameters and svm classification. *BMC Bioinformatics*, 7(S5), 2006.

[2] A. Blum and P. Langley. Selection of relevant features and examples in machine learning. *Artificial Intelligence*, 97(1-2):245–271, 1997.

[3] C. Burge and S. Karlin. Prediction of complete gene structures in human genomic dna. *Journal of Molecular Biology*, 268(1):78–94, 1997.

[4] S. Degroeve, B. D. Baets, Y. V. de Peer, and P. Rouze. Feature subset selection for splice site prediction. In *Proceedings of the European Conference on Computational Biology (ECCB 2002)*, pages 75–83, 2002.

[5] S. Degroeve, Y. Saeys, B. D. Baets, P. Rouze, and Y. V. D. Peer. Splicemachine: predicting splice sites from high-dimensional local context representations. *Bioinformatics*, 21(8):1332–1338, 2005.

[6] R. Guigo, P. Filcek, J. Abril, A. Reymond, J. Lagarde, F. Denoeud, S. Antonarakis, M. Ashburner, V. Bajic, E. Birney, R. Castelo, E. Eyras, C. Ucla, T. Gingeras, J. Harrow, T. Hubbard, S. Lewis, and M. Reese. Egasp: the human encode genome annotation assessment project. *Genome Biology*, 7(S2), 2006.

[7] R. Islamaj, L. Getoor, and W. J. Wilbur. Feature generation algorithm: Application to splice-site prediction. In *International Workshop on Feature Selection for Data Mining: Interfacing Machine Learning and Statistics*, 2006.

[8] R. Islamaj, L. Getoor, and W. J. Wilbur. A feature generation algorithm for sequences with application to splice-site prediction. In *Proceedings of the 10th European Conference on Principles and Practice of Knowledge Discovery in Databases*, 2006.

[9] R. Kohavi and G. H. John. Wrappers for feature subset selection. *Artificial Intelligence*, 97(1-2):273–324, 1997.

[10] D. Koller and M. Sahami. Toward optimal feature selection. In *International Conference on Machine Learning*, pages 284–292, 1996.

[11] H. Liu and H. Motoda. *Feature Extraction, Construction and Selection:*

A Data Mining Perspective. Kluwer Academic Publishers, Norwell, MA, 1998.

[12] H. Liu and L. Wong. Data mining tools for biological sequences. *Journal of Bioinformatics and Computational Biology*, 1:139–168, 2003.

[13] M. Pertea, X. Lin, and S. L. Salzberg. Genesplicer: a new computational method for splice site prediction. *Nucleic Acids Research*, 29(5):1185–1190, 2001.

[14] K.-M. Schneider. A new feature selection score for multinomial naive bayes text classification based on kl-divergence. In *Meeting of the Association of Computational Linguistics (ACL)*, pages 186–189, 2004.

[15] R. Staden. Computer methods to locate signals in nucleic acid sequences. *Nucleic Acids Research*, 12(1):505–519, 1984.

[16] I. H. Witten, A. Moffat, and T. C. Bell. *Managing gigabytes (2nd ed.): compressing and indexing documents and images.* Morgan Kaufmann Publishers Inc., San Francisco, CA, 1999.

[17] Y. Yang and J. O. Pedersen. A comparative study on feature selection in text categorization. In *Proceedings of the 14th International Conference on Machine Learning*, pages 412–420, 1997.

[18] G. Yeo and C. Burge. Maximum entropy modeling of short sequence motifs with applications to rna splicing signals. In *RECOMB '03: Proceedings of the 17th Annual International Conference on Research in Computational Molecular Biology*, pages 322–331, 2003.

[19] L. Yu and H. Liu. Feature selection for high-dimensional data: A fast correlation-based filter solution. In *Machine Learning, Proceedings of the 20th International Conference*, pages 856–863, 2003.

[20] L. Zhang and L. Luo. Splice site prediction with quadratic discriminant analysis using diversity measure. *Nucleic Acids Research*, 31(21):6214–6220, 2003.

[21] M. Q. Zhang. Statistical features of human exons and their flanking regions. *Human Molecular Genetics*, 7(5):919–932, 1998.

[22] M. Q. Zhang and T. G. Marr. A weight array method for splicing signal analysis. *Computational Applications in Biological Sciences*, 9(5):499–509, 1993.

[23] T. Zhang and F. J. Oles. Text categorization based on regularized linear classification methods. *Information Retrieval*, 4(1):5–31, 2001.

[24] X. H.-F. Zhang, K. A. Heller, I. Hefter, C. S. Leslie, , and L. A. Chasin. Sequence information for the splicing of human pre-mrna identified by support vector machine classification. *Genome Research*, 13(12):2637–

2650, 2003.

Chapter 19

An Ensemble Method for Identifying Robust Features for Biomarker Discovery

Diana Chan

Mississippi State University

Susan M. Bridges

Mississippi State University

Shane C. Burgess

Mississippi State University

19.1 Introduction

Biomarker discovery has become an important area of research with the advent of modern high throughput technologies in biology. A biomarker is a molecule or set of molecules that is found in the blood, other body fluids, or tissues that exhibits a distinct pattern of expression under certain conditions and can be used be used for a diagnostic or prognostic test [14]. Although biomarkers can be proteins, mRNA, or metabolites, we restrict ourselves to protein biomarkers.

Multi-step data mining pipelines, based on a wide range of statistical and machine learning techniques, have been developed for biomarker discovery from both mRNA and protein expression data. Both mRNA and protein expression data are characterized by large numbers of noisy features and limited sample sizes. Such data, where the number of potential features greatly outnumbers the number of samples, is "wide data." Our research, as well as that of others, has shown that it is possible, using different data mining approaches, to identify many distinct sets of features from a single dataset that can provide near perfect classification. However, a major challenge for biomarker research is to locate features that are robust in the sense that they yield highly accurate results in the classification of new datasets. This problem is being addressed from two different aspects. One is to improve the accuracy and reproducibility of the data acquisition technologies. The other, and the one that is the focus of our work, is to improve the data mining process. We have developed a unique ensemble-based approach for feature selection for biomarker discovery that is based on the intuition that features that are selected often and yield accurate classifiers, regardless of the method used, will produce the most robust classifiers. We demonstrate that the features selected by our ensemble method yield accurate classifiers and can be used to build accurate classifiers with new data using two publicly available ovarian cancer datasets [5].

19.2   Biomarker Discovery from Proteome Profiles

A biomarker is one or more proteins whose presence in a sample, at a given level, indicates disease status. A few proteins and metabolites are already used diagnostically in clinical pathology. However, manual identification of one or a few biomarkers often has poor specificity and sensitivity. For example, prostate specific antigen (PSA), the recent biomarker for prostate cancer, mispredicts newplasia 70% of the time. Ideal biomarkers are the optimal subset of features that can be extracted from wide datasets to consistently discriminate between the treatment and control samples (i.e., with high specificity and sensitivity). Specific and sensitive biomarkers discriminate between conditions and therefore can lead to improved medical screening and diagnosis. New high throughput technologies such as cDNA microarrays and mass spectrometry that can rapidly measure large numbers of mRNAs, proteins, or metabolites expressed by a tissue have prompted the use of machine learning methods for biomarker discovery. Petricoin et al. [5, 11] were the first to use machine learning feature selection methods to identify biomarkers when classifying ovarian cancer patients based on patterns of protein in their serum using mass spectrometry. Furthermore, this approach was unique because it focused on identifying a set of peaks in the mass spectrometry profiles that reliably distinguish normal from disease rather than definitive identification of the differentially expressed proteins. Petricoin et al. used genetic algorithms and self-organizing maps to discriminate samples of women with ovarian cancer from women without cancer. Their method was able to perfectly classify all of the cancer samples and classify 95% of healthy samples. Their experiment was based on a single test without cross-validation.

During the past few years, many other groups have used a variety of machine learning and statistical methods for biomarker discovery from proteome profiles using the same publicly available Ovarian Cancer dataset [5]. Lilien et al. [9] used statistical feature selection with principal component analysis (PCA) for dimensionality reduction and linear discriminate analysis (LDA) for classification. Their method achieved perfect classification for all of the samples if the number items in the training sets was greater than 75% of the total sample size. Wu et al. [18] used t-statistics and random forests (RF) as their feature selection method for the same dataset. The t-statistic was used first to rank the features according to the relevance of each single feature to the dataset. Feature subsets were then selected to classify the data using different classification algorithms including support vector machines (SVM), random forests (RF), linear discriminant analysis (LDA), quadratic discriminant analysis (QDA), k-nearest neighbors (KNN), and bagged/boosted decision trees. Wu et al. concluded their experiments by stating the best performance was achieved by using RF as both feature selection and classification algorithms.

Surface-enhanced laser desorption/ionization (SELDI) and matrix-assisted laser desorption/ionization (MALDI) and time-of-flight (TOF) mass spectrometry have been used most commonly for biomarker identification. Only a brief review of each will be given here. SELDI-TOF and MALDI-TOF are similar in that they use mass spectrometers to record patterns of proteins ionized from the surface of a plate; have a relatively high tolerance for salt (into the millimolar range); may be sensitive to the fmole range required of biological samples; can be used to measure carbohydrates, oligonucleotides, small polar molecules, as well as peptides, proteins, and their post-translationally-modified forms such as glyco- and phosphor-proteins; and are versatile, convenient, and can rapidly produce lists of many protein or peptide "peaks" that may be altered in their concentration in a biological sample as a result of a disease state. Both SELDI- and MALDI-TOF record patterns of masses divided by the charge that they carry (m/z; generally +1 or +2), with m/z ranging from less than 1000 Da (small peptides) to a maximum of 300 kDa. These mass spectral patterns can be used to differentiate classes of samples without actual identification of the proteins. There are main two differences between SELDI and MALDI. First, in SELDI, the ionization requires a thin metal "chip"(Ciphergen Biosystems, Fremont, CA), which has an affinity for either all proteins in a particular sample or proteins with particular biophysical characteristics (e.g., acidic, basic, hydrophobic, or specific antigenicity) depending on the nature of the chip surface. After sample application and washing, ionization occurs directly from the plate (i.e., "surface enhanced"). In contrast, MALDI plates are not selective and chromatography is done "off-line." The second difference is that, in MALDI, before the samples are applied to the metal MALDI plate, they are mixed with energy-absorbing compounds called a chemical "matrix," which contain small chromophores that absorb light at a given wavelength. Commonly used matrix chemicals are -cyano-4-hydroxycinnamic acid (CHCA), 3,5-dimethoxy-4-hydroxycinnamic (sinapinic) acid, and 2,5-dihydroxybenzoic (gentisic) acid. After spotting to the metal plate, evaporation of water and solvents from the mixture results in sample proteins embedded in a crystalline lattice made up of matrix molecules.

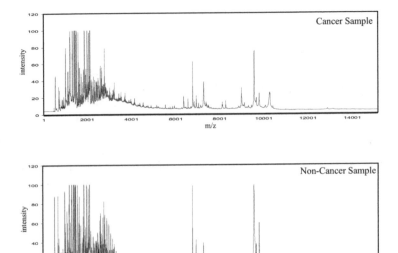

FIGURE 19.1: Spectra of cancer sample and non-cancer sample from Ovarian Cancer Dataset 8-7-02.

The datasets generated by SELDI-TOF and MALDI-TOF consist of tens of thousands of m/z measurements or "features." Because of expense and limited availability of samples, the total number of available instances is generally limited to tens to hundreds of samples. Figure 19.1 shows example spectra from a normal and a cancer sample from Ovarian Cancer Dataset 8-7-02 from the work of Petrocoin et al. [11]. The data was downloaded from the Clinical Proteomics Program Databank Website [5]. The m/z values located on the x-axis are the features that data-mining methods must choose from to find features that consistently distinguish between treatment and control samples. In this dataset, there are 15,154 distinct m/z values or 15,154 potential features. The dataset consists of 91 normal and 162 cancer samples.

19.3   Challenges of Biomarker Identification

Subsequent research by Petricoin's group and others has demonstrated the challenges of reproducibility of feature selection for biomarker discovery. A recent commentary by Ransohoff [12] and articles by Petricoin [10] and Baggerly [3] provide a nice overview of the issues involved. Researchers have found that it is not possible to identify a single m/z value that will accurately distinguish between treatment and control samples for the entire serum profile [6]. Therefore, feature selection algorithms have been widely used to select a subset of features as biomarkers to distinguish between treatment and control samples.

A number of different machine learning and data mining techniques have been shown to be effective in locating biomarkers for analysis of mass spectrometry profiles from wide datasets. Although these methods are able to classify a single dataset with high accuracy, it has proven to be very difficult to reproduce the results across new independent datasets. Problems with reproducibility result from 1) "the lack of a common standard operating procedure between different labs, reproducibility between different machines, and the variation of the research instruments" [7], and 2) selection of feature subsets that are not effective with new data [3]. This chapter describes an ensemble-based feature selection technique that selects a set of robust features that are able to provide reproducibly high accuracy with new datasets. The ensemble method is used to maximize the robustness of the features by using several different methods for each step of the pre-processing, feature selection, and classification steps and then selecting those features that appear often in accurate classifiers. We use Petricoin's ovarian cancer dataset [5] to demonstrate that the features selected by our ensemble method yield accurate classifiers and can be used to build an accurate classifier with new data.

19.4 Ensemble Method for Feature Selection

Data mining procedures for analyzing mass spectrometry profiles generally involve a number of steps including data preprocessing, feature selection, and classification model building. The choices used for different aspects of the data mining procedure have a significant impact on the features selected and the performance of the resulting classifier. We propose a framework for feature selection for wide data that is based on the intuition that features that are selected often under varying preprocessing steps, feature selection methods, and classification algorithms are more likely to be robust. The ensemble method rewards features that are selected frequently and that result in accurate classification. The major steps in the process of the ensemble method for feature selection are

1. Establish a general data mining process.

2. Use a voting procedure for features that rewards features that occur in many accurate classifiers.

3. Build and test a classifier with the features accruing the most votes.

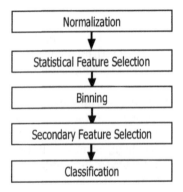

FIGURE 19.2: Data mining process for biomarker selection.

TABLE 19.1:   Options for each step in the data mining process.

Data Mining Step	Options
Normalization	None NV Z-score
Statistical feature pre-Selection	Wilcoxon test
Binning	Minimum p-value Maximum average intensity
Secondary feature selection	None CFS with greedy search CFS with BFS search CFS with genetic search Wrapper with greedy search Wrapper with BFS search Wrapper with genetic search Principal component analysis
Classifier	Back propagation neural networks Naive Bayes Decision tree Support vector machines

A number of different data mining procedures for biomarker identification have been described in the literature (e.g., [8, 9, 11, 16, 18, 19]). The overall data mining process that we have used for this study is similar to that used by Sorace and Zhan [13] (see Figure 19.2), but our ensemble-based approach can easily be applied with any data mining procedure where a number of options are available for each step. Table 19.1 summarizes the different options that we have considered for steps in the data mining process that we have used.

We briefly discuss the options used for each step in the data mining process shown in Table 19.1. Two normalization procedures were used for this study: z-score and normalized value (NV) [4]. Intensity values without normalization are used as a third alternative for this step. In situations where the number of features is huge compared to the number of samples, statistical tests can be used to select fewer features by eliminating redundant or irrelevant features to distinguish cancer from non-cancer samples. The two-sided Wilcoxon test was used for the first level of feature selection to compare the intensity at each of the m/z values for all samples and to identify the m/z values that are most discriminative between cancer and control samples. The 100 m/z values with the lowest p-values were selected as the initial feature subset for later feature selection. Many of the m/z values that pass the Wilcoxon filter represent the same peak. A binning procedure similar to that used by Sorace and Zhan [13] was used to combine these values into bins if their values were separated by less than 1 m/z. Two different methods were used to choose a representative m/z value for each bin. For the first method, the m/z value with the lowest p-value in each bin was selected. This m/z value indicates that it is the most discriminating value between the cancer and control samples for the bin. The second method selects the m/z value with the highest average intensity value across all samples as the representative m/z value.

The first three pre-processing steps provide six different combinations of features in terms of normalization and binning. A second round of selection was used to further reduce the number of features used to train the classifier. Feature selection at this stage is a search through the space of possible combinations of features and is driven by two procedures: attribute evaluation and the search procedure. The attribute evaluator is used to determine the quality of the individual feature for classification. The search procedure determines how the search space of possible features is explored. Different combinations of attribute evaluation and search procedures were used to generate different feature subsets for the ensemble method. Three different search procedures (greedy, best first, and genetic) were used with the CFS and wrapper attribute selection methods yielding a total of six options. Principal components analysis with ranking selection and no secondary feature selection (none) were used as two additional options, giving a total of eight methods for secondary feature selection. Four different options were used to construct ensemble classifiers for feature selection: backpropagation neural networks, naive Bayes, decision trees, and support vector machines.

19.5   Feature Selection Ensemble

Ensemble approaches are typically used to build robust classifiers where a number of different classifiers vote to provide the class for a new sample. However, in our case, the ensemble of classifiers is used to vote for features rather than class labels.

Ensembles are based on the idea of combining multiple classifiers by taking a linear combination of the learners via voting. Ensemble methods require both a method for generating members of the ensemble and a voting procedure.

We have generated the members of the ensemble by using all combinations of options for the data mining procedure that are shown in Table 19.1. The product of the number of options at each step gives the total number of classifiers generated (192). The 10-fold cross validation accuracy of each classifier and the set of features used by each classifier were recorded. A total of 47 unique features were selected by at least one classifier and were evaluated by the voting procedure.

The voting procedure for our ensemble methods works as follows. For each feature (m/z value) that was selected for use by at least one classifier, both a feature score and a weighted feature score were computed. Note that m/z values within 1 were considered to be the same. Features with higher weighted feature scores will be selected over other features. The feature score for a feature f_j is the sum of the accuracy values for all classifiers that included the feature. This score rewards features that are selected often by accurate classifiers. The weighted score is a modification of the feature score where the accuracy for each classifier is divided by the number of features selected. This scoring method favors frequently selected features that are members of small feature sets that yield accurate classifiers.

More formally, the feature score, $s(f_j)$, and weighted feature score, $ws(f_j)$, for feature f_j are defined as follows:

$$s(f_j) = \sum_{i-1}^{N} e_{ij} a_i \tag{19.1}$$

$$ws(f_j) = \sum_{i-1}^{N} (\frac{1}{F_i}) e_{ij} a_i \tag{19.2}$$

where N is the number of classifiers, $e_{ij} = 1$ if f_j is a feature selected for classifier i, a_i is the accuracy of classifier i, and F_i is the number of features for classifier i. After all features were scored, a classifier was constructed using the highest scoring features.

19.6　Results and Discussion

We have used the publicly available Ovarian Cancer Dataset to test the ensemble-based feature selection approach. Two datasets, the SELDI 8-7-02 and 4-3-02 Ovarian Datasets, were downloaded from the Clinical Proteomics Program Databank Website [5]. We first tested the capabilities of the features identified using the ensemble method for the first dataset, 8-7-02. It is also important to determine if the features selected by the ensemble method can perform better in classifying another dataset, which was collected at another time. The second dataset, Ovarian Cancer dataset 4-3-02, was used to test the robustness of the features selected using the ensemble approach from the first dataset.

The 8-7-02 dataset includes serum profiles of 91 non-cancer controls and 162 cancer subjects. Each spectrum consists of 15,154 distinct m/z values (potential features). The 4-3-02 dataset consists of samples from 50 unaffected women and another 50 patients with ovarian cancer [5] and has the same number of m/z values for each sample. This dataset was used as a validation step to test the accuracy of the features selected using the 8-7-02 dataset to build a classifier for the 4-3-02 dataset.

In order to test the ensemble approach for feature selection, we generated the ensemble for each dataset as previously discussed. We then used the voting procedure to find a robust set of features and then constructed a single classifier to test the ensemble-based feature set. We have chosen not to use an ensemble method such as AdaBoost for the final classifier in order to simplify the analysis.

The ensemble feature selection method provides both a feature score s and a weighted feature score ws for each feature selected by any classifier in the ensemble. Preliminary experimental results indicated that ws scoring outperforms s scoring. The results reported are based on ws scoring. When building the final classifier, the m/z with the highest ws was first added to the feature subset for classification. Features were added in decreased order of ws until both classification accuracy and relative absolute error (RAE) did not improve. When used with the 8-7-02 dataset, the six features with the highest ws are (in decreasing order of ws): m/z = 245.24466, 261.88643, 435.07512, 2.8548732, 433.90794, and 222.41828. These features were used to build four different classifiers using neural net, naive Bayes, decision tree, and support vector machine models. The accuracy and relative absolute error (RAE) for all classifiers are shown in Figures 19.3 and 19.4. The graphs plot the classification accuracy and RAE as each additional feature was added to the final feature subset. RAE is the relative absolute error computed by comparing the absolute error with the one obtained if the prediction had been the mean of the class values [17]. Ten-fold cross validation was used to compute classifier accuracy for all experiments.

Perfect classification accuracy was achieved for the first four features with the neural net, decision tree, and SVM classifiers. The SVM had a perfect RAE of zero with these four features. The neural net and decision tree models required the addition of two more features to reach a minimum RAE and never had an RAE as low as the SVM. The naive Bayes model had a higher RAE and lower accuracy than the other three models.

As an additional validation step, the set of six features selected using the 8-7-02 dataset was subsequently used to build a classifier for a different dataset, the 4-3-02 Ovarian Dataset [5]. Ten-fold cross validation with the 4-3-02 dataset using the six features selected with the 8-7-02 dataset resulted in a neural net classifier with a classification accuracy of 88%. However, these

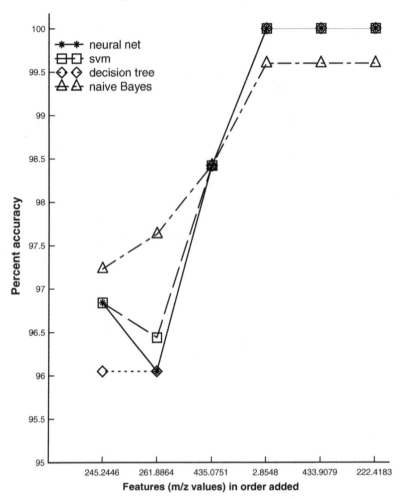

FIGURE 19.3: Classification accuracy (10-fold cross validation) with features added sequentially in decreasing order of weight for four types of classifiers

six features did not yield accurate classifiers for the naive Bayes, decision trees, or support vector machine models. If the top 13 features selected using the 8-7-02 dataset are used, classification accuracy increases for all classifiers. Table 19.2 summarizes the results of using the feature subsets selected by the ensemble method for dataset 8-7-02 and used to build classifiers for the 4-3-02 dataset. Results are based on 10-fold cross validation. Classifiers based on neural nets and decision trees have better classification performance than naïve Bayes and support vector machines for dataset 4-3-02.

Baggerly et al. [2] were not able to reproduce results from these two datasets using features selected by genetic algorithms partly due to calibration problems. Table 19.3 shows the features selected by other researchers for the 8-7-02 dataset. Two m/z values are considered the same if the difference between the two values is less than 1 Th. Some of the features we selected overlap with features selected by others. To measure the robustness of the features selected by other researchers, the seven features selected by Sorace and Zhan [13] based on the 8-07-02 dataset were used to classify dataset 4-3-02. These features are listed in Table 19.3. The last column of Table 19.2 gives the accuracy rates of different types of classifiers that were constructed with the seven features selected by Sorace and Zhan. The accuracy rates of these classifiers are substantially less than those of the classifiers based on our features. In

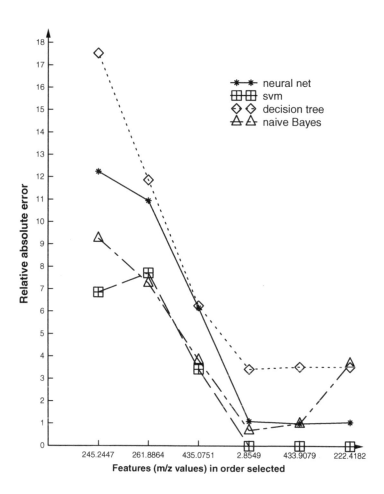

FIGURE 19.4: Relative absolute error (RAE) (10-fold cross validation) with features added sequentially in decreasing order of weight for four types of classifier

TABLE 19.2: Classification accuracies with different 8-7-02 feature subsets for dataset 4-3-02.

	Ensemble feature subset: 6 features	Ensemble feature subset: 13 features	Sorace and Zhan [13] features: 7 features
Neural nets	88%	90%	73%
Naïve Bayes	82%	83%	70%
Decision trees	79%	92%	73%
Support vector machines	79%	82%	72%

TABLE 19.3: Comparison of the features selected by other researchers and those selected by our method. Features (m/z values) shown in bold were selected by our ensemble method.

Alex at al. [1]	**245.8296**, **261.88643**, 336.6502, 418.8773, **435.46452**, 437.0239, 465.97198, 687.38131, 4004.826
Sorace and Zhan [13]	2.7921, **245.53704**, **261.8864**, 418.1136, **435.0751**, 464.3617, 4003.645 (these features were selected by stepwise discriminant analysis according to Rule 1)
Vannucci at al. [15]	**245.3**, **433.2**, **434.6**, 243.9, 430.6, 241.3, 437.2, 605.2, 431.9
Our work	**2.8549**, **222.4183**, **245.2447**, **2661.8864**, **433.9079**, **435.0751**

addition, the ensemble method ranks all of the features and offers the flexibility of allowing the researcher to include additional features in the feature subset to classify another dataset in order to improve performance.

We have also conducted an experiment comparing the performance of neural net classifiers based on a single path through the data mining procedure (one option at each step) and the performance of a neural net classifier based on the features selected using our ensemble method for dataset 4-3-02. The options used to build the "single option" classifiers and the ensemble methods are shown in Table 19.4.

Figure 19.5 shows that, although the feature sets generated from these single option feature selection methods are able to classify the first dataset 8-7-02 with almost 100% accuracy, the performance degrades substantially if these features are used to classify another dataset, 4-3-02. The feature set selected by the ensemble method achieves much higher accuracy with the new dataset.

Our experiments do not offer evidence to show which normalization method, binning method, feature selection method, and search method can improve classification with a new dataset. Feature sets B, C, and D are able to achieve

TABLE 19.4:   Data mining options used to select features using a single option for feature selection (A–F) and the ensemble method.

Feature Set	Normalization	Binning	Feature Selection Method	Feature Selection Search Method	Number of Features
A	None	MaxAvgInt	CFS	Greedy	2
B	NV	MaxAvgInt	CFS	BFS	7
C	None	MinP	None	None	24
D	None	MaxAvgInt	None	None	24
E	NV	MaxAvgInt	PC	Ranker	6
F	Zscore	MinP	Wrapper	Greedy	2
Ensemble	All	All	All	All	10

perfect classification for dataset 8-7-02, but the accuracy drops by more than 10% when the feature sets are used to classify dataset 4-3-02. In addition, there is no evidence to show the number of features has an impact on improving the classification performance of another dataset. Feature set A, consisting of only two features, classifies the first dataset, 8-7-02, with a high accuracy of 96.84%. However, performance drops to 73% when the feature sets are used to classify dataset 4-3-02. The feature sets consisting of many features do not show good performance on dataset 4-3-02. There is a perfect classification when feature set C is used to classify dataset 8-7-03, but the accuracy drops to 76% when all 24 features are used to classify dataset 4-3-02.

19.7   Conclusion

Many different feature selection approaches can be effectively used to locate disease biomarkers. Each approach has advantages and disadvantages. However, what is very consistent is the inconsistency of replicating biomarker selection using different feature selection methods, and this is due to the multifactorial nature of the features [2]. It is also difficult to use features selected from one dataset for classification of another dataset due to the size and nosiness of the data, the variation from one biological sample to another, and the variation due to differences in experimental procedures, between machines and between laboratories. Here we present an ensemble framework for feature selection for building classifiers with wide data. Features that are selected often, that result in accurate classifiers, and that are part of small feature sets are considered more robust and are favored by the weighted voting function. Using widely-studied ovarian cancer datasets, we show that the features selected

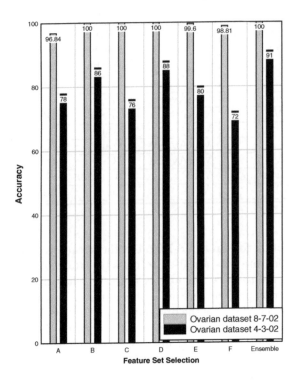

FIGURE 19.5: Comparison of performance of classifiers from single option feature selection and ensemble feature selection methods. Accuracy results are given for the dataset used for feature selection (8-7-02) and a second dataset (4-3-02).

by our method give very high classification rates and reproducible results with new data. A general data mining process with different options has been used for each step to demonstrate the effectiveness of the ensemble approach. This general approach can be easily adapted for use with different options for each data mining step or with a different data mining process. Although our work is proteomics-based, our ensemble method is generally applicable to any wide dataset including mRNA expression data.

References

[1] G. Alexe, S. Alexe, L. Liotta, E. Petricoin, M. Reiss, and P. Hammer. Ovarian cancer detection by logical analysis of proteomic data. *Proteomics*, 4:766–783, 2004.

[2] K. Baggerly, J. Morris, and K. Coombes. Reproducibility of seldi-tof protein patterns in serum: Comparing datasets from different experiments. *Bioinformatics*, 20:777–785, 2004.

[3] K. Baggerly, J. Morris, S. Edmonson, and K. Coombes. Signal in noise: Evaluating reported reproducibility of serum proteomics tests for ovarian cancer. *Journal of the National Cancer Institute*, 97:307–309, 2005.

[4] K. Baggerly, J. Morris, J. Wang, D. Gold, L. Xiao, and K. Coombes. A comprehensive approach to the analysis of matrix-assisted laser desorption/ionization-time of flight proteomics spectra from serum samples. *Proteomics*, 3:1667–1672, 2003.

[5] Clinical proteomics program database detailed explanation of proteome quest for data analysis. Web site, Aug 2005. http://home.ccr.cancer.gov/ncifdaproteomics/ppatterns.asp.

[6] T. Conrads, M. Zhou, E. Petricoin, L. Liotta, and T. Veenstra. Cancer diagnosis using proteomic patterns. *Expert Reviews in Molecular Diagnostic*, 3:411–420, 2003.

[7] A. Constans. Rethinking clinical proteomics. *Scientist*, 19, 2005.

[8] J. Li, Z. Zhang, J. Rosenweig, Y. Wang, and D. Chan. Proteomics and bioinformatics approaches for identification of serum biomarkers to detect breast cancer. *Clinical Chemistry*, 48:1296–1304, 2002.

[9] R. Lilien, H. Farid, and B. Donald. Probabilistic disease classification of expression-dependent proteomic data from mass spectrometry of human serum. *Journal of Computational Biology*, 10:925–946, 2003.

[10] L. Liotta, M. Lowenthal, T. Conrads, T. Veenstra, D. Fishman, and E. Petricoin. Misinformation generated by lack of communication between producers and consumers of publicly available experimental data. *Journal of the National Cancer Institute*, 97:310–314, 2005.

[11] E. Petricoin, A. Ardekani, B. Hitt, P. Levine, V. Fusaro, S. Steinberg, G. Mills, C. Simone, D. Fishman, and E. Kohn. Use of proteomic patterns in serum to identify ovarian cancer. *Lancet*, 359:572–577, 2002.

[12] D. Ransohof. Lessons from controversy: ovarian cancer screening and serum proteomics. *Journal of the National Cancer Institute*, 97:315–319, 2005.

[13] J. Sorace and M. Zhan. A data review and re-assessment of ovarian cancer serum proteomic profiling. *BMC Bioinformatics*, 4, 2003.

[14] Medical terminology and drug database. Web site, December 2005. http://www.stjude.org/glossary.

[15] M. Vannucci, N. Sha, and P. Brown. Nir and mass spectra classification:

Bayesian methods for wavelet-based feature selection. *Chemometrics and Intelligent Laboratory Systems*, 77:139–148, 2005.

[16] M. Wagner, D. Naik, and A. Pothen. Protocols for disease classification from mass spectrometry data. *Proteomics*, 3:1692–1698, 2003.

[17] I. H. Witten and F. Eibe. *Data Mining: Practical Machine Learning Tools and Techniques (Second Edition)*. Morgan Kaufmann, San Francisco, CA, 2005.

[18] B. Wu, T. Abbott, D. Fishman, W. McMurray, G. Mor, K. Stone, D. Ward, K. Williams, and H. Zhao. Comparison of statistical methods for classification of ovarian cancer using mass spectrometry data. *BioInformatics*, 19, 2003.

[19] H. Zhu, C. Yu, and H. Zhang. Tree-based disease classification using protein data. *Proteomics*, 3:1673–1677, 2003.

Chapter 20

Model Building and Feature Selection with Genomic Data

Hui Zou

University of Minnesota

Trevor Hastie

Stanford University

20.1 Introduction

Feature selection is fundamental in statistical modeling. When the number of predictors is large, it is crucial to identify a few important variables that can well explain the response. A sparse model is much more interpretable than the full model using all predictors, and feature selection can often improve the prediction accuracy of the model. Traditional model selection methods combine best-subset selection with some model selection criteria such as AIC and BIC. This approach has two fundamental drawbacks. First, the best-subset selection is not computationally feasible for high-dimensional data. The number of subset models increases exponentially. Second, the best-subset selection is very unstable in the sense that a small perturbation on the data yields a very different model [2]. Modern methods in high-throughput biology such as gene expression arrays produce massive high-dimensional data that traditional variable selection approaches are not capable of handling.

Recently, penalization-based variable selection methods have attracted a lot of attention. Regularization is crucial in order to build a predictive model with high accuracy, especially when there are a huge number of predictors. A non-regularized model is guaranteed to overfit the data. The L_2 penalty has been widely used in various learning problems such as smoothing splines, the support vector machine, and neural networks, as summarized in [13]. The L_2 penalty is good at controlling the model complexity by shrinking all coefficients toward zero, but they all stay in the model. On the other hand, the best-subset selection is able to reduce the model size, but it results in an unconstrained fit using the chosen variables. The lasso [20] was proposed as a compromise between the L_2 regularization and the best-subset selection. By imposing an L_1 constraint on the parameters, the lasso does both shrinkage and variable selection simultaneously. The L_1 penalization opens a new door to variable selection. However, the lasso may produce unsatisfactory results in some scenarios: (1) The number of predictors (greatly) exceeds the number of observations; and (2) the predictors are highly correlated and form *groups*. These two issues are very common with genomic data. A typical example is the gene selection problem in microarray analysis. When the number of predictors greatly exceeds the number of observations, the *grouped variables* situation is a particularly important concern, which has been addressed a number of times in the literature. Tree harvesting [11] uses supervised learning methods to select groups of predictive genes found by hierarchical clustering. Using an algorithmic approach, the authors of [3] performed the clustering and supervised learning together. A careful study in [18] strongly motivates the use of a regularized regression procedure to find the grouped genes. With gene expression and similar data, a desirable method should have the following properties:

1. Variable selection is performed via continuous shrinkage and is built into the procedure so that the irrelevant variables are automatically removed from the model.

2. Variable selection is not limited by the fact that the number of predictors is much larger than the sample size.

3. It should be able to select groups of significant variables.

The first property implies the method does variable selection in a fashion similar to the lasso, while the last two properties fix the drawbacks of the lasso for high-dimensional data.

In this chapter we introduce a new regularization technique called the elastic net for simultaneous modeling and feature selection. The elastic net addresses many of the problems encountered in model building and feature selection with high-dimensional data. This chapter describes how the elastic net can be used for regression, classification, and sparse eigen-gene analysis.

20.2   Ridge Regression, Lasso, and Bridge

We first review the ridge regression, lasso, and bridge regression. Analysis of the strengths and weaknesses of these methods motivates the development of a new variable selection method that addresses their drawbacks with high-dimensional data.

Consider the linear model. Given N predictors $\mathbf{x}_1, \cdots, \mathbf{x}_N$, the response $\mathbf{Y}$ is predicted by

$$\hat{\mathbf{Y}} = \hat{\beta}_0 + \mathbf{x}_1\hat{\beta}_1 + \cdots + \mathbf{x}_N\hat{\beta}_N \tag{20.1}$$

Suppose the dataset has M observations with N predictors. Let $\mathbf{Y} = (y_1, \ldots, y_M)^T$ be the response and $\mathbf{X} = [\mathbf{x}_1 | \cdots | \mathbf{x}_N]$ be the model matrix, where $\mathbf{x}_j = (x_{1j}, \ldots, x_{Mj})^T, j = 1, \ldots, N$ are the predictors. After a location and scale transformation, we can assume the response is centered and the predictors are standardized,

$$\sum_{i=1}^{M} y_i = 0, \quad \sum_{i=1}^{M} x_{ij} = 0, \text{ and } \sum_{i=1}^{M} x_{ij}^2 = 1, \text{ for } j = 1, 2, \ldots, N \tag{20.2}$$

It is well known that the ordinary least squares (OLS) estimator often does poorly in both prediction and interpretation. Ridge regression [14] is often used to improve the prediction of least squares. The ridge regression model is defined as

$$\hat{\beta}(\text{ridge}) = \arg\min_{\beta} \|\mathbf{Y} - \mathbf{X}\beta\|^2 + \lambda\|\beta\|^2 \tag{20.3}$$

where

$$\|\beta\|^2 = \sum_{j=1}^{N} \beta_j^2$$

As a continuous shrinkage method, ridge regression achieves its better prediction performance through a bias-variance trade-off. However, ridge regression cannot produce a parsimonious model, for it always keeps all the predictors in the model.

A promising technique called the lasso was proposed in [20]. The lasso is a penalized least squares method imposing an L_1 penalty on the regression coefficients:

$$\hat{\beta}(\text{lasso}) = \arg\min_{\beta} \|\mathbf{Y} - \mathbf{X}\beta\|^2 + \lambda\|\beta\|_1 \tag{20.4}$$

where

$$\|\beta\|_1 = \sum_{j=1}^{N} |\beta_j|$$

The L_1 penalty ($\|\beta\|_1$) is not differentiable at zero. Due to this property, the L_1 penalty continuously shrinks the OLS estimates toward zero and some components will be exactly zero if λ is large enough. Thus the lasso does both continuous shrinkage and automatic variable selection simultaneously.

A class of penalized least squares methods called bridge regression using the L_q penalty was considered in [6]:

$$\widehat{\beta}(\text{bridge}) = \arg\min_{\beta} \|\mathbf{Y} - \mathbf{X}\beta\|^2 + \lambda\|\beta\|_q \qquad (20.5)$$

where

$$\|\beta\|_q = \sum_{j=1}^{N} |\beta_j|^q$$

Ridge regression is bridge with $q = 2$ while the lasso amounts to using $q = 1$ in bridge. For $q > 1$, the bridge solution contains all predictors. However, bridge estimates with $q \leq 1$ enjoy a sparse representation. But there is another very subtle point. If $q < 1$, the bridge solution is not continuous, because the penalty function is concave. Therefore, the lasso is very unique in the bridge family, for it is the only L_q estimator that simultaneously enjoys sparsity and continuity.

20.3 Drawbacks of the Lasso

Note that like OLS, the lasso solution is not uniquely defined when $M > N$. This is the first obvious drawback of the lasso. In contrast, ridge regression can easily handle the $M > n$ case. In terms of prediction, it has been observed that if predictors are highly correlated, the prediction performance of the lasso is dominated by ridge regression [20]. With high-dimensional data it is often true that many predictors are highly correlated. Thus the lasso is not the best prediction method for high-dimensional data.

The biggest advantage of the lasso over ridge regression is its ability to do automatic variable selection. However, with high-dimensional data, the lasso selection is not satisfactory either. When $N > M$, the lasso can select at most M variables before it saturates, due to the nature of the L_1 optimization. This seems to be a limiting feature for a variable selection method. Consider the problem of identifying genes that affect the survival time of cancer patients based on the measurements of the gene expression level of 5000 genes. If 20 patients are enrolled in the study, the lasso can only select 20 or fewer genes. The biological truth might be that 100 genes are related to the survival time.

The other drawback of the lasso selection is that the lasso tends to under-select when there is a group of variables among which the pairwise correlations are very high, because the lasso tends to select only one variable from the group and does not care which one is selected [5].

These drawbacks of the lasso make it an inappropriate variable selection method in modeling high-dimensional data. We illustrate our points by considering the gene selection problem in microarray data analysis. A typical microarray dataset has many thousands of predictors (genes) and often less than a hundred samples. For those genes sharing the same biological pathway, the correlations among them can be high [18]. We think of those genes as forming a group. The ideal gene selection method should be able to do two things: eliminate the trivial genes, and automatically include whole groups into the model once one gene among them is selected (*grouped selection*). The lasso cannot perform grouped selection. The number of selected genes is artificially bounded by the sample size.

20.4 The Elastic Net

With the advantages and drawbacks of the lasso in mind, we want to invent a new method that works as well as the lasso whenever the lasso does the best, and can fix its fundamental drawbacks. To be specific, the new method should be able to select more than M variables in the $N \gg M$ problems, if it is necessary to select more variables than the samples. In addition, the new method should select groups of correlated variables via continuous shrinkage.

20.4.1 Definition

For that purpose, the authors of [24] proposed the *elastic net* method. For any fixed non-negative λ_1 and λ_2, we define the naive elastic net criterion

$$L(\lambda_1, \lambda_2, \boldsymbol{\beta}) = \|\mathbf{Y} - \mathbf{X}\boldsymbol{\beta}\|^2 + \lambda_2\|\boldsymbol{\beta}\|^2 + \lambda_1\|\boldsymbol{\beta}\|_1 \qquad (20.6)$$

where

$$\|\boldsymbol{\beta}\|^2 = \sum_{j=1}^{N} \beta_j^2 \text{ and } \|\boldsymbol{\beta}\|_1 = \sum_{j=1}^{N} |\beta_j|$$

The naive elastic net estimator $\hat{\boldsymbol{\beta}}$ is the minimizer of (20.6):

$$\hat{\boldsymbol{\beta}}(\text{naive elastic net}) = \arg\min_{\boldsymbol{\beta}} L(\lambda_1, \lambda_2, \boldsymbol{\beta}) \qquad (20.7)$$

The penalty function $\lambda_2\|\boldsymbol{\beta}\|^2 + \lambda_1\|\boldsymbol{\beta}\|_1$ is named the elastic net penalty [24]. It is a linear combination of the ridge and lasso penalties. When $\lambda_2 = 0$, the elastic net penalty reduces to the lasso penalty. However, we consider using $\lambda_2 > 0$. A positive λ_2 leads to fundamental differences between the elastic net and the lasso.

First of all, the elastic net penalty function is strictly convex with $\lambda_2 > 0$, whereas the lasso penalty is convex but not strictly convex. Thus, when using a positive λ_2, the elastic net is well defined even when $N \gg M$. The elastic net fixes the first drawback of the lasso.

Meanwhile, the L_1 component of the elastic net penalty makes the penalty singular at zero, which allows the elastic net to do automatic variable selection. To easily understand the thresholding property, let us consider an orthogonal design with orthogonal predictors. It is easy to show that with parameters (λ_1, λ_2), the naive elastic net solution is

$$\hat{\beta}_i(\text{naive elastic net}) = \frac{\left(\left| \hat{\beta}_i(\text{ols}) \right| - \frac{\lambda_1}{2} \right)_+}{1 + \lambda_2} \text{sgn}\left(\hat{\beta}_i(\text{ols}) \right) \qquad (20.8)$$

where $\hat{\beta}(\text{ols}) = \mathbf{X}^T \mathbf{Y}$ and z_+ denotes the positive part, which is z if $z > 0$, else 0. The solution of ridge regression with parameter λ_2 is given by $\hat{\beta}(\text{ridge}) = \hat{\beta}(\text{ols})/(1 + \lambda_2)$, and the lasso solution with parameter λ_1 is $\hat{\beta}_i(\text{lasso}) = \left(\left| \hat{\beta}_i(\text{ols}) \right| - \frac{\lambda_1}{2} \right)_+ \text{sgn}\left(\hat{\beta}_i(\text{ols}) \right)$.

Including the L_2 penalty in the lasso introduces a double amount of shrinkage. Both L_1 and L_2 components shrink OLS estimates. Double shrinkage does not help to reduce the variances much and introduces unnecessary extra bias. The double shrinkage is evident in the orthogonal design. The authors of [24] suggested using the corrected elastic net estimates that are defined as follows:

$$\hat{\beta}(\text{elastic net}) = (1 + \lambda_2)\hat{\beta}(\text{naive elastic net}) \qquad (20.9)$$

Hence the elastic net estimator is a rescaled naive elastic net estimator. Such a scaling transformation preserves the variable-selection property of the naive elastic net and is the simplest way to undo shrinkage. Why use $(1 + \lambda_2)$ as the scaling factor? Consider the exact solution of the naive elastic net when the predictors are orthogonal. The lasso is known to be minimax optimal [4] in this case, which implies the naive elastic net is not optimal. After scaling by $1 + \lambda_2$, the elastic net automatically achieves minimax optimality.

From now on, let $\hat{\beta}$ stand for $\hat{\beta}$ (elastic net) for convenience. We can view the elastic net as a regularized lasso. Reference [24] showed the following fact. The elastic net estimates $\hat{\beta}$ are equivalently defined as

$$\hat{\beta} = \arg\min_{\beta} \beta^T \left(\frac{\mathbf{X}^T \mathbf{X} + \lambda_2 \mathbf{I}}{1 + \lambda_2} \right) \beta - 2\mathbf{Y}^T \mathbf{X} \beta + \lambda_1 \|\beta\|_1 \qquad (20.10)$$

It is easy to see that

$$\hat{\beta}(\text{lasso}) = \arg\min_{\beta} \beta^T (\mathbf{X}^T \mathbf{X}) \beta - 2\mathbf{Y}^T \mathbf{X} \beta + \lambda_1 \|\beta\|_1 \qquad (20.11)$$

Hence we can interpret the elastic net as a stabilized version of the lasso. Note that $\hat{\mathbf{\Sigma}} = \mathbf{X}^T \mathbf{X}$ is a sample version of the correlation matrix ($\mathbf{\Sigma}$) and $\frac{\mathbf{X}^T \mathbf{X} + \lambda_2 \mathbf{I}}{1 + \lambda_2} = (1 - \gamma)\hat{\mathbf{\Sigma}} + \gamma \mathbf{I}$ with $\gamma = \frac{\lambda_2}{1 + \lambda_2}$ shrinks $\hat{\mathbf{\Sigma}}$ toward the identity matrix. Together (20.10) and (20.11) say that rescaling after the elastic net penalization is mathematically equivalent to replacing $\widehat{\Sigma}$ with its shrunk version in the lasso. In linear discriminant analysis, prediction accuracy can often be improved by replacing $\widehat{\Sigma}$ by a shrunken estimate [7, 13]. Likewise, we improve the lasso by regularizing $\widehat{\Sigma}$ in (20.11).

Finally, we show here that the elastic net encourages a grouping effect in the sense that correlated variables tend to enter or leave the model together. Consider an extreme scenario where some predictors are exactly identical. The authors of [24] showed that strict convexity of the elastic net or the ridge penalty guarantees a unique solution that assigns identical coefficients to the identical variables. In contrast, the lasso does not even have a unique solution. In general, the elastic net coefficients of highly correlated variables are shrunk toward each other. In [24], it was proved that for any pair of variables $(\mathbf{x}_i, \mathbf{x}_j)$,

$$\frac{1}{\|\mathbf{Y}\|_2} \left| \hat{\beta}_i - \hat{\beta}_j \right| \leq \frac{1 + \lambda_2}{\lambda_2} \sqrt{2(1 - \rho)} \qquad (20.12)$$

and

$$\frac{1}{\|\mathbf{Y}\|_2} \left| \hat{\beta}_i + \hat{\beta}_j \right| \leq \frac{1 + \lambda_2}{\lambda_2} \sqrt{2(1 + \rho)} \qquad (20.13)$$

where ρ is the sample correlation between $\mathbf{x}_i$ and $\mathbf{x}_j$. Thus, if $\rho \doteq 1$, $\hat{\beta}_i \approx \hat{\beta}_j$, and if $\rho \doteq -1$, $\hat{\beta}_i \approx -\hat{\beta}_j$. These two inequalities provide a quantitative description for the grouping effect of the elastic net.

The lasso does not possess the grouping effect. For a simple illustration, let us consider the linear model with $p = 2$. The authors of [20] gave the explicit expression for $(\hat{\beta}_1, \hat{\beta}_2)$, from which we easily get $|\hat{\beta}_1 - \hat{\beta}_2| = |cos(\theta)|$, where θ is the angle between $\mathbf{Y}$ and $\mathbf{x}_1 - \mathbf{x}_2$. It is easy to construct examples such that $\rho = cor(\mathbf{x}_1, \mathbf{x}_2) \to 1$ but $cos(\theta)$ does not vanish.

20.4.2 A Stylized Example

We now present an idealized example showing the important differences between the elastic net and the lasso. Let Z_1 and Z_2 be two independent $Unif(0, 20)$ variables. The response $\mathbf{Y}$ is generated as $N(Z_1 + 0.1 \cdot Z_2, 1)$. The predictors are generated as follows:

$$\mathbf{x}_1 = Z_1 + \epsilon_1, \ \mathbf{x}_2 = -Z_1 + \epsilon_2, \ \mathbf{x}_3 = Z_1 + \epsilon_3$$
$$\mathbf{x}_4 = Z_2 + \epsilon_4, \ \mathbf{x}_5 = -Z_2 + \epsilon_5, \ \mathbf{x}_6 = Z_2 + \epsilon_6$$

where ϵ_i are i.i.d. $N(0, \frac{1}{16})$. One hundred observations were generated from this model. The variables $\mathbf{x}_1, \mathbf{x}_2, \mathbf{x}_3$ form a group whose underlying factor is Z_1, and $\mathbf{x}_4, \mathbf{x}_5, \mathbf{x}_6$ form a second group whose underlying factor is Z_2. The within-group correlations are almost 1 and the between-group correlations are almost 0. An oracle would identify the Z_1 group as the important variables. This simulation model tries to mimic the biological pathways. We can think of Z_1 and Z_2 as pathways. $\mathbf{x}_1, \mathbf{x}_2, \mathbf{x}_3$ are the genes that function through the pathway Z_1, and $\mathbf{x}_4, \mathbf{x}_5, \mathbf{x}_6$ are the genes that function through the pathway Z_2. The response is mainly affected by pathway Z_1.

Figure 20.1 compares the solution paths of the lasso and the elastic net on two simulated datasets from the above simple model. We can see that the lasso solution paths can be very different with a different dataset generated from the same model. Thus the lasso paths are unstable. Furthermore, the lasso plots do not reveal any correlation information by itself. In contrast, the elastic net has much smoother and more stable solution paths while clearly showing the grouped selection: $\mathbf{x}_1, \mathbf{x}_2, \mathbf{x}_3$ are in one significant group and $\mathbf{x}_4, \mathbf{x}_5, \mathbf{x}_6$ are in the other trivial group.

20.4.3 Computation and Tuning

The elastic net is a computationally efficient method. First, we can see that, by its definition, both the lasso and the elastic net can be recast as a quadratic programming problem. Thus, standard quadratic programming software can be directly used to solve the lasso/elastic net. The authors of [5] invented the LARS algorithm for computing the entire lasso solution path. The authors of [24] showed that the elastic net solution paths are also piecewise linear functions of λ_1 for each fixed λ_2. This fact implies that the elastic net can be solved by a path algorithm similar to the LARS. The authors of [24] proposed the LARS-EN algorithm for efficiently computing the entire elastic net solution paths. Starting from zero, the LARS-EN algorithm sequentially updates the elastic net fits. In the $N \gg M$ case, such as with microarray data, it is not necessary to run the LARS-EN algorithm to the end (early stopping). The optimal results are achieved at an early stage of the LARS-EN algorithm [24]. If we stop the algorithm after m steps, then it requires $O(m^3 + Nm^2)$ operations.

We now discuss how to choose the type and value of the tuning parameter in the elastic net. In the lasso, the conventional tuning parameter is the L_1 norm of the coefficients (t) or the fraction of the L_1 norm (s). The advantage of using (λ_2, s) is that s is always valued within $[0, 1]$. In the $N \gg M$ problem it is better to use (λ_2, k) as the tuning parameters where k is the number of the LARS-EN steps. There are well-established methods for choosing such tuning parameters [13, Chapter 7]. If only training data are available, 10-fold cross validation is a popular method for estimating the prediction error and comparing different models, and we use it here. Note that there are two tuning parameters in the elastic net, so we need to cross validate on a two-dimensional surface. Typically we first pick a (relatively small) grid of values for λ_2, say $(0, 0.01, 0.1, 1, 10, 100)$. Then, for each λ_2, the other tuning parameter (s or k) is selected by 10-fold CV. The chosen λ_2 is the one giving the smallest CV error. It is worth mentioning that for each λ_2, the computational cost of a 10-fold CV is the same as ten OLS fits. Thus, the two-dimensional CV is computationally thrifty in the usual $M > N$ setting. In the $N \gg M$ case, the

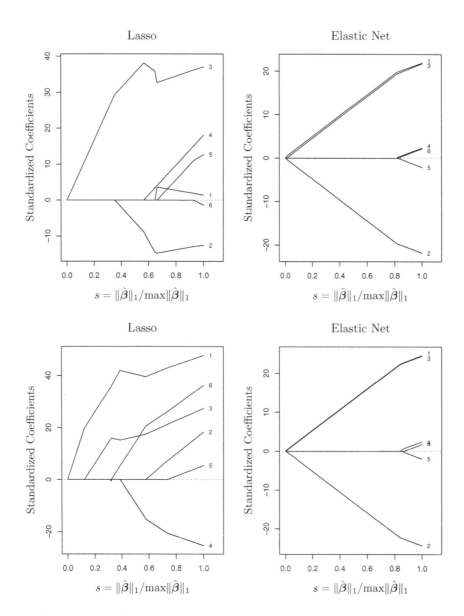

FIGURE 20.1: The left and right panels show the lasso and the elastic net ($\lambda_2 =$ 0.5) solution paths respectively. We fitted the lasso and the elastic net using two independent datasets simulated from the same model.

cost grows linearly with N, and is still manageable. Practically, early stopping is used to ease the computational burden. For example, if $M = 30, N = 5,000$, and we do not want to use more than 200 variables in the final model, we may stop the LARS-EN algorithm after 500 steps.

We use the prostate cancer data to illustrate the computations with the elastic net. The data in this example come from a study of prostate cancer [19]. The predictors are eight clinical measures: log cancer volume (lcavol), log prostate weight (lweight), age, log of the amount of benign prostatic hyperplasia (lbph), seminal vesicle invasion (svi), log capsular penetration (lcp), Gleason score (gleason), and percentage Gleason score 4 or 5 (pgg45). The response is the log of prostate specific antigen (lpsa). The prostate cancer data were divided into two parts: a training set with 67 observations, and a test set with 30 observations. Model fitting and tuning parameter selection by 10-fold cross validation were carried out on the training data. Figure 20.2 displays the lasso and the elastic net solution paths, in which the vertical lines indicate the selected model. The lasso includes lcavol, lweight lbph, svi, and pgg45 in the final model, while the elastic net selects lcavol, lweight, svi, lcp, and pgg45. The prediction error of the elastic net is about 24% lower than that of the lasso.

20.4.4 Analyzing the Cardiomypathy Data

We apply the elastic net to analyze the cardiomypathy data. The data have 30 observations and 6319 predictors. The response variable in this study is a G protein-coupled receptor, designated Ro1. When the receptor is overexpressed in the heart of adult mice, the mice develop a lethal dilated cardiomyopathy that has many hallmarks of human disease. The mice recover when the expression of the receptor is turned off [18]. The goal of the study is to investigate the association between the changes in gene expression and the expression of Ro1. Thirty-two mice were tested in the study [17]. To determine which changes in gene expression were due to the expression of the Ro1 transgene, the authors of [18] suggested identifying the genes that correlate with the Ro1 expression profile. Genes that explain this expression profile are potential candidates to provide additional therapeutic targets and clues to the mechanism of disease.

The lasso model selects 21 genes. We fit the elastic net model by the LARS-EN algorithm with early stopping after 100 steps. The optimal tuning parameters for the elastic net are $\lambda = 0.1, k = 47$, where k is the number of steps in the LARS-EN algorithm. In the elastic net model, 44 genes are selected. Note that the size of the training set is 30, so the lasso can at most select 30 genes. In contrast, the number of genes selected by the elastic net is 44, greater than the sample size. Figure 20.3 displays the elastic net solution paths as a function of k.

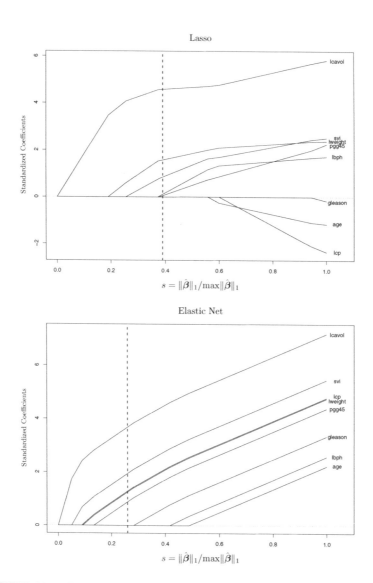

FIGURE 20.2: Prostate cancer data. The top panel shows the lasso estimates as a function of s, and the bottom panel shows the elastic net estimates as a function of s. Both of them are piecewise linear, which is a key property of our efficient algorithm. In both plots the vertical dotted line indicates the selected final model.

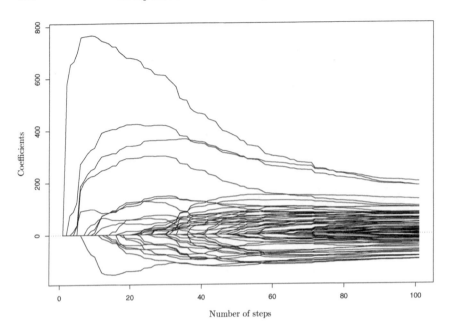

FIGURE 20.3: Cardiomypathy data: the elastic net coefficients paths.

20.5 The Elastic-Net Penalized SVM

We have discussed the elastic net method in the linear regression model. The elastic net penalty can be used in classification problems, too. The resulting classifier should also retain the nice properties of the elastic net in regression. The support vector machine (SVM) [21] is now a very popular classification tool with numerous applications. We focus on the use of the elastic net in SVMs.

20.5.1 Support Vector Machines

In a standard two-class classification problem, the response y is coded as $\in \{1, -1\}$. The goal is to find a classification rule from the training data, so that when given a new input $\mathbf{x}$, we can assign a class label to it. The SVM has been a popular tool for the two-class classification problem in the machine learning field. Recently, it has also gained increasing attention from the statistics community. Readers are referred to [13] for a complete statistical approach to the SVM.

It turns out that the SVM is also equivalent to a regularized function fitting problem. With $f(x) = \beta_0 + \mathbf{x}^T\boldsymbol{\beta}$, consider the optimization problem:

$$\min_{\beta_0,\boldsymbol{\beta}} \sum_{i=1}^{n} [1 - y_i f(x_i)]_+ + \lambda\|\boldsymbol{\beta}\|_2^2, \tag{20.14}$$

where the subscript "+" indicates the positive part and λ is a tuning parameter. Note that (20.14) has the form *loss + penalty*, which is a familiar paradigm to statisticians in function estimation. The loss function $(1-yf)_+$ is called the *hinge* loss. The penalty is the L_2-norm of the coefficient vector, the same as that used in the ridge regression. The role of the ridge penalty in the SVM is the same as in linear regression. The ridge penalty shrinks the fitted coefficients towards zero to control the variance of fitted coefficients, hence possibly achieving a better bias-variance trade-off, especially when there are many highly correlated variables. Although the SVM enjoys a sparse representation due to the support vectors, it cannot select significant variables. Often people combine the SVM with some external feature selection method such as the recursive feature elimination (RFE) [9].

The 1-norm SVM was used in [23] to perform automatic feature selection in the SVM. With $f(x) = \beta_0 + \mathbf{x}^T\boldsymbol{\beta}$, the 1-norm SVM solves

$$\min_{\beta_0,\boldsymbol{\beta}} \sum_{i=1}^{n} [1 - y_i f(x_i)]_+ + \lambda\|\boldsymbol{\beta}\|_1 \tag{20.15}$$

The 1-norm SVM shares many of the nice properties of the lasso. The L_1 (lasso) penalty encourages some of the coefficients to be shrunken to exact zero if λ is appropriately chosen. Hence the 1-norm SVM performs feature selection through regularization. The 1-norm SVM has significant advantages over the 2-norm SVM when there are many noise variables.

20.5.2 A New SVM Classifier

We have seen that in regression problems the lasso penalty has some fundamental drawbacks and the elastic net penalty fixes these limitations. The authors of [22] applied the elastic net penalty to the SVM. Consider the following doubly regularized SVM, which is referred to as the DrSVM:

$$\min_{\beta_0,\boldsymbol{\beta}} \sum_{i=1}^{n} \left[1 - y_i(\beta_0 + \mathbf{x}_i^T\boldsymbol{\beta})\right]_+ + \lambda_2\|\boldsymbol{\beta}\|_2^2 + \lambda_1\|\boldsymbol{\beta}\|_1 \tag{20.16}$$

where both λ_1 and λ_2 are tuning parameters. The role of the L_1 penalty is to allow automatic variable selection, and the role of the L_2 penalty is to help groups of correlated variables get selected together.

The grouping effect also shows in the DrSVM. The following inequalities were proven in [22]:

$$|\beta_i - \beta_j| \leq \frac{\sqrt{n}}{\lambda_2}\sqrt{2(1-\rho)} \qquad (20.17)$$

and

$$|\beta_i + \beta_j| \leq \frac{\sqrt{n}}{\lambda_2}\sqrt{2(1+\rho)} \qquad (20.18)$$

where ρ is the sample correlation between $\mathbf{x}_i$ and $\mathbf{x}_j$. Thus, if $\rho \doteq 1$, $\hat{\beta}_i \approx \hat{\beta}_j$, and if $\rho \doteq -1$, $\hat{\beta}_i \approx -\hat{\beta}_j$. The 1-norm SVM does not possess the grouping effect.

The DrSVM is computationally efficient. It is interesting to note that for each fixed λ_1, the DrSVM solution is a piecewise linear function of $\frac{1}{\lambda_2}$. This is similar to a result in [10] which showed that the 2-norm SVM solution is a piecewise linear function of $\frac{1}{\lambda_2}$. The authors of [10] developed a path-following algorithm for computing the entire 2-norm SVM paths. The authors of [22] developed a similar path algorithm for computing the DrSVM solution for all λ_2 with a fixed λ_1. On the other hand, the DrSVM solution is a piecewise linear function of λ_1 for each fixed λ_2. This is similar to a result in [23] which showed that the 1-norm SVM solution is a piecewise linear function of λ. The authors of [23] developed a path-following algorithm for computing the entire 1-norm SVM paths. The authors of [22] developed a similar path algorithm for computing the DrSVM solution for all λ_1 with a fixed λ_2. Hence we have two path-following algorithms to compute the DrSVM. The readers are referred to [22] for the technical details of the two path algorithms.

To illustrate the piecewise linearity property of the DrSVM, we compute its solution paths using a small simulated dataset. We generate 8 training data in each of two classes. Each input $\mathbf{x}_i$ is a $p = 5$ dimensional vector. For the "+1" class, x_i has a normal distribution with mean $\mu_+ = (1,0,0,0,0)^T$ and the diagonal elements of the covariance matrix are 1, and the off-diagonal elements are all equal to $\rho = 0.8$. The "−1" class has a similar distribution, except that the mean is $\mu_- = (-1,0,0,0,0)^T$. The solution paths are displayed in Figure 20.4, where any segment between two adjacent vertical lines is linear.

We demonstrate the use of the elastic-net penalized SVM in microarrays classification and gene selection on the leukemia data consisting of 7129 genes and 72 samples [8]. In the training dataset, there are 38 samples, among which 27 are type 1 leukemia (ALL) and 11 are type 2 leukemia (AML). The goal is to construct a diagnostic rule based on the expression level of those 7219 genes to predict the type of leukemia. The remaining 34 samples are used to test the prediction accuracy of the diagnostic rule. We compared the three types of SVMs: the SVM, the 1-norm SVM, and the DrSVM. The tuning parameters are chosen according to 10-fold cross validation, then the final model is fitted on all the training data and evaluated on the test data. The results are summarized in Table 20.1. The SVM uses recursive feature elimination to select genes. As we can see, the DrSVM seems to have the best prediction performance. However, notice this is a very small dataset, so the

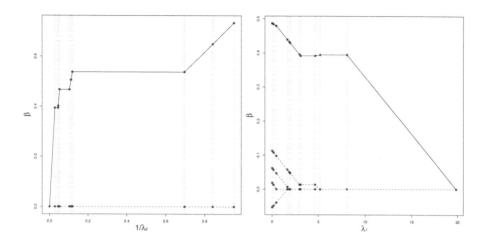

FIGURE 20.4: The solid line corresponds to β_1, the dashed lines correspond to $\beta_2, \ldots, \beta_5$. The right panel is for $\boldsymbol{\beta}(\lambda_1)$ (with $\lambda_2 = 30$), and the left panel is for $\boldsymbol{\beta}(\lambda_2)$ (with $\lambda_1 = 6$).

TABLE 20.1:   Summary of leukemia classification results

Method	10-fold CV error	Test error	No. of genes
SVM + RFE	2/38	1/34	31
1-norm SVM	2/38	1/34	22
DrSVM	0/38	0/34	78

difference may not be significant. It is also worth noting that the 22 genes selected by the L_1-norm SVM are a subset of the 78 genes selected by the DrSVM.

20.6   Sparse Eigen-Genes

We have seen that the elastic net is very useful in supervised learning problems (regression and classification). It turns out that the elastic net can also be used to performance variable selection in un-supervised learning problems. The authors of [25] considered using the elastic net to obtain principal components with sparse loadings. This property has very useful applications in gene expression data analyses. Principal components of gene expression arrays are called eigen-genes. If the sparse eigen-genes can explain a large part of the total variance of gene expression levels, then the subset of genes representing the sparse eigen-genes is considered important.

20.6.1 PCA and Eigen-Genes

Principal component analysis (PCA) [15] is a popular un-supervised learning and dimension reduction technique. Recently PCA has been used in gene expression data analysis [1]. The authors of [12] proposed the so-called *gene shaving* techniques using PCA to cluster highly variable and coherent genes in microarray datasets. PCA seeks the linear combinations of the original variables such that the derived variables capture maximal variance. PCA can be computed via the singular value decomposition (SVD) of the data matrix. In detail, let the data $\mathbf{X}$ be an $M \times N$ matrix, where n and p are the number of observations and the number of variables, respectively. Without loss of generality, assume the column means of $\mathbf{X}$ are all 0. Let the SVD of $\mathbf{X}$ be

$$\mathbf{X} = \mathbf{U}\mathbf{D}\mathbf{V}^T \tag{20.19}$$

$\mathbf{Z} = \mathbf{U}\mathbf{D}$ are the principal components (PCs), and the columns of $\mathbf{V}$ are the corresponding loadings of the principal components. The sample variance of the i-th PC is $\mathbf{D}_{ii}^2/n$. In gene expression data the standardized PCs $\mathbf{U}$ are called the *eigen-arrays* and $\mathbf{V}$ are the *eigen-genes* [1]. Usually the first q ($q \ll \min(n,p)$) PCs are chosen to represent the data, thus a great dimensionality reduction is achieved. An obvious drawback of PCA is that each PC is a linear combination of all p variables and the loadings are typically nonzero. This often makes it difficult to interpret the derived PCs. It is desirable not only to achieve the dimensionality reduction but also to reduce the number of explicitly used variables. An ad hoc way to sparsity in PCA is to artificially set the loadings with absolute values smaller than a threshold to zero. We prefer a principled approach to deriving a sparse PCA.

20.6.2 Sparse Principal Component Analysis

The theory and algorithm of sparse principal component analysis (SPCA) were developed in [25]. To focus on the main idea, we introduce SPCA for the leading principal component.

SPCA starts with an equivalent formulation of PCA. Note that for any $\lambda > 0$, we let

$$\min_{\boldsymbol{\alpha},\boldsymbol{\beta}} \sum_{i=1}^{M} \|\mathbf{x}_i - \boldsymbol{\alpha}\boldsymbol{\beta}^T \mathbf{x}_i\|^2 + \lambda \|\boldsymbol{\beta}\|^2$$
$$\text{subject to} \quad \|\boldsymbol{\alpha}\|^2 = 1 \tag{20.20}$$

Then the solution $\boldsymbol{\beta}$ is proportional to the first principal component. Hence PCA is identical to a regression-type problem. This fact suggests that one could borrow the sparse modeling techniques from regression to produce sparse principal components. It is important to note that the positive λ is necessary when $N \gg M$ [25].

The following SPCA method was proposed in [25]:

$$\min_{\alpha,\beta} \sum_{i=1}^{M} \|\mathbf{x}_i - \alpha\beta^T\mathbf{x}_i\|^2 + \lambda\|\beta\|^2 + \lambda_1\|\beta\|_1$$

$$\text{subject to} \quad \|\alpha\|^2 = 1 \tag{20.21}$$

We see that $\lambda\|\beta\|^2 + \lambda_1\|\beta\|_1$ is the elastic net penalty. Its L_1 part will shrink some components of β to exact zero, just like the lasso shrinkage in regression and the SVM. The zero components in β correspond to zero loadings in the principal component. The empirical results in [25] suggest that the solution is not sensitive to the choice of λ as long as $\lambda > 0$. Hence the L_2 part of the elastic net penalty is primarily used to ensure the sparse principal component is identical to the ordinary principal component when we do not need the sparsity (using $\lambda_1 = 0$).

The optimization problem in SPCA is nonconvex. The authors of [25] proposed an alternating algorithm for solving SPCA. Note that one can easily solve β for a given α. It is an elastic net regression problem. On the other hand, if we fix β, solving α can be found by reduced rank Procrustes rotation [25]. We can start with the ordinary PCA and iterate between these two steps until convergence.

We illustrate the sparse PC selection method on Ramaswamy's data [16], which has 16063 ($p = 16063$) genes and 144 ($n = 144$) samples. Its first principal component (eigen-gene) explains 46% of the total variance. Note that all 16063 genes are used in the first eigen-gene. To derive a sparse eigen-gene, we applied SPCA to find the leading sparse PC. We found that as few as 2.5% of these 16063 genes can sufficiently construct the leading principal component with an affordable loss of explained variance (from 46% to 40%).

20.7 Summary

The elastic net is a novel shrinkage and selection method for producing a sparse model with good prediction accuracy. The elastic net encourages the grouping effect and elegantly handles the high-dimensionality. The elastic net also enjoys great computational efficiency with the help of efficient path algorithms. In many ways the elastic net is a more appropriate tool for variable selection with high-dimensional data than the lasso. We have seen the applications of the elastic net in regression and classification problems. The elastic net penalty can also be used in principal components, leading to a sparse version of PCA that automatically omits unimportant variables from the PCA directions. This method can be used to find sparse eigen-genes. There are other statistical models that are used in modeling genomic data.

For instance, the Cox proportional hazard model is the standard model for modeling censored survival data. The elastic net can be directly used in those models.

References

[1] O. Alter, P. Brown, and D. Botstein. Singular value decomposition for genome-wide expression data processing and modeling. *Proceedings of the National Academy of Sciences*, 97:10101–10106, 2000.

[2] L. Breiman. Heuristics of instability and stabilization in model selection. *The Annals of Statistics*, 24:2350–2383, 1996.

[3] M. Dettling and P. Bühlmann. Finding predictive gene groups from microarray data. *Journal of Multivariate Analysis*, 90:106–131, 2004.

[4] D. Donoho, I. Johnstone, G. Kerkyacharian, and D. Picard. Wavelet shrinkage: asymptopia? (with discussion). *Journla of the Royal Statistical Society: Series B*, 57:301–337, 1995.

[5] B. Efron, T. Hastie, I. Johnstone, and R. Tibshirani. Least angle regression. *The Annals of Statistics*, 32(2):407–499, 2004.

[6] I. Frank and J. Friedman. A statistical view of some chemometrics regression tools. *Technometrics*, 35:109–148, 1993.

[7] J. Friedman. Regularized discriminant analysis. *Journal of the American Statistical Association*, 84:249–266, 1989.

[8] T. Golub, D. Slonim, P. Tamayo, C. Huard, M. Gaasenbeek, J. Mesirov, H. Coller, M. Loh, J. Downing, and M. Caligiuri. Molecular classification of cancer: class discovery and class prediction by gene expression monitoring. *Science*, 286:513–536, 1999.

[9] I. Guyon, J. Weston, S. Barnhill, and V. Vapnik. Gene selection for cancer classification using support vector machines. *Machine Learning*, 46:389–422, 2002.

[10] T. Hastie, S. Rosset, R. Tibshirani, and J. Zhu. The entire regularization path of the support vector machine. *Journal of Machine Learning Research*, pages 1391–1415, 2004.

[11] T. Hastie, R. Tibshirani, D. Botstein, and P. Brown. Supervised harvesting of expression trees. *Genome Biology*, 2:0003.1–0003.12, 2003.

[12] T. Hastie, R. Tibshirani, M. Eisen, P. Brown, D. Ross, U. Scherf, J. Weinstein, A. Alizadeh, L. Staudt, and D. Botstein. "Gene Shaving" as

a method for identifying distinct sets of genes with similar expression patterns. *Genome Biology*, 1:1–21, 2000.

[13] T. Hastie, R. Tibshirani, and J. Friedman. *The Elements of Statistical Learning; Data mining, Inference and Prediction.* Springer-Verlag, New York, 2001.

[14] A. Hoerl and R. Kennard. Ridge regression. In *Encyclopedia of Statistical Sciences*, volume 8, pages 129–136. Wiley, New York, 1988.

[15] I. Jolliffe. *Principal component analysis.* Springer Verlag, New York, 1986.

[16] S. Ramaswamy, P. Tamayo, R. Rifkin, S. Mukheriee, C. Yeang, M. Angelo, C. Ladd, M. Reich, E. Latulippe, J. Mesirov, T. Poggio, W. Gerald, M. Loda, E. Lander, and T. Golub. Multiclass cancer diagnosis using tumor gene expression signature. *Proceedings of the National Academy of Sciences*, 98:15149–15154, 2001.

[17] C. Redfern, M. Degtyarev, A. Kwa, N. Salomonis, N. Cotte, T. Nanevicz, N. Fidelman, K. Desai, K. Vranizan, E. Lee, P. Coward, N. Shah, J. Warrington, G. Fishman, D. Bernstein, A. Baker, and B. Conklin. Conditional expression of a gi-coupled receptor causes ventricular conduction delay and a lethal cardiomyopathy. *PNAS*, 97:4826–4831, 2000.

[18] M. Segal, K. Dahlquist, and B. Conklin. Regression approach for microarray data analysis. *Journal of Computational Biology*, 10:961–980, 2003.

[19] T. Stamey, J. Kabalin, J. Mcneal, F. F. Johnstone, I., E. Redwine, and N. Yang. Prostate specific antigen in the diagnosis and treatment of adenocarcinoma of the prostate, ii: Radical prostatectomy treated patients. *Journal of Urology.*, 16:1076–1083, 1989.

[20] R. Tibshirani. Regression shrinkage and selection via the lasso. *Journal of the Royal Statistical Society, B*, 58:267–288, 1996.

[21] V. Vapnik. *The Nature of Statistical Learning Theory.* Springer-Verlag, New York, 1995.

[22] L. Wang, J. Zhu, and H. Zou. The doubly regularized support vector machine. *Statistica Sinica*, 16:589–616, 2006.

[23] J. Zhu, S. Rosset, T. Hastie, and R. Tibshirani. 1-norm svms. *Advances in Neural Information Processing Systems 16*, 2003.

[24] H. Zou and T. Hastie. Regression and variable election via the elastic net. *Journal of the Royal Statistical Society, Series B*, 67:301–320, 2005.

[25] H. Zou, T. Hastie, and R. Tibshirani. Sparse principal component analysis. *Journal of Computational and Graphical Statistics*, 15:265–286, 2006.

Index

T - #0080 - 101024 - C0 - 234/156/23 [25] - CB - 9781584888789 - Gloss Lamination